# The Neuroendocrine-Immune Network


Editor

**Serem Freier, M.B., B.S.**

Chairman
Department of Pediatrics
Shaare Zedek Medical Center
Jerusalem, Israel

Associate Professor
Hebrew University of Jerusalem
Jerusalem, Israel

Director of Research
Institute of Gastroenterology
Shaare Zedek Medical Center
Jerusalem, Israel



CRC Press
Taylor & Francis Group
Boca Raton London New York

CRC Press is an imprint of the
Taylor & Francis Group, an **informa** business

First published 1990 by CRC Press
Taylor & Francis Group
6000 Broken Sound Parkway NW, Suite 300
Boca Raton, FL 33487-2742

Reissued 2018 by CRC Press



**Library of Congress Cataloging-in-Publication Data**

The Neuroendocrine-immune network/editor, Serem Freier.
p. cm.
Includes bibliographies and index.
ISBN 0-8493-4625-8
1. Neuroendocrinology. 2. Neuroimmunology. I. Freier, S.
[DNLM; 1. Cell Communication. 2. Neurosecretion—drug effects.
3. Neurosecretory Systems—immunology. 4. Neurosecretory Systems—physiology. WL 102 N494532]
QP356.4.N4772 1990
612.8—dc20
DNLM/DLC 89-1014
for Library of Congress

A Library of Congress record exists under LC control number: 89001014

Publisher's Note
The publisher has gone to great lengths to ensure the quality of this reprint but points out that some imperfections in the original copies may be apparent.

Disclaimer
The publisher has made every effort to trace copyright holders and welcomes correspondence from those they have been unable to contact.

ISBN 13: 978-1-315-89809-4 (hbk)
ISBN 13: 978-1-351-07719-4 (ebk)

Visit the Taylor & Francis Web site at http://www.taylorandfrancis.com and the
CRC Press Web site at http://www.crcpress.com

# PREFACE

The present explosion of knowledge has brought in its wake the need for stepping up the rate of disseminating the vast store of new information. Ever more books are being written to keep up with new developments, and people who have never had any ambitions to edit or write books find themselves unexpectedly becoming editors or authors. I am one of them.

The subject of this book exemplifies one of the changes in our concept of the function of the human body. Although we always paid lip service to the fact that the human body is a functional unit, in practice, the various specialists such as immunologists and endocrinologists each went their own way. This book reminds us that the body is fully integrated with all the so-called systems interacting with each other. Rather than think in terms of systems or even organs, we have to think in terms of interactions. Thus, for instance, the gut, which is best known as an organ of absorption, is also the greatest collection of lymphoid tissue in the human body and in addition, secretes a greater variety of hormones than any of the classical endocrine glands. We cannot, therefore, discuss absorption without taking into consideration the immune and endocrine changes which are part of the process of absorption. This book discusses these neuroendocrine-immune interactions from a number of aspects; developmental, anatomical, and histological approaches are used in some chapters. In others, the multiple effects of peptide hormones come under consideration, while again in others, the interactions of the various "systems" are described. Certain aspects have been omitted, notably the fascinating field of the interaction of mental disease and the immune system.

The task of being an editor begins by approaching suitable contributors. I have been fortunate in that most of the potential authors who have been approached did agree to participate. First, I would like to thank all the co-authors for giving their time and energy to review the fields of their expertise. Second, I would like to thank Mrs. Hana Hoffman of "Teum Congresses" who undertook the correspondence with all the would be authors at the outset of this venture. Third, this is an opportunity to thank the many friends who have supported the work of my laboratory throughout the years, in particular Mrs. Adelaide Kennedy-Leigh of London and her late husband Michael, the trustees of his estate, as well as Mr. and Mrs. Abe Fish of Toronto, and my brother Mr. Shalheveth Freier of Rehovoth. Last but not least, I owe a great debt of gratitude to my wife, Miriam, and my children, Shlomit, Micah, Etan, Joseph, and Tamar, who for years have put up with my preoccupations and my absent-mindedness. To all of them, this book is dedicated.

**Serem Freier**

# EDITOR

**Serem Freier, M.B., B.S., D.C.H.,** is Chairman of the Department of Pediatrics and Director of Research of the Institute of Gastroenterology at the Shaare Zedek Medical Center, Jerusalem, Israel. He is Associate Professor of Pediatrics at the Hebrew University of Jerusalem.

Dr. Freier studied medicine at St. Bartholomew's Hospital Medical College at the University of London and graduated in 1949. He obtained his M.B., B.S. degree in 1950 and his D.C.H., from the Conjoint Board of Examiners of England and Wales, in the year 1951.

Dr. Freier is a past president of the Israel Pediatric Association. He is a member of the European Society of Pediatric Gastroenterology and Nutrition, the Israel Society of Pediatric Gastroenterology and Nutrition, the Israel Society of Immunology, the Israel Society of Allergy and Clinical Immunology, and the Israel Society of Clinical Pediatrics.

He has organized two international symposia. One on Breast Feeding in Tel-Aviv in February 1980 and a second one on the Immunology of the Gastrointestinal Tract and the Liver in Jerusalem in March 1987. He was the recipient of the Buswell Award from the Dean of the School of Medicine SUNY at Buffalo, New York in 1980. He was visiting professor at the Universities of Adelaide and Perth in Australia in 1973 and the University of New York at Buffalo in 1981. He has received research grants from the U.S. Department of Agriculture, the U.S. Agency for International Development-Co-operative Developmental Research, the Israel Cancer Fund, the Dysautonomia Foundation Inc., and from industry.

Dr. Freier is the author of over 90 papers mostly dealing with his main field of interest, the immunology of the gastrointestinal tract, and has edited a book on breast feeding. His present main interest is the impact of the peptide hormones on the immune system in the gut, and, more specifically, the role of these peptides in regulating the delivery of the immunoglobulins to the lumen of the intestine.

# CONTRIBUTORS

**Kurt D. Ackerman, M.D.**
Department of Neurobiology and Anatomy
University of Rochester
School of Medicine
Rochester, New York

**Chana Altaratz, M.Sc.**
Department of Embryology and Teratology
Tel Aviv University Medical School
Tel Hashomer, Israel

**Zvi Bar-Shavit, Ph.D.**
Senior Lecturer
Department of Experimental Medicine and Cancer Research
Hebrew University Hadassah Medical School
Jerusalem, Israel

**Dean Befus, Ph.D.**
Professor
Department of Microbiology and Infectious Diseases
University of Calgary Health Science Center
Calgary, Alberta, Canada

**Benzion Beilin, M.D.**
Head
Department of Anesthesiology
Golda Medical Center
Hasharon Hospital
Petah Tigvah, Israel

**Denise L. Bellinger, Ph.D.**
Research Associate
Department of Neurobiology and Anatomy
University of Rochester
School of Medicine
Rochester, New York

**John Bienenstock, M.D.**
Professor
Department of Pathology
McMaster University
Hamilton, Ontario, Canada

**Michael G. Blennerhassett, Ph.D.**
Postdoctoral Fellow
Departments of Pathology and Neurosciencs
McMaster University
Hamilton, Ontario, Canada

**Dale E. Bockman, Ph.D.**
Professor and Chairman
Department of Anatomy
Medical College of Georgia
Augusta, Georgia

**William H. Brooks, M.D.**
Clinical Professor
Department of Surgery
University of Kentucky Medical Center
Lexington, Kentucky

**Sonia L. Carlson, Ph.D.**
Postdoctoral Fellow
Department of Medical Microbiology and Immunology
University of Kentucky Medical Center
Lexington, Kentucky

**Richard J. Cross, Ph.D.**
Assistant Research Professor
Department of Microbiology and Immunology
University of Kentucky Medical Center
Lexington, Kentucky

**Josef Donnerer, M.D.**
Assistant Professor
Department of Experimental and Clinical Pharmacology
University of Graz
Graz, Austria

**Anthony Eglezos, Ph.D.**
Research Assistant
National Research Institute of Gerontology and Geriatric Medicine
Mount Royal Hospital
Parkville, Victoria, Australia

**Yitzhak S. Engelberg, M.D.**
Department of Pathology
Tel Hashomer Hospital
Ramat-Gan, Israel

**David L. Felten, Ph.D.**
Professor
Department of Neurobiology and Anatomy
University of Rochester School of Medicine
Rochester, New York

**Suzanne Y. Felten, Ph.D.**
Associate Professor
Department of Neurobiology and Anatomy
University of Rochester
School of Medicine
Rochester, New York

**Serem Freier, D.C.H.**
Chairman
Department of Pediatrics
Director of Research
Institute of Gastroenterology
Shaare Zedek Medical Center
Jerusalem, Israel

**Vithal K. Ghanta, Ph.D.**
Associate Professor
Department of Biology and Microbiology
University of Alabama at Birmingham
Birmingham, Alabama

**Rachel Goldman, Ph.D.**
Senior Scientist
Department of Membrane Research
Weizmann Institute of Science
Rehovot, Israel

**Deborah V. Harbour, Ph.D.**
Assistant Professor
Department of Psychiatry and Behavioral Sciences
University of Texas Medical Branch
Galveston, Texas

**Robert D. Helme, M.D.**
Professor
National Research Institute of Gerontology and Geriatric Medicine
Mount Royal Hospital
Parkville, Victoria, Australia

**Raymond N. Hiramoto, Ph.D.**
Professor
Department of Microbiology
University of Alabama at Birmingham
Birmingham, Alabama

**Yasunori Kakuta, M.D.**
Postdoctoral Fellow
Department of Pathology
McMaster University
Hamilton, Ontario, Canada

**Margaret L. Kirby, Ph.D.**
Professor
Department of Anatomy and Petiatrics
Medical College of Georgia
Augusta, Georgia

**Emanuel Lebenthal, M.D.**
Chief
Division of Gastroenterology and Nutrition
Children's Hospital
Professor of Pediatrics
State University of New York
Buffalo, New York

**Shmuel Livnat, Ph.D.**
Assistant Professor
Departments of Psychiatry and Microbiology and Immunology
University of Rochester
School of Medicine
Rochester, New York

**Kelly S. Madden, Ph.D.**
Postdoctoral Fellow
Department of Psychiatry and Microbiology and Immunology
University of Rochester
School of Medicine
Rochester, New York

**Laslo Nebel, M.D.**
Department of Embryology and Teratology
Tel Aviv University Medical School
Tel Hashomer, Israel

**M. Sue O'Dorisio, M.D., Ph.D.**
Professor
Department of Pediatrics and Immunology
Ohio State University
Columbus, Ohio

**Mary H. Perdue, Ph.D.**
Assistant Professor
Department of Pathology
McMaster University
Hamilton, Ontario, Canada

**Marek Rola-Pleszczynski, M.D.**
Director
Immunology Division
Department of Pediatrics
University of Sherbrooke
Sherbrooke, Quebec, Canada

**Thomas L. Roszman, Ph.D.**
Professor
Department of Microbiology and Immunology
University of Kentucky Medical Center
Lexington, Kentucky

**Shraga Segal, Ph.D.**
Department of Microbiology and Immunology
Faculty of Health Sciences
Ben Gurion University of Negev
Beer Sheva, Israel

**Piersante Sestini, M.D.**
Assistant Professor
Institute of Respiratory Disease
University of Siena
Siena, Italy

**Yehuda Shavit, Ph.D.**
Senior Lecturer
Department of Psychology
Hebrew University of Jerusalem
Jerusalem, Israel

**Eric M. Smith, Ph.D.**
Associate Professor
Departments of Microbiology and Psychiatry and Behavioral Sciences
University of Texas Medical Branch
Galveston, Texas

**H. Brent Solvason, Ph.D.**
Department of Microbiology
University of Alabama at Birmingham
Birmingham, Alabama

**Ron H. Stead, F.I.M.L.S.**
Lecturer
Department of Pathology
McMaster University
Hamilton, Ontario, Canada

**David A. Sullivan, Ph.D.**
Assistant Professor
Department of Ophthamology
Harvard Medical School
and
Associate Scientist
Immunology Unit
Eye Research Institute
Boston, Massachusetts

**Boris Tartakovsky, Ph.D.**
Department of Chemical Immunology
Weizmann Institute of Science
Rehovot, Israel

**Raz Yirmiya, Ph.D.**
Postdoctoral Fellow
Department of Anatomy and Psychiatry
University of California
Los Angeles, California

# TABLE OF CONTENTS

Chapter 1

# THE ROLE OF THE NEURAL CREST IN THE DEVELOPMENT OF THE IMMUNE SYSTEM AND ENDOCRINE ORGANS

**Dale E. Bockman and Margaret L. Kirby**

## TABLE OF CONTENTS

# I. INTRODUCTION

Development of the immune system is dependent upon proper development of the thymus. Proper development of the thymus, in turn, is dependent upon contributions from the neural crest. Like the thymus, certain endocrine organs are dependent upon neural crest derivatives for differentiation. Furthermore, neural crest plays a key role in the formation of the neural network which serves lymphoid organs, including their blood supply.

The development of these interdependent elements is accomplished through an exquisitely timed series of differentiative steps, migrations, and interactions.

This chapter will first discuss the salient features of the differentiation of these elements, including the interactive features which control them. Next, it will consider clinical observations which have been interpreted as the result of failures in proper development of the neural crest, and experimental studies in which developmental defects have been induced as a means of trying to understand the interactive mechanisms involved.

# II. CENTRAL ROLE OF THYMUS IN DEVELOPMENT OF THE IMMUNE SYSTEM

The thymus is a central lymphoid organ[1] which is necessary for the full development of the immune system. Within its specialized microenvironment, lymphocytes proliferate and differentiate by a process which is not antigen driven. Differentiated cells migrate from the thymus to populate peripheral lymphoid organs and tissues and to provide them with the thymus-dependent component of the immune system. Here, antigen-driven proliferation and differentiation constitute immune reactivity; failure of the thymus to provide a population of reactive cells to the peripheral lymphoid organs causes severe deficiencies in immune capability.

Thymic lymphoid development is initiated when the thymic primordium attracts lymphoid stem cells which arise in other places and circulate in the blood.[2,3] At this stage, the thymus is an epithelial bud with surrounding mesenchyme (Plate 1).* Lymphoid stem cells enter the circulation from the yolk sac and fetal liver, and later from the bone marrow. They migrate from the blood vessels and enter the thymic primordium, where they proliferate and differentiate in the spaces between thymic epithelial cells.

Several subpopulations of lymphocytes are generated within the thymus;[4-6] most die there. A small percentage of cells, carrying marker antigens which indicate their differentiative fate, leave the epithelial framework of the thymus to enter the bloodstream, and seed spleen, lymph nodes, and other peripheral lymphoid organs.

# III. DEVELOPMENT OF THE THYMUS, THYROID, AND PARATHYROIDS

## A. THE PRIMITIVE PHARYNGEAL REGION

As the early embryo assumes a cylindrical shape, a blind-ending tube of epithelium derived from endoderm is folded into the head region. This primitive foregut develops into the upper part of the digestive system. The pharyngeal part of the foregut is elaborated rapidly. The primitive pharyngeal cavity is limited in extent from floor to ceiling, but is expanded laterally, where it forms four or five pouchlike projections on each side. Pouch five may be considered a posterior extension of pouch four.[7]

The primitive pharynx gives rise to a number of important structures. Most of these lose their connections with the pharynx as they migrate or change their relative positions, retaining

* Plate 1 appears following page 6.

no evidence of their pharyngeal origin. The structures derived from the primitive pharynx, which are pertinent to the present discussion, include the thyroid and parathyroid glands, the ultimobrancial bodies, and the thymus.

### B. DERIVATIVES OF PHARYNGEAL EPITHELIUM

Regional specializations of the epithelium lining the primitive pharyngeal cavity mark the earliest primordia of the glands in question. The thyroid appears first as a ventral bulge in the midline of the floor of the foregut. As elaboration of epithelial cells continues, it forms a distinct mass which is located between the first pair of pharyngeal pouches. After losing continuity with the pharynx, it assumes a progressively more posterior relative location due to the forward growth of the pharyngeal area. A connective tissue capsule is formed around the developing thyroid from local mesenchyme. The connective tissue of the thyroid is derived from the neural crest, as will be described later. The epithelial cells organize into the follicles which are typical of the definitive thyroid and which elaborate the thyroid hormones.

The ultimobranchial bodies are epithelial elaborations from the endoderm of the fifth (or posterior fourth) pharyngeal pouches. They receive mesenchymal contributions from the neural crest. Their hormone-producing cells may be derived from the neural crest.[8] As the thyroid assumes a relatively more posterior position, the ultimobranchial bodies fuse with it. The hormone-producing cells derived from the ultimobranchial bodies become the C cells or parafollicular cells of the thyroid, which produce calcitonin (thyrocalcitonin).

The parathyroid glands develop as epithelial elaborations from the dorsal portion of each of the third and fourth pharyngeal pouches. They acquire connective tissue septa and capsules from surrounding mesenchyme and separate from the pharynx. They become attached to the thyroid, but remain four separate organs.

The thymic primordia, derived from the ventral portion of the third pharyngeal pouches, are at first hollow epithelial structures. The lumina of the thymic primordia disappear as cell proliferation continues. The primitive condition in which thymic spaces are continuous with the pharyngeal cavity is therefore not maintained, and eventually, the primordia lose all contact with the pharynx. The epithelial component of the thymus is modified to form a closed reticular framework as lymphocytes proliferate and differentiate within the interstitial spaces (Plate 1). The left and right lobes of the thymus unite with each other in the midline, forming the definitive single organ; migration and relative growth of the thymus leads to its final position in the upper part of the thoracic cavity in most mammals (multiple cervical lobes are present in birds). The connective tissue of the thymus is acquired from neural crest-derived mesenchyme which surrounds the primitive pharynx (Plate 1).

## IV. DEVELOPMENT AND CONTRIBUTIONS OF THE NEURAL CREST

### A. FORMATION

The neural crest is a transient structure which is present during the early formation of the nervous system and which subsequently expands to produce much of the peripheral nervous system and a large number of definitive nonnervous structures in the head and neck region. It produces a population of mesenchymal cells which interact with epithelial structures to induce their differentiation.

The primitive neural tube is formed when a thickened region of ectoderm (the neural plate) curves upward until its edges meet in the midline and fuse. The edges of the neural plate are continuous with unthickened ectoderm. This region projects upward as the plate rolls into a tube, forming the neural folds. The neural crest forms from the neural folds. As the tops of the neural tube meet and fuse, cells in the region where the ectoderm and neural tube meet divide rapidly. The neural crest thus formed projects laterally; its cells migrate from the edges, through extracellular matrix, to supply progeny to all parts of the body.[7,9]

### B. NEURAL CONTRIBUTIONS

The neural crest, with contributions from neural placodes, provides much of the peripheral nervous system.[7,9] This includes sensory ganglia of the brain and spinal cord and the autonomic nervous system. The motor fibers of the autonomic nervous system supply smooth muscle and glands. They supply the lymphoid organs, particularly their blood vessels. Sensory fibers for the lymphoid organs come from cell bodies in the cranial and spinal ganglia; these cells are derived from the neural crest.

The neural crest is involved directly with formation of the adrenal gland. Neural crest cells migrate from a particular region in the trunk, form aggregates of cells which become associated with the primordial adrenal cortex, then differentiate into the hormone-producing cells of the adrenal medulla.

### C. MESENCHYMAL CONTRIBUTIONS

The neural crest in the head region of the embryo and back to the level of the fifth somite gives rise to mesenchymal cells as well as neural components.[9,10] These cells comprise the ectomesenchyme, so called because it is mesenchyme which arises from ectoderm rather than mesoderm. Ectomesenchymal cells form much of the cartilage, bone, and general connective tissue in the head and neck, as well as the walls of the large arteries conducting blood from the heart. Ectomesenchymal cells are involved in the septation of the outflow tract of the heart.[11]

The ectomesenchymal cells which are most pertinent to the present discussion are those which interact with the epithelial promordia of thymus, thyroid, and parathyroids. Cells from the cranial neural crest migrate through and proliferate within the pharyngeal arches, some coming into intimate contact with the epithelial buds which are capable of forming the pharyngeal-derived glands. The inductive interaction of these two components is necessary for normal differentiation to occur. Neither the epithelial nor the mesenchymal component can develop in isolation; the combination is necessary for lobulation and lymphoid proliferation to occur.[3,12]

Thus, it is clear that for normal development of the thymus (and by extension the immune system), cells derived from cranial neural crest must interact with epithelial primordia in order to allow the thymus to support proliferation and differentiation of lymphoid cells (Plate 1). Furthermore, the interaction of these elements is time dependent. In the chick, seeding of the thymic primordium by lymphoid stem cells is limited to a 36-h period of receptivity.[3]

## V. DEFECTIVE DEVELOPMENT OF THE THYMUS AND ENDOCRINE ORGANS

### A. CLINICAL DATA

The DiGeorge syndrome is characterized in part by immune deficiency. T-cell function is impaired because of agenesis or hypoplasia of the thymus. In addition, hypocalcemia is present because of defective parathyroid function. Frequently, anomalies of the heart and great vessels, thyroid, and craniofacial structure are present. Burke et al.[13] recently reported that only three of 11 DiGeorge patients had thyrocalcitonin-containing cells in their thyroid. The pattern of defects associated with the DiGeorge syndrome is what would be expected if its etiology were based on the failure of the cranial neural crest to develop normally[14-18] because they are multiple and because the neural crest is directly or indirectly involved in the development of each structure involved.

Clusters of defects thought to be due to defective neural crest development have been referred to as neurocristopathies.[19,20] Thymic deficiency is prominent in the DiGeorge syndrome, less so in others.[14,16] There are various patterns in different described clusters of anomalies. Part of the variation is probably due to real differences in the exact nature of the etiological mechanism. Some of the variation may be due to failure to examine the structure and function of all the organs which could possibly be affected.

Some clinical syndromes which are associated with fetal exposure to certain agents display a pattern of anomalies like those of the DiGeorge syndrome. Lammer et al.[21] recently have described the effects of fetal exposure to retinoic acid, an analog of vitamin A, which was administered primarily for acne treatment. In malformed infants, there was a characteristic pattern of malformation involving thymic, cardiac, craniofacial, and central nervous system structures. Abnormalities of thymic morphogenesis included ectopia, hypoplasia, and aplasia. No follow-up studies on immune function after fetal exposure to retinoic acid were completed at the time of this report.

Similar combinations of defects are characteristic of the fetal alcohol syndrome, caused by exposure of the fetus to high levels of alcohol, as discussed by Ammann et al.[22]

## B. EXPERIMENTAL DATA

Direct evidence of the importance of the neural crest in the development of thymus, thyroid, and parathyroids has come from experimental studies using chick embryos.[15] In these studies, appropriate regions of the cranial neural crest were ablated before migration to the pharyngeal area had occurred, and the thymus was examined later to determine the effect. At the stage in development when neural folds (containing the neural crest progenitors) are present over the region corresponding to the first five somites, the neural folds were removed bilaterally with a vibrating needle or by microcautery. The eggs were then sealed and incubated until incubation days 13 through 16, at which time thymic development was compared with controls whose neural folds were left intact. The development of the thymus was inhibited markedly by neural crest ablation. In some experimental animals, no thymus could be found by microdissection under a dissecting microscope. In others, the mass of thymic tissue was markedly reduced. Mean weight of thymus in the experimental group was significantly less than in controls. Histological examination of the thymus in experimental animals, when present, sometimes showed distinct signs of delayed development, such as remnants of the primitive pharyngeal pouch in thymic tissue. Lymphocytes were present within the thymus of experimental animals, however, and there was the beginning of differentiation of cortical from medullary regions. Thyroids and parathyroids also were affected. They frequently could not be located by microdissection, or were reduced in size on one or both sides. Anomalies of the outflow tract of the heart were present in the experimental animals in this study, confirming the demonstration by Kirby et al.[23] that proper aorticopulmonary septation is dependent upon contributions from the neural crest. Thus ablation of the cranial neural crest produces not only thymic defects, but also defective development of a cluster of organs which are affected in the DiGeorge syndrome.

Other experimental methods have been demonstrated to be capable of producing similar clusters of defects, including defective thymic development. Fertilysin ($N,N^1$-*bis*[dichloroacetyl]-1,8-octamethyl diamine, also referred to commonly as *bis*-diamine), has been shown to cause defective development of thymus and the cardiovascular system, as well as craniofacial structures.[24] Porter and Schmidt[25] recently have demonstrated that oral administration of Fertilysin to pregnant mice early in gestation produces an overall reduction in the size of the thymus, and cortical hypoplasia, when the offspring are examined 36 d after birth. Morphology of the spleen is also affected. Furthermore, the functional responsiveness of splenic lymphocytes in the postnatal period is reduced by exposure to Fertilysin during gestation.

Experimental studies also have confirmed the clusters of defects which have been ob-

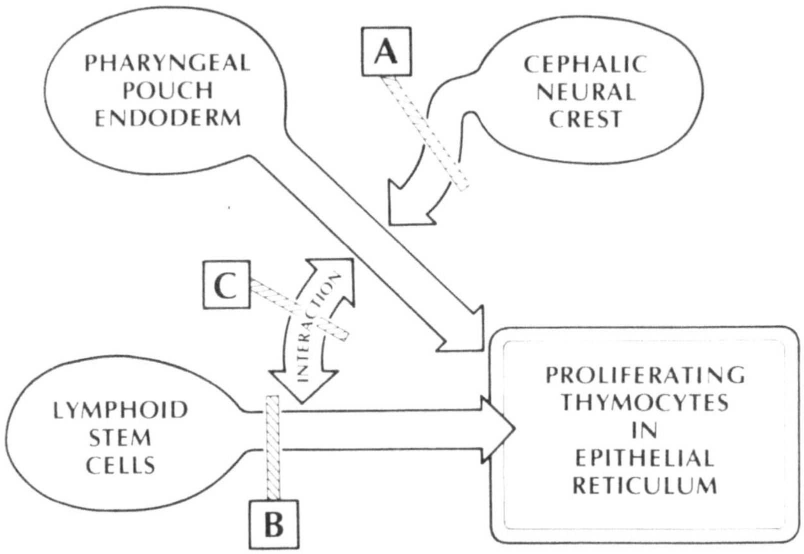


FIGURE 1. Conceptual representation of contributions and interactions necessary for the development of a functional thymus. Enoderm from the third and fourth pharyngeal pouches is induced by migrating mesenchymal elements derived from the neural crest. As induced epithelial cells proliferate, they interact with blood-borne lymphoid stem cells, inducing thymocyte differentiation and proliferation in their reticulum. Thymic aplasia or dysplasia, as produced experimentally by neural crest ablation and in the DiGeorge syndrome, would result from an incomplete contribution by the neural crest (A). The "epithelial" thymus of severe combined immunodeficiency, lacking in lymphocytes, would result from a lack of lymphoid stem cells (B). The epithelial thymus of other lymphoid deficiencies could result from a lack of interaction between epithelium and stem cells (C) or a deficiency in a subpopulation of stem cells. (From Bockman, D. E. and Kirby, M. L., Dependence of thymus development on derivatives of the neural crest, *Science*, 223, 498, 1984. Copyright 1984 by AAAS.)

served in clinical cases by exposure to other agents, and have strengthened the possibilities that an effect on the neural crest is involved in the etiology. Administration of retinoic acid to mice during the early somite stages of development causes characteristic malformations of the thymus, face, and great vessels.[26] Histological evidence shows that this is due to interference with neural crest development. Similar defects may be induced in mice by the administration of alcohol to the mother.[27,28] An important area for further study in these experimental approaches is an assessment of the functional capability of the immune system after treatments which impair, but do not destroy, the thymus.

## VI. CONCLUSION

The neural crest has an important role in the development of the neuroendocrine-immune network. It is necessary in the formation of the nerves which serve the lymphoid organs, particularly their blood vessels. It participates (directly or indirectly) in the formation of endocrine organs, particularly adrenal medulla, thyroid, and parathyroids. It is necessary for the normal development of the thymus, a central lymphoid organ upon which the immune system depends for development of its full capabilities. Experimental evidence indicates that depriving the developing embryo of some contributions from the neural crest leads to improper development of the thymus (Figure 1), thyroids, parathyroids, heart, and facies. Clusters of congenital defects, including the DiGeorge syndrome, and syndromes resulting

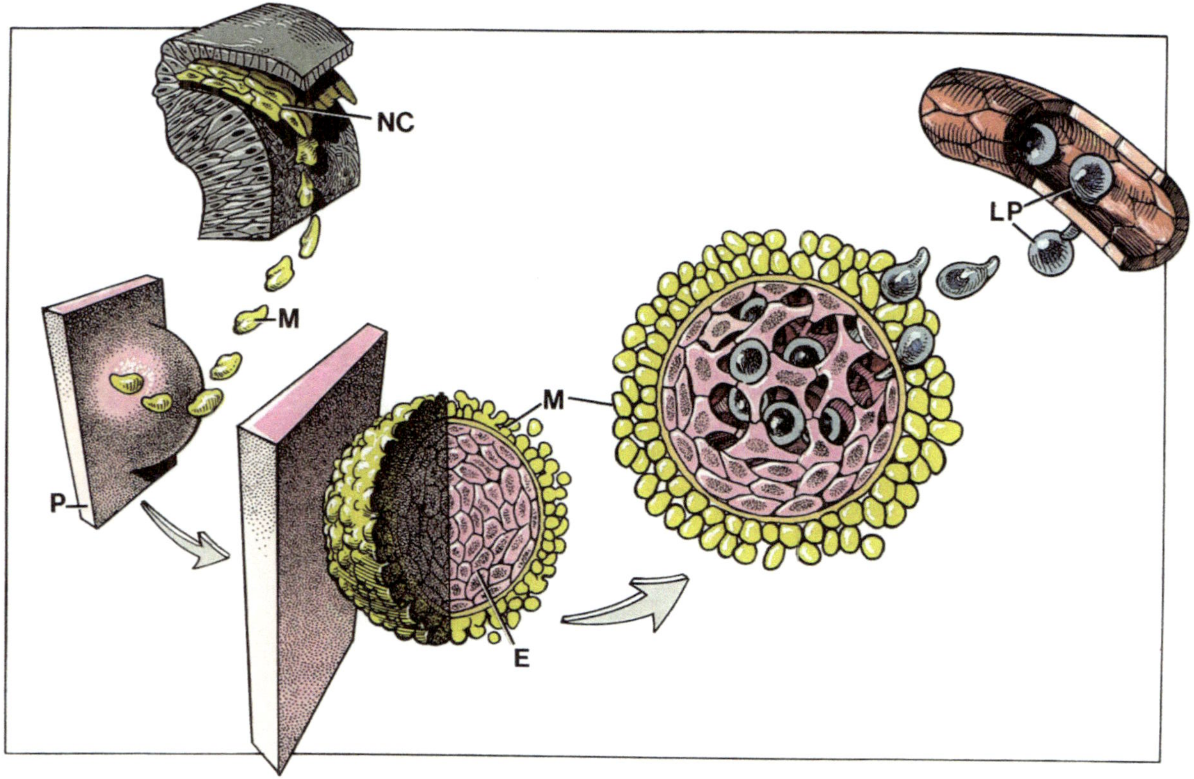


PLATE 1. Diagram showing the contribution to thymic development from three sources: epithelium (E), mesenchyme (M), and lymphoid precursor cells (LP). The epithelial primordium of the thymus derives from an outpouching of the wall of the primitive pharynx (P). The epithelium is surrounded by inductive mesenchyme derived from the neural crest (NC). The thymocytes which begin to differentiate within the epithelial network are derived from lymphoid precursor cells which originate in the fetal liver or bone marrow and are carried to the region of the thymus by blood vessels. (From Reif, A. E. and Schlesinger, M., Eds., *Thy 1: Immunology, Neurology, and Therapeutic Applications*, Marcel Dekker, New York, 1988. With permission.)

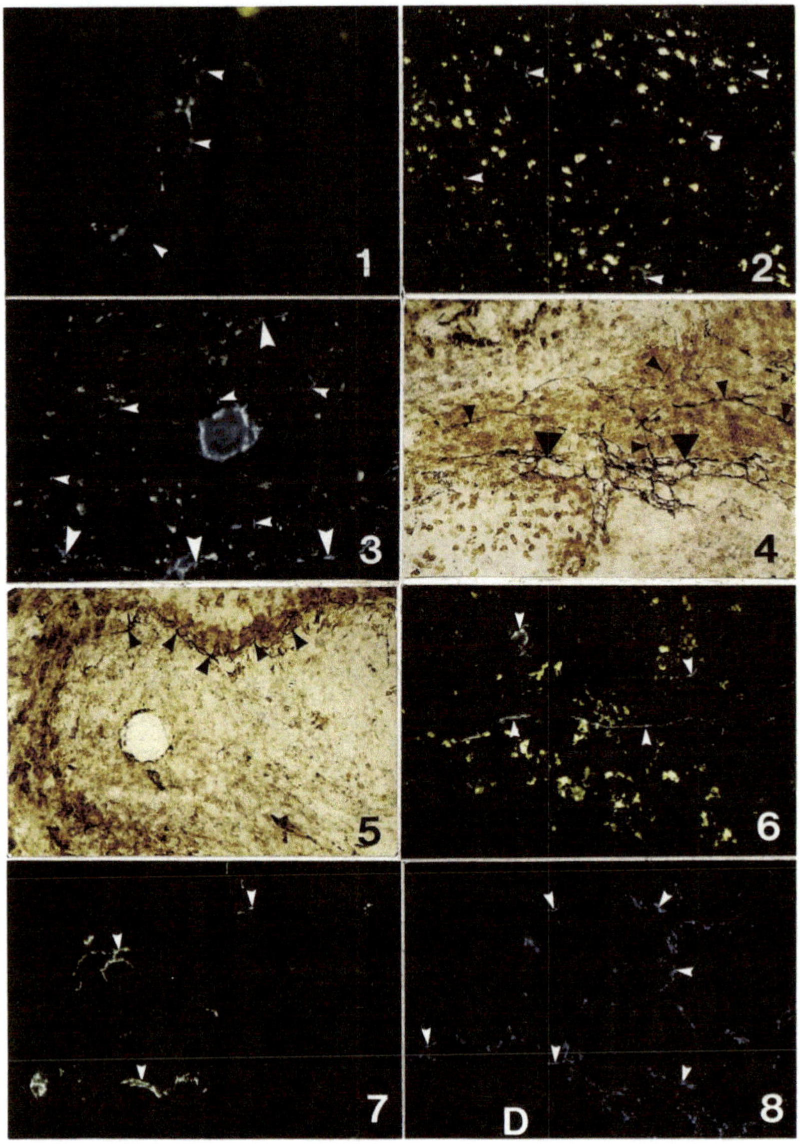


PLATE 2. (1) Noradrenergic nerve fibers (arrowheads) adjacent to cells in the bone marrow. Fluorescence histochemistry. Fischer 344 rat, 3 months (Magnification × 250). (2) Noradrenergic nerve fibers (arrowheads) in the cortex of the thymus. Large yellow autofluorescent cells of unknown subtype are present in this region in addition to thymocytes. Fluorescence histochemistry. Fischer 344 rat, 8 months. (Magnification × 125). (3) Noradrenergic nerve fibers in the white pulp of the spleen. A plexus of fibers surrounds a central arteriole. Some fibers are present in the parenchyma of the periarteriolar lymphatic sheath (PALS) (small arrowheads) among lymphocytes. Additional linear profiles of nerve fibers run along the marginal sinus (large arrowheads). Fluorescence histochemistry. Fischer 344 rat, 4 months (Magnification × 125). (4) Tyrosine hydroxylase (TH)-positive nerve fibers (black) in the white pulp of the spleen. Some of these fibers extend along a central arteriole in longitudinal section (large arrowheads), while other fibers arborize among T lymphocytes in the PALS. The nerve fibers are stained with an antibody to TH (nickel enhanced, black), while the T lymphocytes are stained with antibody to OX-19 (brown), a pan-T cell marker. Double label immunocytochemistry. Fischer 344 rat, 3 months (Magnification × 125). (5) TH-positive nerve fibers (black) along the marginal sinus in the white pulp of the spleen. These fibers run adjacent to antigen-presenting macrophages, stained brown with antibody to ED-3. Double-label immunocytochemistry. Fischer 344 rat, 12 months. (Magnification × 125). (6) Noradrenergic nerve fibers traversing a cortical region of a cervical lymph node, near the capsule. These fibers are present among T lymphocytes in this region. Fluorescence histochemistry. C57BL/6 mouse, 3 months (Magnification × 125). (7) Noradrenergic nerve fibers in a medullary cord of mesenteric lymph node. These fibers are present among macrophages, lymphocytes, and other cell types in this region. Fluorescence histochemisty. NZW mouse, $1^1/_2$ months. (Magnification × 125). (8) Noradrenergic nerve fibers in the lamina propria of GALT in the rabbit appendix. These fibers branch throughout the lamina propria, among numerous cell types of the immune system, but remain separated from the nodular region of the dome (D) by a moat. Fluorescence histochemistry. Rabbit, adult. (Magnification × 125).

from fetal exposure to a vitamin A analog and alcohol, are similar to experimentally produced deficiencies. A reasonable hypothesis for the congenital defects is the failure of neural crest-derived ectomesenchyme to reach the proper embryonic primordia in the proper quantities at the proper time to induce their development in a normal fashion. It must be kept in mind, however, that the interactions during this period are intricately interrelated and exquisitely timed, and that multiple components are involved. It is necessary, therefore, to realize that alternate mechanisms may be capable of blocking the development of this network.

## REFERENCES


1. **Cooper, M. D., Gabrielsen, A. E., and Good, R. A.,** Role of the thymus and other central lymphoid tissues in immunological disease, *Annu. Rev. Med.*, 18, 113, 1967.
2. **Moore, M. A. S. and Owen, J. J. T.,** Experimental studies on the development of the thymus, *J. Exp. Med.*, 126, 715, 1967.
3. **Le Douarin, N. and Jotereau, F. V.,** Tracing of cells of the avian thymus through embryonic life in interspecific chimeras, *J. Exp. Med.*, 142, 17, 1975.
4. **Tidman, N., Janossy, G., Bodger, M., Granger, S., Kung, P. C., and Goldstein, G.,** Delineation of human thymocyte differentiation pathways utilizing double-staining techniques with monoclonal antibodies, *Clin. Exp. Immunol.*, 45, 457, 1981.
5. **Asma, G. E. M., Van Den Bergh, R. L., and Vossen, J. M.,** Use of monoclonal antibodies in a study of the development of T lymphocytes in the human fetus, *Clin. Exp. Immunol.*, 53, 429, 1983.
6. **Rothenberg, E. and Lugo, J. P.,** Differentiation and cell division in the mammalian thymus, *Dev. Biol.*, 112, 1, 1985.
7. **Arey, L. B.,** *Developmental Anatomy*, W. B. Saunders, Philadelphia, 1965, 213.
8. **Le Douarin, N., Fontaine, J., and Le Lievre, C.,** New studies on the neural crest origin of the avian ultimobranchial glandular cells — interspecific combinations and cytochemical characterization of C cells based on the uptake of biogenic amine precursors, *Histochemistry*, 38, 297, 1974.
9. **Le Douarin, N.,** *The Neural Crest*, Cambridge University Press, Cambridge, 1982, 1.
10. **Le Lievre, C. S. and Le Douarin, N. M.,** Mesenchymal derivatives of the neural crest: analysis of chimaeric quail and chick embryos, *J. Embryol. Exp. Morphol.*, 34, 125, 1975.
11. **Phillips, M. T., Kirby, M. L., and Forbes, G.,** Analysis of cranial neural crest distribution in the developing heart using quail-chick chimeras, *Circ. Res.*, 60, 27, 1987.
12. **Auerbach, R.,** Morphogenetic interactions in the development of the mouse thymus gland, *Dev. Biol.*, 2, 271, 1960.
13. **Burke, B. A., Johnson, D., Gilbert, E. F., Drut, R. M., Ludwig, J., and Wick, M. R.,** Thyrocalcitonin-containing cells in the DiGeorge anomaly, *Hum. Pathol.*, 18, 355, 1987.
14. **Couly, G., Lagrue, A., and Griscelli, C.,** Le syndrome de DiGeorge, neurocristopathie rhombencephalique exemplaire, *Rev. Stomatol. Chir. Maxillofac.*, 84, 103, 1983.
15. **Bockman, D. E. and Kirby, M. L.,** Dependence of thymus development on derivatives of the neural crest, *Science*, 223, 498, 1984.
16. **Kirby, M. L. and Bockman, D. E.,** Neural crest and normal development: a new perspective, *Anat. Rec.*, 209, 1, 1984.
17. **Bockman, D. E. and Kirby, M. L.,** Neural crest interactions in the development of the immune system, *J. Immunol.*, 135, 766s, 1985.
18. **Van Mierop, L. H. and Kutsche, L. M.,** Cardiovascular anomalies in Di George syndrome and importance of neural crest as a possible pathogenetic factor, *Am. J. Cardiol.*, 58, 133, 1986.
19. **Bolande, R. P.,** The neurocristopathies. A unifying concept of disease arising in neural crest maldevelopment, *Hum. Pathol.*, 5, 409, 1974.
20. **Johnston, M. C., Vig, K. W., and Ambrose, L. J.,** Neurocristopathy as a unifying concept: clinical considerations, *Adv. Neurol.*, 29, 97, 1981.
21. **Lammer, E. J., Chen, D. T., Hoar, R. M., Agnish, N. D., Benke, P. J., Braun, J. T., Curry, C. J., Fernhoff, P. M., Grix, A. W., Jr., Lott, I. T., Richard, J. M., and Sun, S. C.,** Retinoic acid embryopathy, *N. Engl. J. Med.*, 313, 837, 1985.
22. **Ammann, A. J., Wara, D. W., Cowan, M. J., Barrett, D. J., and Stiehm, E. R.,** The DiGeorge syndrome and the fetal alcohol syndrome, *Am. J. Dis. Child.*, 136, 906, 1982.

23. **Kirby, M. L., Gale, T. F., and Stewart, D. E.,** Neural crest cells contribute to normal aorticopulmonary septation, *Science,* 220, 1059, 1983.
24. **Taleporos, P., Salgo, M. P., and Oster, G.,** Teratogenic action of bis(dichloroacetyl)diamine, *Teratology,* 18, 5, 1978.
25. **Porter, J. F. and Schmidt, R. R.,** Developmental toxicity of $N,N^1$-bis(dichloroacetyl)-1,8-octamethylene diamine: effects of in utero exposure on the postnatal murine immune system, *Biol. Neonate,* 50, 221, 1986.
26. **Webster, W. S., Johnston, M. C., Lammer, E. J., and Sulik, K. K.,** Isotretinoin embryopathy and the cranial neural crest: an in vivo and in vitro study, *J. Craniofac. Genet. Dev. Biol.,* 6, 211, 1986.
27. **Sulik, K. K., Johnston, M. C., and Webb, M. A.,** Fetal alcohol syndrome: embryogenesis in a mouse model, *Science,* 214, 936, 1981.
28. **Daft, P. A., Johnston, M. S., and Sulik, K. K.,** Abnormal heart and great vessel development following acute ethanol exposure in mice, *Teratology,* 33, 93, 1986.

Chapter 2

# PERIPHERAL INNERVATION OF LYMPHOID TISSUE

**David L. Felten, Suzanne Y. Felten, Kurt D. Ackerman, Denise L. Bellinger, Kelly S. Madden, Sonia L. Carlson, and Shmuel Livnat**

## TABLE OF CONTENTS

## I. EVIDENCE FOR NEURAL LINKS WITH THE IMMUNE SYSTEM

Evidence from several fields of investigation indicates a bidirectional link between the nervous system and the immune system. The link from the nervous system to the immune system is mediated by neuroendocrine outflow, autonomic outflow, or both the major routes available to the central nervous system (CNS) by which it can influence the viscera. As discussed in several chapters of this volume and elsewhere,[1-4] psychosocial factors such as bereavement, depression, marital separation, and stress of examinations in medical students are associated with suppressed immune functions. Experimental studies in rodents demonstrate a causal relationship between a variety of stressors and subsequent immune alterations, or diminished resistance to tumors or pathogens. While these studies indicate the involvement of neural circuitry in the modulation of immune reactivity, they do not reveal the specific identity of that circuitry.

A search for the identification of specific circuitry involved in neural-immune interactions was undertaken with lesion studies in the CNS.[5-10] Discrete lesions in the hypothalamus (particularly the preoptic/anterior area), limbic forebrain structures (amygdala, hippocampal formation, septum), brain stem autonomic and reticular regions (raphe nuclei, parabrachial nuclei, reticular formation), and the cerebral cortex were associated with alterations in immune reactivity, either suppression or enhancement, depending upon the site and extent of the lesion. These same sites also showed altered neural electrical activity or monoamine metabolism in response to immunization. This combined evidence suggests an integrated circuitry of cerebral cortex, limbic forebrain, hypothalamus, and brain stem autonomic nuclei which would, in turn, regulate both neuroendocrine outflow and autonomic outflow.

Both the neuroendocrine and efferent autonomic systems produce biologically active molecules that can interact with cells of the immune system through cell surface or intracellular receptors. Although neuroendocrine circuitry (such as the CRF-ACTH-glucocorticoid axis) is implicated in immunomodulation, its activity cannot explain all immunological responses to neurally derived signals. For example, behaviorally conditioned immunosuppression does not appear to depend upon adrenal cortical mechanisms,[11] and immunosuppression in rodents resulting from inescapable shock can occur in the absence of adrenal glands[12] or even the pituitary. Such findings strongly suggest that autonomic outflow may participate in the translation of central communication to the immune system. This chapter examines briefly the evidence for direct autonomic innervation of lymphoid organs, particularly by the noradrenergic postganglionic sympathetic nerves. Since the early 1970s the existence of adrenergic receptors on lymphocyte, monocytes, and neutrophils was known; but only recently has it become evident that noradrenergic nerves abundantly innervate the parenchyma of organs of the immune system (Plate 2, Parts 1 to 8)* and establish direct synaptic-like contacts with immunocytes (Figure 1).

## II. GENERAL PRINCIPLES AND PATTERNS OF PERIPHERAL INNERVATION OF LYMPHOID ORGANS

Both primary lymphoid organs (bone marrow, thymus) and secondary lymphoid organs (spleen, lymph nodes, gut-associated lymphoid tissue [GALT]) are innervated by noradrenergic nerves derived from ganglia of the sympathetic nervous system (Plate 2, Parts 1 to 8). This innervation shows similarities in the patterns of fiber distribution and in principles of general organization in these organs. The noradrenergic fibers are derived from extrinsic ganglia of either the sympathetic chain or collateral ganglia. The fibers distribute with the

* Plate 2, Parts 1 to 8, appear following page 6.

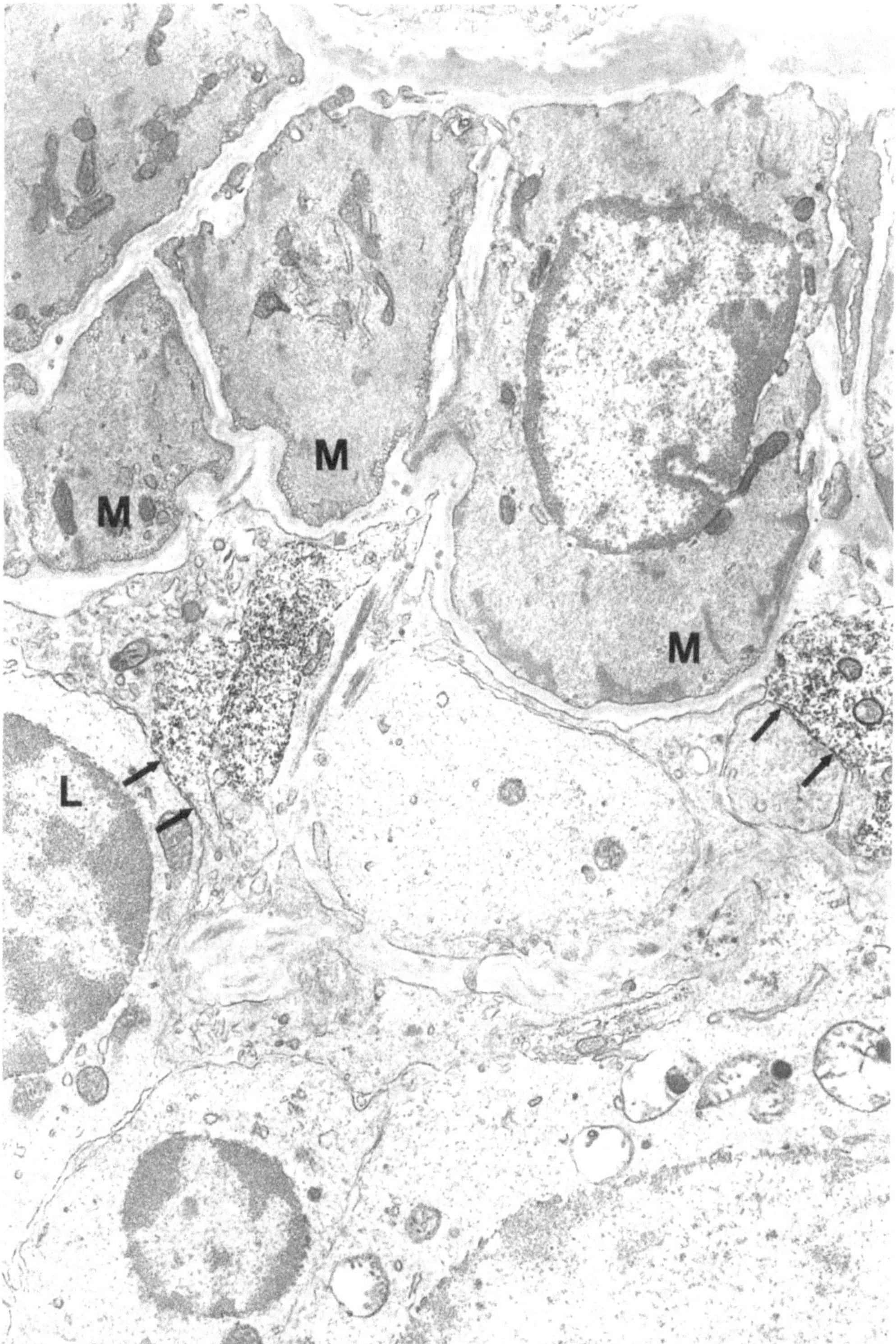


FIGURE 1. Tyrosine hydroxylase positive nerve profiles (arrows) adjacent to smooth muscle cells (M) of the central arteriole and lymphocytes (L) in the periarteriolar lymphatic sheath. Immunoelectron microscopy. Rat spleen, adult. (Magnification × 10,600.)

vasculature, innervate smooth muscle of the capsular and trabecular systems of these organs, and also distribute among immunocytes in the parenchyma. In the secondary organs, the fibers show a distinct compartmentation, with distribution within T lymphocyte and macrophage zones, and alongside B lymphocyte areas. It is likely that the sympathetic neurotransmitter norepinephrine (NE) is made available for interaction with these cells both by paracrine secretion and local release in a synaptic-like region of contact between nerve terminals and lymphocytes (Figure 1) or other cell types.

Norepinephrine is not the only neurotransmitter available for interaction with cells of the immune system. We have found nerve profiles with substance P immunoreactivity, with neuropeptide Y immunoreactivity, and possibly other neuropeptides. These neuropeptides may be colocalized with NE in postganglionic sympathetic nerves, as we suspect in the case of neuropeptide Y, or may be found in independent nerve fibers, as would be expected for sensory fibers. These possibilities have not been resolved; their resolution is of great importance for defining the physiologically relevant ligands for potential interaction with neurotransmitter receptors on lymphocytes, macrophages, and accessory cells. The presence of muscarinic cholinergic receptors on lymphocytes, coupled with the presence of acetylcholinesterase (AChE)-positive profiles in some lymphoid organs (clearly not an exclusive cholinergic marker), has led to the proposal that acetylcholine (ACh) acts as a neurotransmitter interacting with cells of the immune system. However, thus far, evidence for the presence of acetylcholine, choline acetyltransferase (ChAT) activity, and connections from cholinergic circuitry are either negative or contradictory. Further careful studies are needed for conclusion about possible cholinergic innervation of immune organs.

For acceptance of a compound as a bona fide neurotransmitter acting on specific immune cells, it must fulfill the basic criteria for neurotransmission required of all innervated structures, whether they be central neurons or periperal targets of the autonomics: (1) nerve fibers that possess or synthesize the neurotransmitter must be present in the region of the target cells, and must be compartmentalized in such a manner to make the neurotransmitter potentially available; (2) the neurotransmitter must be released from the nerve fibers; (3) the target cells must possess receptors for the neurotransmitter and should have demonstrable second messenger links or intracellular responses to the neurotransmitter; and (4) the target cell should demonstrate predictable responses to manipulation of the neurotransmitter and its receptors. These criteria appear to be fulfilled, at least in a minimal fashion, for NE as a neurotransmitter interacting with lymphocytes or other cell types in lymphoid tissue (see Felten et al.[13] for a detailed discussion of these criteria). Such criteria are only partially fulfilled for other neurotransmitters. For example, some immune cells possess receptors for a neurotransmitter that is not present or released in the region where that cell usually resides (e.g., muscarinic and dopaminergic receptors on splenic or peripheral blood lymphocytes). This may reflect an actual interaction of neurotransmitter and receptor only in certain sites to which the cell must migrate, or only at certain times during ontogeny or other periods in the life of the immune cell. In other cases, lymphocytes may respond *in vitro* to a neurotransmitter, but no classical neurotransmitter receptors can be identified on the target cells (e.g., serotonin). Experimental work has shown the presence of receptors and the modulation of specific immune functions to some neuropeptides, such as substance P,[14-16] somatostatin,[14] and VIP.[17-22] Ottaway et al.[22] have shown vasoactive intestinal peptides (VIP) fibers near postcapillary venules in the gut, a possible source for VIP interacting with receptors on lymphocytes migrating to this tissue. The demonstration of release and availability is established less solidly. While it is highly likely that these neuropeptides will fulfill the criteria for neurotransmission in specific sites in the immune system and also participate in inflammatory responses, the systematic demonstration of the fulfillment of these criteria for neurotransmission is a major challenge in the field of neural-immune interactions. In the case of putative neurotransmitters where the sources, availability, and specific receptor interactions are unclear or are lacking (e.g., some of the opioid peptides, ACh), utmost caution is needed.

# III. PERIPHERAL INNERVATION OF PRIMARY LYMPHOID ORGANS

Noradrenergic nerve fibers innervate the bone marrow and thymus. Calvo[23,24] demonstrated nerve fibers in the substance of the bone marrow, with terminals ending among hemopoietic elements, using neurohistological and electron microscopic observations. Using fluorescence histochemistry for catecholamines, we found plexuses of noradrenergic fibers (Plate 2, Part 1) traveling with the vasculature in bone marrow, with some fibers distributing into the parenchyma.[25,26] It is not clear whether all of the fibers observed by Calvo were noradrenergic; it is possible that other neurotransmitters may be either colocalized in the noradrenergic terminals or may be present in independent nerve fibers. Scattered evidence suggests that key functions in bone marrow, such as proliferation, differentiation, and migration of mature elements may be subject to noradrenergic modulation (see Felten et al.[13] for discussion).

Noradrenergic fibers derived from the superior cervical ganglion[27,28] innervate the thymus gland. Fibers enter the thymus with the vasculature, distribute with the capsular/septal system, and enter the thymic cortex both in association with the vasculature and as free nerve fibers.[25-27,29-31] Noradrenergic fibers distribute throughout the cortex (Plate 2, Part 2) and associate with the vessels near the cortico-medullary junction. We have observed the conservation of this compartmentation in the face of involution with age. The noradrenergic fibers remain only in the cortex in aged rats, but appear to be more densely clustered than in their younger counterparts, suggesting a maintenance of noradrenergic innervation in a diminished volume of tissue. Some functional studies, such as those by Singh and Owen,[32] suggest that proliferation, differentiation, and emigration of thymocytes may be subject to sympathetic noradrenergic modulation (see Felten et al.[13] for discussion).

# IV. PERIPHERAL INNERVATION OF SECONDARY LYMPHOID ORGANS

## A. INNERVATION OF THE SPLEEN

Noradrenergic fibers arising from the superior mesenteric/coeliac ganglion, which is supplied by preganglionic fibers from the T6 to T12 intermediolateral cell column of the spinal cord, innervate the splenic white pulp.[25,26,29,30,33-39] The noradrenergic fibers travel with the splenic nerve into the hilar region of the spleen, distribute with the vasculature and with the capsular/trabecular system of smooth muscle, and innervate specific compartments of the white pulp. In the white pulp, fibers follow the central arteriolar system, and further branch into the parenchyma (Plate 2, Part 3). We used fluorescence histochemistry and neurochemical analysis to identify the noradrenergic innervation, and then utilized double-label immunocytochemical methods to identify tyrosine hydroxylase (TH)-positive (noradrenergic in the periphery) fibers intermingled with specific immune cell types in the splenic white pulp. The TH-positive fibers were found in the following compartments: (1) in the periarteriolar lymphatic sheath (PALS) among T lymphocytes of both major subtypes, T-helper and T-suppressor cells (Plate 2, Part 4); (2) along the marginal sinus, where antigen-presenting (ED-3 positive) macrophages are found (Plate 2, Part 5); (3) within the marginal zone itself, populated by macrophages and B lymphocytes; and (4) in parafollicular sites along the outer edge of clusters of B lymphocytes. In addition, some fibers may distribute within the follicles, although the abundance of these fibers may vary with the state of maturation during ontogeny, or with the degree of activation in adults.

Felten and Olschowka[34] examined the TH-positive terminals in the PALS using electron microscopic immunocytochemistry. As expected, terminals were found near smooth muscle cells of the central arterioles, separated by a basement membrane and often by cell processes. However, additional TH-positive nerve terminals were found directly contacting lymphocytes (Figure 1), both near the smooth muscle cells and in deeper parts of the PALS, distant from smooth muscle. These terminals formed smooth contacts, sometimes indented into the lymphocyte membrane, separated by 6 nm. Our previous finding of micromolar concentration of NE in the spleen from *in vivo* dialysis measurements[40] suggested the paracrine presence of NE in the splenic microenvironment. These direct synaptic-like contacts between TH-positive nerve terminals and lymphocytes suggest the possibility of an additional directed synaptic interaction.

We have immunocytochemical evidence for the presence of substance P and neuropeptide Y-containing fibers in the spleen. The distribution of neuropeptide Y closely matches the distribution of TH (and NE with fluorescence histochemistry), raising the possibility of colocalization which has been shown elsewhere in the periphery. Substance P-positive fibers are distributed along the central arteriolar system and in the PALS. Substance P has been identified as a neurotransmitter in primary sensory neurons, particularly small unmyelinated nociceptive fibers; this possibility in lymphoid organs must be substantiated with tracing studies and ablation studies. As is the case with most neuropeptides, additional experimental evidence for presence and compartmentation in nerve fibers is needed, but has not been easy to obtain due to difficulty in interpreting immunocytochemical findings in the spleen. Dendritic cells and other immune cells cross-react with many antibodies to neuropeptides, producing a pattern of staining deceptively similar to nerve profiles, and the presence of endogenous peroxidase prevents the use of simple immunocytochemical approaches that usually are used in brain. Thus, confirmation of even the most obvious immunocytochemical findings, using tracing studies, ablation studies, and neurochemistry, are needed for conclusive acceptance of neuropeptide presence and compartmentation of in nerve fibers within lymphoid organs.

## B. INNERVATION OF LYMPH NODES

Noradrenergic nerve fibers derived from ganglion cells in the sympathetic chain or in collateral sympathetic ganglia travel into the hilar region of lymph nodes with the vasculature, distribute with the vasculature, and arborize within specific lymphoid compartments.[25,26,41,42] The fibers travel through the medullary cords alongside the medullary sinuses (Plate 2, Part 6), extend past the cortico-medullary junctions, and arborize within the paracortical region of T lymphocytes. Additional noradrenergic fibers distribute with a capsular/subcapsular system and arborize within the cortical region of T lymphocytes (Plate 2, Part 7). This pattern of innervation is similar to its counterpart in the spleen, with innervation of the following functional zones: (1) sites of lymphocyte entry (marginal zone of spleen, cortico-medullary junction of lymph nodes (LN)); (2) site of antigen capture and processing (marginal zone and sinus of spleen, subcapsular sinus of LN); (3) site of antigen presentation and activation of T lymphocytes (PALS of spleen, paracortex of LN); (4) site of antigen presentation and activation of B lymphocytes (parafollicular zone and marginal zone of spleen, medullary cords of LN); and (5) site of lymphocyte egress (outer marginal zone of spleen, medullary sinus of LN). We know little about neuropeptides in lymph nodes. Functional evidence for effects of NE and some peptides has been obtained, but evidence for other criteria for neurotransmission is sorely lacking.

## C. INNERVATION OF GUT-ASSOCIATED LYMPHOID TISSUE (GALT)

Noradrenergic fibers arising from mesenteric ganglia distribute to the gut and follow a specific pattern of innervation of GALT. The most extensive studies of NE innervation were carried out in the rabbit appendix[43] and related tissue (sacculus rotundus, Peyer's patches). The fibers enter the gut at the mesentary, travel to an internodular region with the muscular layers, then turn radially and travel toward the mucosal surface in a narrow plexus between the nodules. No fibers are observed to distribute into the nodules along this course. The noradrenergic fibers travel directly through the T-cell zones and enter the lamina propria, where they arborize extensively among lymphocytes, macrophages, enterochromaffin cells, mast cells, and plasma cells (Plate 2, Part 8).

The enteric nervous system is innervated abundantly by neuropeptide-containing neurons;[44] however, the presence of these peptides in GALT, and potential interactions with immunocytes, remains largely unknown. Ottaway et al.[22] found VIP-containing nerve fibers adjacent to the postcapillary venules, sites where T lymphocytes enter the GALT. The presence of VIP receptors on those lymphocytes appears to be important to their migration into GALT.[20] Stead and colleagues[45] found a close association between substance P and calcitonin gene-related peptide-positive nerve fibers in the gut and mast cells, suggesting a possible neural interaction with these cells of the immune system.

## V. A POTENTIAL ROLE FOR NOREPINEPHRINE IN LYMPHOID TISSUE

NE is present in nerves in specific compartments of both primary and secondary lymphoid organs. Micromolar concentrations of NE have been detected by *in vivo* dialysis,[40] supporting the earlier demonstration of release of NE from the splenic nerve.[46] Numerous studies have reported adrenergic receptors on lymphocytes, monocytes, and neutrophils.[13,47,48] Typically, beta$_2$-adrenergic receptors are found on B and T lymphocytes; the existence of alpha-adrenergic receptors on these cells has not been demonstrated conclusively. Catecholamine interactions with these receptors activates the production of cyclic AMP, a classic second messenger system known to participate in a variety of lymphocyte functions.[48] Thus, a functional role for NE in the regulation of immunological processes is highly likely.

Work from Livnat's laboratory,[13,25,26,41,49,50] utilizing the experimental approach of chemical sympathectomy, has documented extensive functional alterations in immune responses following denervation. In several strains of mice, sympathectomy diminished primary antibody responses by as much as 80% and 97% in spleens and lymph nodes, respectively, and suppressed the secondary antibody response as well. T-cell-mediated responses, such as delayed hypersensitivity to epicutaneous immunization and cytotoxic T-lymphocyte responses to alloantigens, were reduced by 50% or more in denervated mice. In contrast, proliferation, and possibly differentiation, of lymphocytes (mainly B cells) in lymph nodes, in the absence of immunization, was markedly stimulated by sympathectomy. Furthermore, natural killer (NK) cell activity in the spleen and lungs was augmented following denervation. These findings belie simple interpretation. The apparent permissive role of the sympathetic nervous system for responses to antigen is supported by the *in vitro* observations of a stimulatory effect of NE (and beta$_2$-agonists) on primary cytotoxic T-lymphocyte[49-51] and primary antibody responses.[52] However, this interpretation does not support the observation that surgical sympathectomy is followed by enhancement of the immune response.[53-55] The use of surgical denervation introduces additional complications, such as the removal of all ganglion cells (including noradrenergic and peptidergic) and the introduction of surgical stress in both experimental and sham-operated controls that grossly alters the neuroendocrine status of the animal. A more detailed discussion of these issues can be found in Livnat et al.[26] and Felten et al.[13]

NE interacts directly with some cells of the immune system via binding to beta- or alpha-adrenergic receptors. NE may have indirect effects by activating or inhibiting the function and secretory products of other cell types. NE can regulate other physiological activities in the immune system such as lymphocyte traffic, vascular permeability, and blood flow — adding another layer of complexity to its role as an immunomodulator. The response of target cells in the immune system also is complex, with activation, proliferation, differentiation, secretion, or locomotion potentially affected. Since one action of NE in the brain is to modulate the excitability or sensitivity of some neurons to other neurotransmitters, an analogous role can be envisioned in the immune system. NE might be expected to affect the expression of receptors to other immunoregulatory molecules, endocrine hormones, or neurally derived mediators, or to modify postreceptor events via intersecting second messenger systems.

At the very least, NE and other molecules, traditionally known as neurotransmitters, appear to constitute yet another class of immunoregulatory agents. In a manner analogous to neuron-neuron communication in the brain, the peripheral neurotransmitters provide the immune system with a highly sophisticated class of modulatory factors that interact in a complex and integrative fashion with other communication molecules and act as and are influenced by messages from the CNS. Along with pituitary and target organ hormones, the neurotransmitters provide a link that integrates the brain, behavior, and immunity, translating psychosocial and environmental stimuli into chemical signals that impact on immunological function, which we know has a major influence on the health of the organism.

## ACKNOWLEDGMENTS

This work was supported by N00014-84-K-0488 from the Office of Naval Research; R01 NS25223, T32 GM07356, and F32 NS07980 from NIH; R01 MH42073, K01 MH00572, and F31 MH09356 from NIMH; and a MacArthur Foundation Prize Fellowship to D.L.F. Antiserum to ED-3 macrophage antigen was generously supplied by Dr. C. D. Dikjstra, Free University, Amsterdam, The Netherlands. We thank John Housel, Sarah Kennedy, Kristin Gallo, Jackie Quay, Dorothy Herrera, and Nancy Dimmick for excellent technical assistance.

## REFERENCES

1. **Ader, R., Ed.,** *Psychoneuroimmunology,* Academic Press, New York, 1981.
2. **Jankovic, B. D., Markovic, B. M., and Spector, N. H., Eds.,** Neuroimmune interactions, *Ann. N.Y. Acad. Sci.,* 1987, 496.
3. **Guillemin, R., Cohn, M., and Melnechuk, T., Eds.,** *Neural Modulation of Immunity,* Raven Press, New York, 1985.
4. **Goetzl, E. J., Ed.,** Supplement on neuromodulation of immunity and hypersensitivity, *J. Immunol.,* 135(2), 1985.
5. **Brooks, W. H., Cross, R. J., Roszman, T. L., and Markesbery, W. R.,** Neuroimmunomodulation: neural anatomical basis for impairment and facilitation, *Ann. Neurol.,* 12, 56, 1982.
6. **Cross, R. J., Roszman, T. L., and Markesbery, W. R.,** Hypothalamic-immune interactions. Effect of hypophysectomy on neuroimmunomodulation, *J.Neurol. Sci.,* 53, 557, 1982.
7. **Roszman, T. L., Cross, R. J., Brooks, W. H., and Markesbery, W. R.,** Hypothalamic-immune interactions. II. The effect of hypothalamic lesions on the ability of adherent spleen cells to limit lymphocyte blastogenesis, *Immunology,* 45, 737, 1982.
8. **Masek, K., Kadlecova, O., and Petrovicky, P.,** The effect of brain stem lesions on the immune response, in *Advances in Immunopharmacology,* Hadden, J. W., Chedid, L., Dukor, P., Spreafico, F., and Willoughby, D., Eds., Pergamon Press, Elmsford, NY, 1983.
9. **Stein, M., Schleifer, S J., and Keller, S. E.,** Hypothalamic influences on immune responses, in *Psychoneuroimmunology,* Ader, R., Ed., Academic Press, New York, 1981, 429.
10. **Renous, G., Biziere, K., Renoux, M., Guillamin, J. M., and Degenne, D.,** A balanced brain asymmetry modulates T cell-mediated events, *J. Neuroimmunol.,* 5, 227, 1983.

11. **Ader, R., Cohen, N.,and Grota, L. J.**, Adrenal involvement in conditioned immunosuppression, *Int. J. Immunopharmacol.*, 1, 141, 1979.
12. **Keller, S. E., Weiss, J. M., Schleifer, S. J., Miller, N. E., and Stein, M.**, Stress-induced suppression of immunity in adrenalectomized rats, *Science*, 221, 1301, 1983.
13. **Felten, D. L., Felten, S. Y., Bellinger, D. L., Carlson, S. L., Ackerman, K. D., Madden, K. S., Olschowska, J. A., and Livnat, S.**, Noradrenergic sympathetic neural interactions with the immune system: structure and function, *Immunol. Rev.*, 100, 225, 1987.
14. **Payan, D. G. and Goetzl, E. J.**, Modulation of lymphocyte function by sensory neuropeptides, *J. Immunol.*, 135, 783s, 1985.
15. **Stanisz, A. M., Scicchitano, R., Dazin, P., Bienenstock, J., and Payan, D. G.**, Distribution of substance P receptors on murine spleen and Peyer's patch T and B cells, *J. Immunol.*, 139, 749, 1987.
16. **McGillis, J. P., Organist, M. L., and Payan, D. G.**, Substance P and immunoregulation, *Fed. Proc.*, 46, 196, 1987.
17. **Danek, A., O'Dorisio, M. S., O'Dorisio, T. M., and George, J. M.**, Specific binding sites for vasoactive intestinal polypeptide on nonadherent peripheral blood lymphocytes, *J. Immunol.*, 131, 1173, 1983.
18. **O'Dorisio, M. S., Hermina, N. S., O'Dorisio, T. M., and Balcerzak, S. P.**, Vasoactive intestinal polypeptide modulation of lymphocyte adenylate cyclase, *J. Immunol.*, 127, 2551, 1981.
19. **O'Dorisio, M. S., Wood, C. L., and O'Dorisio, T. M.**, Vasoactive intestinal peptide and neuropeptide modulation of the immune response, *J. Immunol.*, 135, 792s, 1985.
20. **Ottaway, C. A.**, Evidence for local neuromodulation of T cell migration in vivo, *Adv. Exp. Med.*, 186, 637, 1985.
21. **Ottaway, C. A. and Greenberg, G. R.**, Interaction of vasoactive intestinal peptide with mouse lymphocytes: specific binding and the modulation of mitogen responses, *J. Immunol.*, 132, 417, 1984.
22. **Ottaway, C. A., Lewis, D. L., and Asa, S. L.**, Vasoactive intestinal peptide-containing nerves in Peyer's patches, *Brain Behav. Immun.*, 1, 148, 1987.
23. **Calvo, W.**, The innervation of the bone marrow in laboratory animals, *Am. J. Anat.*, 123, 315, 1968.
24. **Calvo, W. and Forteza-Vila, J.**, On the development of bone marrow innervation in newborn rats as studied with silver impregnation and electron microscopy, *Am. J. Anat.*, 126, 355, 1969.
25. **Felten, D. L., Felten, S. Y., Carlson, S. L., Olschowka, J. A., and Livnat, S.**, Noradrenergic and peptidergic innervation of lymphoid tissue, *J. Immunol.*, 135, 755s, 1985.
26. **Livnat, S., Felten, S. Y., Carlson, S. L., Bellinger, D. L., and Felten, D. L.**, Involvement of peripheral and central catecholamine systems in neural-immune interactions, *J. Neuroimmunol.*, 10, 5, 1985.
27. **Bulloch, K. and Pomerantz, W.**, Autonomic nervous system innervation of thymic-related lymphoid tissue in wild-type and nude mice, *J. Comp. Neurol.*, 228, 57, 1984.
28. **Nance, D. M., Hopkins, D. A., and Bieger, D.**, Re-investigation of the innervation of the thymus gland in mice and rats, *Brain Behav. Immun.*, 1, 134, 1987.
29. **Williams, J. M., Peterson, R. G., Shea, P. A., Schmedtje, J. F., Bauer, D. C., and Felten, D. L.**, Sympathetic innervation of murine thymus and spleen: evidence for a functional link between the nervous and immune systems, *Brain Res. Bull.*, 6, 83, 1981.
30. **Williams, J. M. and Felten, D. L.**, Sympathetic innervation of murine thymus and spleen: a comparative histofluorescence study, *Anat. Rec.*, 199, 531, 1981.
31. **Bulloch, K.**, Neuroanatomy of lymphoid tissue: a review, in *Neural Modulation of Immunity*, Guillemin, R., Cohn, M., and Melnechuk, T., Eds., Raven Press, New York, 1985, 111.
32. **Singh, U. and Owen, J. J. T.**, Studies on the maturation of thymus stem cells. The effects of catecholamines, histamine, and peptide hormones on the expression of T alloantigens, *Eur. J. Immunol.*, 6, 59, 1976.
33. **Felten, D. L., Ackerman, K. D., Wiegand, S. J., and Felten, S. Y.**, Noradrenergic sympathetic innervation of the spleen. I. Nerve fibers associate with lymphocytes and macrophages in specific compartments of the splenic white pulp, *J. Neurosci. Res.*, 18, 28 — 36 and 118 — 121 (color plates), 1987.
34. **Felten, S. Y. and Olschowka, J. A.**, Noradrenergic sympathetic innervation of the spleen. II. Tyrosine hydroxylase (TH)-positive nerve terminals form synaptic-like contacts on lymphocytes in the splenic white pulp, *J. Neurosci. Res.*, 18, 37, 1987.
35. **Ackerman, K. D., Felten, S. Y., Bellinger, D. L., and Felten, D. L.**, Noradrenergic sympathetic innervation of the spleen. III. Development of innervation in the rat spleen, *J. Neurosci. Res.*, 18, 49 — 54 and 122 — 125 (color plates), 1987.
36. **Bellinger, D. L., Felten, S. Y., Collier, T. J., and Felten, D. L.**, Noradrenergic sympathetic innervation of the spleen. IV. Morphometric analysis in adult and aged F344 rats, *J. Neurosci. Res.*, 18, 55 — 63 and 126 — 129 (color plates), 1987.
37. **Carlson, S. L., Felten, D. L., Livnat, S., and Felten, S. Y.**, Noradrenergic sympathetic innervation of the spleen: V. Acute drug-induced depletion of lymphocytes in the target fields of innervation results in redistribution of noradrenergic fibers but maintenance of compartmentation, *J. Neurosci. Res.*, 18, 64 — 69 and 130 — 131 (color plates), 1987.

38. **Felten, S. Y., Bellinger, D. L., Collier, T. J., Coleman, P. D., and Felten, D. L.,** Decreased sympathetic innervation of spleen in aged Fischer 344 rats, *Neurobiol. Aging,* 8, 159, 1987.
39. **Ackerman, K. D., Felten, S. Y., Bellinger, D. L., Livnat, S., and Felten, D. L.,** Noradrenergic sympathetic innervation of spleen and lymph nodes in relation to specific cellular compartments, *Prog. Immunol.,* 6, 588, 1987.
40. **Felten, S. Y., Housel, J., and Felten, D. L.,** Use of in vivo dialysis for evaluation of splenic norepinephrine and serotonin, *Soc. Neurosci. Abstr.,* 12, 1065, 1986.
41. **Felten, D. L., Livnat, S., Felten, S. Y., Carlson, S. L., Bellinger, D. L., and Yeh, P.,** Sympathetic innervation of lymph nodes in mice, *Brain Res. Bull.,* 13, 693, 1984.
42. **Giron, L. T., Crutcher, K. A., and Davis, J. N.,** Lymph nodes — a possible site for sympathetic neuronal regulation of immune responses, *Ann. Neurol.,* 8, 520, 1980.
43. **Felten, D. L., Overhage, J. M., Felten, S. Y., and Schmedtje, J. F.,** Noradrenergic sympathetic innervation of lymphoid tissue in the rabbit appendix — further evidence for a link between the nervous and immune systems, *Brain Res. Bull.,* 7, 595, 1981.
44. **Gershon, M. D.,** The enteric nervous system, *Annu. Rev. Neurosci.,* 4, 227, 1981.
45. **Stead, R. H., Tomioka, M., Quinonez, G., Simon, G., Felten, S. Y., and Bienenstock, J.,** Intestinal mucosal mast cells in normal and nematode-infected rat intestines are in intimate contact with peptidergic nerves, *Proc. Natl. Acad. Sci. U.S.A.,* 84, 2975, 1987.
46. **von Euler, U. S.,** The presence of a substance with sympathin E properties in spleen extracts, *Acta Physiol. Scand.,* 11, 168, 1946.
47. **Hadden, J. W., Hadden, E. M., and Middleton, E., Jr.,** Lymphocyte blast transformation. I. Demonstration of adrenergic receptors in human peripheral lymphocytes, *Cell. Immunol.,* 1, 583, 1970.
48. **Strom, T. D., Lundin, A. P., and Carpenter, C. B.,** The role of cyclic nucleotides in lymphocyte activation and function, *Prog. Clin. Immunol.,* 3, 115, 1977.
49. **Livnat, S., Madden, K. S., Felten, D. L., and Felten, S. Y.,** Regulation of the immune system by sympathetic neural mechanisms, *Prog. Neuro-psychopharmacol. Biol. Psychiatr.,* 11, 145, 1987.
50. **Livnat, S., Eisen, J., Felten, D. L., Felten, S. Y., Irwin, J., Madden, K. S., and Sundaresan, P. J.,** Behavioral and sympathetic neural modulation of immune function, in *Progress in Catecholamine Research: Part A. Basic Aspects and Peripheral Mechanisms,* Dahlstrom, A., Belmaker, R., and Sandler, M., Eds., Alan R. Liss, New York, 1989, 539—546.
51. **Hatfield, S. M., DiMicco, J. A., and Petersen, B. A.,** Adrenergic receptor modulation of the generation of cytotoxic T lymphocytes in vitro, *Proc. 6th Int. Congr. Immunol.,* 475, 1986.
52. **Sanders, V. M. and Munson, A. E.,** Beta-adrenoceptor mediation of the enhancing effect of norepinephrine on the murine primary antibody response in vitro, *J. Pharmacol. Exp. Ther.,* 230, 183, 1984.
53. **Besedovsky, H. O., del Rey, A. E., Sorkin, E., Da Prada, M., and Keller, H. H.,** Immunoregulation mediated by the sympathetic nervous system, *Cell. Immunol.,* 48, 346, 1979.
54. **Braun, M., Alito, A., Baler, R., Romeo, H., and Cardinali, D.,** Effect of the autonomic nervous system on immune responses, *Proc. 6th Int. Congr. Immunol.,* 479, 1986.
55. **Alito, A. E., Carlomagno, M. A., Cardinali, D., and Braun, M.,** Effect of regional sympathetic denervation on local immune reactions, *Fed. Proc.,* 44, 564, 1985.

Chapter 3

# THE INNERVATION OF MAST CELLS

**Ron H. Stead, Mary H. Perdue, Michael G. Blennerhassett, Yasunori Kakuta, Piersante Sestini, and John Bienenstock**

## TABLE OF CONTENTS

# I. INTRODUCTION

In this chapter we hope to persuade the reader that many, if not all mast cells are innervated, i.e., mast cells are closely apposed to varicose axons and that the interchange of chemical messengers occurs at these sites. This may not be surprising to anatomists and morphologists who have documented the microanatomical association of numerous mast cells with nerve bundles for many years.[1,2] In spite of this well described relationship, the potential for direct "neurocrine" communication between individual axons and mast cells has not received much attention, although an "axon-reflex" pathway involving mast cells has been proposed in the skin.[3,4]

As yet, a "normal" function for mast cells has not been identified; they are generally thought of in terms of their involvement in hypersensitivity and allergic states, when recognition of foreign antigens by membrane-bound IgE results in mediator release.[5,6] However, it is noteworthy that Pearse originally included mast cells in the APUD (amine precursor uptake and decarboxylation) system.[7] The biogenic amine-handling capacity of mast cells and their acknowledged paracrine and endocrine functions certainly support their categorization as APUD cells and membership of the diffuse neuroendocrine system.[7,8] This implies a normal homeostatic function for mast cells, like other cells of that system.

Pearse's original hypothesis involved the neural-crest derivation of APUD cells,[9] but this has since been shown to apply only to some members of the family such as adrenal medulla cells and melanocytes; others, such as those in the gastrointestinal tract and pancreas, are now thought not to be neural-crest derived.[10] Notwithstanding this statement, the extensive experimentation showing that mast cells can be derived from the bone marrow and form an integral part of the immune system, has detracted from their candidature as APUD cells.[6] It must, however, be stressed that the enteroendocrine cells of the gastrointestinal epithelium have been shown to be derived from a progenitor which also gives rise to enterocytes per se,[11-13] and the former are still considered to be a major part of the diffuse neuroendocrine system.[14,15] Furthermore, the derivation of mast cells from the bone marrow does not mean that their progenitors do not exist in other tissues. In fact, the gastrointestinal intraepithelial leukocyte population has been shown to have a greater frequency of mast cell precursors than bone marrow, at least in the mouse (Ernst, P. B. and Bienenstock, J., unpublished observations).[16]

Cells of the APUD or diffuse neuroendocrine system are found predominantly in epithelial tissues where they exert their effects through paracrine, endocrine, and neurocrine interactions.[10,15,17,18] The enteroendocrine cells, for example, are thought to sample luminal contents and communicate with adjacent epithelial cells as well as the subjacent capillary and nerve network in the lamina propria.[18] Enteroendocrine cells have been demonstrated in intimate association with subepithelial nerve varicosities and project filopodia between adjacent epithelial cells,[18-20] yet they are also known to migrate through the crypt and villus epithelium.[21] It does not seem unreasonable to us, therefore, for mast cells (which are generally considered to be migratory) to also associate with peripheral nerves and for there to be neurocrine interactions at these sites. In fact, we have demonstrated an intimate microanatomical association of the presumably bone-marrow-derived mucosal mast cell with subepithelial nerves in the gastrointestinal lamina propria, and have suggested an axon-reflex pathway similar to that proposed in the skin.[22]

In the following pages we present a brief review of the evidence to date for the innervation of mast cells in several tissue sites. The majority of this evidence is from the skin and the respiratory and gastrointestinal tracts. We find this particularly intriguing since these are the major body sites at which a challenge with foreign antigens occurs. Thus, we feel that the mast cells in these tissues might act as relays, communicating immunological, mechanical, and chemical stimuli between the outside world and the organism and partaking in the regulation of homeostasis. In this chapter, we also present some of our recent morphological and physiological evidence to support this suggestion in both the gastrointestinal and respiratory tracts. Finally, we describe our most recent experiments involving co-culture of mast cells and explanted nerves and discuss some of our experiments with nerve-growth factor, which has been found to cause both mast-cell degranulation and proliferation.

Before addressing the issue of mast cell/nerve associations, it is important to consider the question of mast-cell heterogeneity, since it is now clear that there are major subclasses of this cell type which respond differently to different secretagogues.

## II. MAST CELL HETEROGENEITY

Since Ehrlich first described a cell distinguished by the presence of metachromatic granules, mast cells have been identified in most tissues and are characterized by their metachromatic staining properties with thiazin dyes, the binding of cationic dyes at low pH, and the presence of high affinity IgE receptors. It is through the binding of IgE to these receptors and the subsequent effects of challenge with antigen, resulting in hypersensitivity reactions and allergy, that these cells are best known.[5,23]

### A. TWO MAJOR SUBCLASSES OF RAT MAST CELLS

Mast cells in different sites are heterogenous with respect to histochemical, biochemical, pharmacological, and functional characteristics.[5,6] Major differences have been demonstrated between mast cells isolated from the peritoneal cavity and those found in the intestinal mucosa in the rat (Table 1). This heterogeneity had been largely overlooked until 1966 when Enerback demonstrated that mucosal mast cells could not be stained following routine formalin fixation.[24] It is now usual to refer to those mast cells found in the intestinal lamina propria as "intestinal mucosal mast cells" (IMMC), although some authors use the term "mucosal mast cells" (MMC), which we feel is an incomplete and inadequate description. The other major subclass of mast cells has been named "connective tissue mast cells" (CTMC), of which the readily isolated peritoneal mast cells (PMC) have been extensively studied. It is clear, however, that mast cells in different connective tissue sites (i.e., skin vs. tongue) need not be the same, nor are the mast cells within any one tissue necessarily homogeneous. It is also worth remembering that IMMC actually reside in the connective tissue element (lamina propria) of the mucosa.

The mucosal mast cells of the rat contain a unique protease, rat mast cell protease II (RMCP-II); whereas the connective tissue mast cells contain rat mast cell protease I (RMCP-I).[5,6] These proteases are preformed and released upon stimulation, along with other preformed or newly synthesized mediators.[5,6] Among the former are histamine and serotonin which have well-described effects on vascular permeability and tissue edema, but are also well known as neurotransmitters. The newly formed products include arachidonic acid metabolites, either the leukotriene derivatives of the lipoxygenase pathway (which predominate in mucosal mast cells) or the prostaglandins derived via the cyclo-oxygenase route (which predominate in connective tissue mast cells).[5,6]

### B. HISTOCHEMICAL HETEROGENEITY

The basis of the histochemical heterogeneity of rat mast cells appears to be the presence of heparin in the connective tissue-type mast cell and its absence, or relative paucity, in the mast cells in the intestinal lamina propria.[25] IMMC have been shown to contain a different proteoglycan, consisting predominantly of chondroitin di-B sulfate in the rat.[26] Consequently, CTMC bind berberine sulfate, while IMMC do not; it is generally considered that CTMC stain red and IMMC blue with a combined Alcian blue and Safranin (AB/S) stain.[27,28]

**TABLE 1**
**Evidence for Rat Mast Cell Heterogeneity[a]**

| | Peritoneal mast cell | Mucosal mast cell |
|---|---|---|
| Morphology | | |
| Size | Large (c. 20μ), uniform | Small (c. 10—12 μ), pleomorphic |
| Nucleus | Unilobed | Uni- or bilobed |
| Granules | Many | Few |
| Granule constituents | | |
| Protease | Type I (RMCPI) | Type II (RMCPII) |
| Proteoglycans | Heparin | Chondroitin di-B Sulfate |
| Histamine | ≥15 pg/cell | <1 pg/cell |
| Serotonin | <1.5 pg/cell[b] | <0.01—0.1 pg/cell[b] |
| IgE | Surface | Surface and cytoplasmic |
| Histochemistry | | |
| Formalin sensitivity | No | Yes |
| Alcian Blue/Safranin | Red[c] | Blue |
| Life span | >40 days | ≤40 days |
| Proliferation | Thymus-independent | Thymus-dependent |
| Secretagogues | | |
| Antigen | + | + |
| Anti-IgE | + | + |
| Compound 48/80 | + + | 0 |
| Bee venom 401 | + + | 0 |
| A23187 | + + | + |
| Con A | + | + |
| Antiallergic compounds | | |
| Cromoglycate | + + | 0 |
| Theophylline | + + | 0 |
| Doxantrazole | + + | + + |
| AH9679 | + + | 0 |
| Enhancement of secretion | | |
| Phosphatidyl serine | + + | 0 |
| Adenosine | + + | + + |

*Note:* Rough indications of the degree of effect of various compounds: + = some effect, + + = greater effect, and 0 = no effect.

[a] Modified from Reference 29.
[b] Serotonin in PMC is age dependent and increases tenfold in MMC during nematode infection. The values for MMC are estimated from Reference 8.
[c] Immature connective tissue mast cells stain blue.

## C. PHARMACOLOGICAL HETEROGENEITY

Pharmacological differences between mucosal and peritoneal mast cells are also well documented in the rat (Table 1).[29] These include differential responsiveness to various secretagogues, such as compound 48/80 and bee venom peptide 401, and to neuropeptides and endorphins. The antiallergic compounds sodium cromoglycate, theophylline, and AH9679 are effective only on connective tissue mast cells and do not stabilize the mucosal cell type.[30] However, doxantrazole, quercetin, and related flavonoids have been shown to be effective in inhibiting histamine release from both cell types.[31]

## D. ACTION OF NEUROPEPTIDES

With regards to the innervation of mast cells, the heterogeneity of this population would seem to be of importance. Experiments conducted by Shanahan et al. in our laboratory[32,33]

**TABLE 2**
**Effect of Neurotransmitters and Nerve Growth Factor on Histamine Release from Rat Mast Cells**

| | Peritoneal mast cells | Intestinal mucosal mast cells |
|---|---|---|
| Somatostatin | ↑ ↑ | – |
| Vasoactive intestinal peptide | ↑ ↑ | – |
| Substance P | ↑ ↑ | ↑ |
| Bradykinin | ↑ ↑ | – |
| Neurotensin | ↑ | – |
| Dynorphin | ↑ ↑ | – |
| β-Endorphin | ↑ | – |
| α-Neoendorphin | ↑ | – |
| Acetylcholine | ↑ ↑ | ? |
| Noradrenaline | ↑ ↓ ? | ? |
| Nerve Growth Factor | ↑ ↑ | (–)[a] |

*Note:* ↑ ↑ = increase in response; ↓ ↓ = decrease in response.

[a] Preliminary experiments in our laboratory have failed to show NGF mediated histamine release from intestinal mucosal mast cells.

and others[34-36] have shown that several neuropeptides are effective connective tissue mast cell secretagogues, but of those tested, only substance P had a significant histamine-releasing effect on mucosal mast cells (Table 2).

## E. ADRENERGIC AND CHOLINERGIC EFFECTS

There is also pharmacological evidence for adrenergic and cholinergic effects on mast cells from different species; however, the data is somewhat conflicting. For example, in the rat, norepinephrine (NE) has been shown to have both enhancing[37,38] and inhibitory[39,40] effects on histamine release. It is assumed that the inhibition is through a β-adrenergic receptor which has been demonstrated on rat peritoneal mast cells;[37,41] although α-adrenergic blockers have been shown to inhibit the NE-induced release of leukotrienes,[38] α-adrenergic receptors have not been shown on mast cells. Contradictory results in this field may represent interspecies and interstrain mast cell heterogeneity or differences in isolation techniques. This has been discussed recently by Botana et al.[37] and Phillips et al.[42] With regard to cholinergic stimulation of mast cells, reports on the probable presence[43-45] and absence[41,46] of cholinergic receptors have been published, and the issue of whether or not acetylcholine (ACh) is a mast cell secretagogue has been debated for some time.[45,47,48] Most recently, Masini et al. demonstrated that the mast cell response to ACh was dependent upon the amount of bound IgE, suggesting that sensitized animals would be far more responsive.[49] This is obviously of importance when considering neurogenic activation of this cell type in an inflammatory state. Conversely, the different mediators produced by mast cell subtypes may also be of significance in terms of their action on adjacent axons or their interaction with other cell types.

## F. GROWTH FACTORS

Mast cell numbers are controlled by a complex series of regulatory molecules.[5,6,50] The evidence in the mouse is that the so-called bone-marrow-derived and mucosal mast cells depend on interleukin-3 (IL-3) secreted by T-cells for their growth and survival.[51] In the rat, IL-3 or an IL-3-like molecule secreted by stimulated T-cells is thought to be responsible for the mucosal mastocytosis following *Nippostrongylus brasiliensis* infection.[52] However, it should be noted that neonatal nude rats and mice contain normal numbers of mucosal mast cells.[53] It is apparent, therefore, that at least in early development, other factors may be involved in the maintenance of this mast cell subtype. Connective tissue mast cells further require other factors such as IL-4 for their growth.[54,55] As yet undefined molecules produced by fibroblasts may also be important in the development of this mast cell subset.[56,57]

Most of the above evidence pertains to two major subclasses of mast cells now well established in the rat. For this reason, we have chosen to perform our experiments on the interactions of mast cells and nerves in this species. It must be realized, however, that the definition of only two mast cell types totally understates the complexity of the metachromatic granulated cell population, which also includes basophils. It is also not acceptable to assume equivalent heterogeneity in other species, although there is now similar evidence in mice and humans.[5,6]

## III. EVIDENCE FOR THE INNERVATION OF MAST CELLS

### A. SKIN

#### 1. Morphological Evidence

The presence of mast cells in peripheral nerves has been known for many years and has been reviewed by Olsson.[1,2] It is also well established that mast cells are present in benign nerve sheath tumors,[58] including neurofibromata,[59] and in amputation neuromas in humans[2] and mice.[60] Mast cells are also found in glomus tumors where they have been observed closely apposed to unmyelinated nerve fibers and sometimes exhibit various stages of degranulation.[61,62] In the study of Weisner-Menzel et al., it was the morphologically immature-appearing mast cells which were closest to the nerves.[62] It is also interesting that abundant substance P-containing nerves have been identified in glomus tumors.[63] Furthermore, mast cells have been described close to nerves in normal skin.[64,65]

#### 2. Functional Evidence

The injection of substance P into skin mimics the response of the local injection of histamine, or the mast cell degranulating agent, compound 48/80.[4,66-69] This effect is inhibited by antihistamines,[4,66,67] but is not blocked by pretreatment with capsaicin.[70] Vasoactive intestinal peptide, neurotensin, and secretin are also effective in causing the wheal and flare reaction in skin and can be blocked by the antihistamine mepyramine.[71] Piotrowski and Foreman recently added somatostatin and calcitonin gene-related peptide to the list of neuropeptides capable of effecting histamine release from skin mast cells *in situ*.[69,72]

Other *in vivo* experiments have shown that antidromic nerve stimulation can cause mast cell degranulation in the skin,[73,74] and that the concomitant neurogenic edema can be reduced by local pretreatment with compound 48/80.[4,70,74,75] Systemic pretreatment with capsaicin has also been shown to be effective in reducing the edema following the topical application of benzene,[74] as well as that resulting from antidromic stimulation,[4,74] implicating substance P-containing nerves in the response. Collectively, this evidence supports the hypothesis proposed by Kiernan[3] and Lembeck and Holzer[4] that mast cells and nerves (probably substance P-containing) are involved in an axon reflex responsible for the flare reaction to noxious stimuli seen in the skin.

### B. GASTROINTESTINAL TRACT

#### 1. Morphological Evidence

Stach[76] and Heine and Forster[64,65] previously noted that mast cells were associated with nerves in the gastrointestinal tract, and Newson et al. provided ultrastructural evidence for bouton formation between mast cells and enteric nerves.[77] Additionally, mast cells have been observed in contact with nerves in the gastric glands of the opossum[78] and in the rat

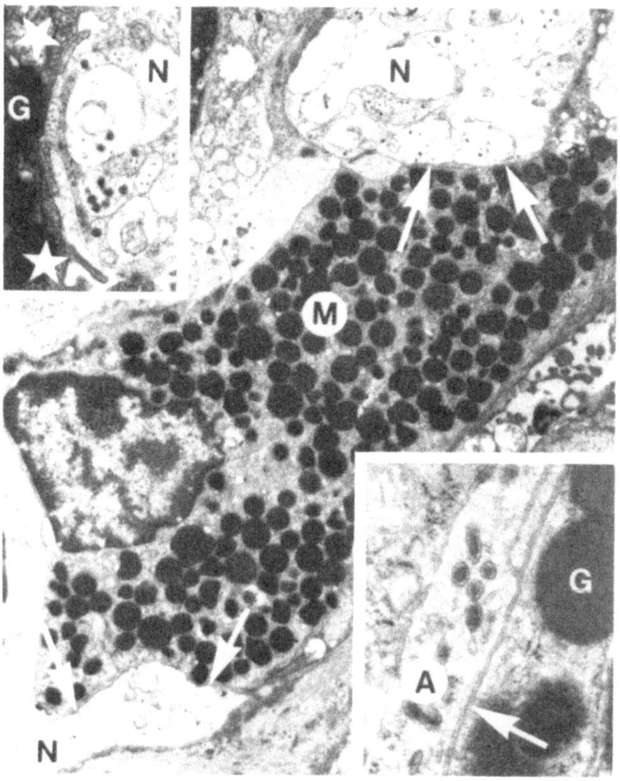


FIGURE 1. The intimate association of a mast cell (M) and two unmyelinated nerves (N) is clearly illustrated in this electron micrograph. (Magnification × 5400.) Upper inset shows an IMMC with its characteristic granules (G) and lamellopodia (★) projecting around a nerve. (Magnification × 14,000.) Lower inset illustrates membrane-membrane contact of an IMMC and an axon (A) containing dense-core vesicles. (Magnification × 27,000.) Arrows indicate mast cell-nerve apposition. (Reproduced with permission from Stead, R. H. et al, *Proc. Natl. Acad. Sci. U.S.A.*, 84, 2975, 1987.)

colon.[79] In the human, Yonei has demonstrated a similar association in the colonic mucosa in ulcerative colitis.[80] In the last study, the mast cells were often degranulated, and the author suggests that excessive cholinergic stimulation, resulting in mast cell activation, might be involved in the exacerbation of the disease.

In our own studies, we have demonstrated that the majority of intestinal mucosal mast cells are associated with subepithelial nerves in the lamina propria of the rat gastrointestinal tract.[22] In animals infected with the nematode *Nippostrongylus brasiliensis*, 67% of the mast cell population were touching nerves, as demonstrated by a combined immunocytochemical technique for neurone specific enolase and Alcian blue staining of mast cells. We also demonstrated that 61% of the mast cell population were in contact with nerves exhibiting substance P-like immunoreactivity, and that some calcitonin gene-related peptide-containing nerves were touching mast cells. We confirmed these findings using the electron microscope, which revealed intimate membrane-membrane contact between 8% of the mast cells in infected rats and unmyelinated axons containing 70 to 170 nm-dense core vesicles, suggesting that they were peptidergic (Figure 1). An additional 30% of the mast cells were within 250 nm of nerves, while other mast cells appeared to embrace nerve bundles with the projection of lamellopodia.

**2. Functional Evidence**

Bani-Sacchi and colleagues have recently shown that field stimulation of the rat ileum results in histamine release and a diminution in mast cell granule metachromasia.[81] They also demonstrated an increased release of ACh and that both atropine and tetrodotoxin depressed these responses, suggesting that ACh released by the field stimulation in turn evoked the secretion of histamine from mast cells.

Other studies by Ganguly and colleagues have shown that mast cells in the rat stomach increase in granularity following sectioning of the vagus nerve;[82] the stress of pylorus ligation causes a decrease in tissue histamine (possibly mast cell derived) which can be prevented by vagotomy.[83] This information suggests that mast cells might be under tonic influence which would support a role for this cell type in the maintenance of homeostasis in the gastrointestinal tract.

Some of our earlier studies, and those of others, have shown that mast cell mediators alter epithelial transport function.[84-88] In Ussing chamber studies, using intestine either from rats sensitized to ovalbumin[85,86] or previously infected with *Nippostrongylus brasiliensis*,[87] the short circuit current (Isc) increased in response to challenge with antigen, indicating changes in net ion transport. Similar experiments using rat or guinea pig intestine from animals infected with *Trichinella spiralis* have demonstrated that mucosal mast cell mediators such as histamine, serotonin, and arachidonic acid metabolites were involved.[88,89] In our studies, the number of stainable mucosal mast cells and histamine levels in the mucosa were also reduced by antigen challenge,[84] and the abnormalities were prevented by doxantrazole,[85,90] which stabilizes both mucosal and connective tissue mast cells,[31] but not by sodium cromoglycate,[90] which has no protective effect on mucosal mast cells.[10] We have also provided evidence that ACh[91] and substance P[91,92] are involved as functional neurotransmitters in the regulation of ion transport in the small intestinal mucosa. More recently, we, and others, have shown that the nerve-blocking drug tetrodotoxin (TTX) reduced the ion transport abnormalities produced in isolated jejunum from rats sensitized and exposed to antigen.[86,88,93] This would suggest that at least some of the changes in epithelial function seen after antigen challenge are mediated via the intestinal nerves.

Additional exciting information has been obtained from our most recent studies.[94,95] Rats were infected with *Nippostrongylus brasiliensis* and studied at various stages of inflammation during the primary infection at the following days: 0 (control), 7 and 10 (acute), and 35 (post). Changes in crypt length and mucosal mast cell numbers were documented. Additionally, Isc responses to electrical transmural stimulation of enteric nerves and to histamine were measured. During acute inflammation at days 7 and 10, the response to transmural stimulation was significantly decreased. At this time the numbers of visible (stained) mucosal mast cells were also decreased, probably indicating that they had degranulated in response to the parasites. A mucosal lesion involving crypt hyperplasia was also evident at day 7, and both crypt hyperplasia and villus atrophy were present at day 10. The reduced responses to transmural stimulation were not due to an inability of the tissue to respond to secretory stimuli, since responses to added histamine (which are not altered in the presence of TTX) were greater during inflammation, which may have reflected the increase in secretory crypt epithelium bulk. This data suggests that mast cell degranulation and reduced responsiveness of enteric nerves are related. Two possibilities exist: (1) neurotransmitters are depleted, possibly by the action of mast cell mediators; and/or (2) nerves have actually been damaged by the same mechanism which caused the mucosal injury. At the postinfection stage (day 35), mucosal mast cell numbers were dramatically increased, and an increased short-circuit current to worm antigen was evident. At that time, the responses to transmural stimulation were also increased. These results may reflect hypertrophy of the intestinal nerves or increased activity.

Overall, the above findings suggest that the interaction of mast cells and nerves in the gastrointestinal mucosa is important in the regulation of epithelial transport function.

## C. RESPIRATORY TRACT

### 1. Morphological Evidence

We are unaware of any published information on the microanatomical association of mast cells and nerves in the respiratory tract. However, we are just completing an extensive study of the rat lung and trachea.[96] In animals given four infections of *Nippostrongylus brasiliensis* to induce mast cell hyperplasia in the lung, we observed approximately one-third of the AB-stained mast cell population in direct association with neurone specific enolase-containing nerves. We also examined noninfected animals and rats given two challenges of the nematode. In the noninfected animals, 22% of the mast cells were closely apposed to nerves, and in the animals killed after secondary infection, 29% of the mast cells were in association with nerves. In the trachea, the percentage of mast cells associated with nerves was lower: in noninfected animals, 15%, and in infected animals, 11% of the intraepithelial mast cells were touching nerves. Notwithstanding these figures, it is clear that many mast cells in the trachea are situated between epithelial cells at which sites both substance P and calcitonin gene-related peptide-containing nerves can be found.[97,98] Clearly, the microanatomical basis for communication exists.

### 2. Functional Evidence

There is considerable functional evidence for the innervation of mast cells in the respiratory tract. First, direct stimulation of the vidian nerve in patients with chronic hypertrophic nonallergic rhinitis causes morphological mast cell changes.[99] In such experiments, Masini et al. have shown that there is a significant decrease in tissue-based histamine and concomitant degranulation of mast cells. This resulted in vasodilation and enhanced nasal secretion.

In the trachea, we have investigated the effects of antigen on epithelial function using *in vitro* studies in an Ussing chamber.[100] Rats were immunized with egg albumin in an adjuvant consisting of pertussis and aluminum hydroxide to promote IgE antibody formation. The subsequent addition of antigen to the luminal surface of the isolated tracheas resulted in an increase in short-circuit current, whereas there was no effect in control, unsensitized animals. The changes were abolished by doxantrazole, but not by sodium cromoglycate, suggesting involvement of mucosal mast cells. To investigate whether substance P-containing nerves were involved, animals were treated neonatally with capsaicin (which depletes these nerves) and subsequently sensitized to ovalbumin using the same procedure. In both, the capsaicin-treated and control animals, substance P produced an increase in short-circuit current of the same magnitude, and capsaicin itself produced an increase in control animals, but not in the capsaicin-treated group. When antigen was added to the luminal side of tracheas from sensitized, capsaicin-treated animals, the change in short-circuit was approximately 50% of the change seen in untreated, sensitized control animals. The levels of IgE antibody were not significantly different between the groups of animals nor were the numbers of mast cells in the tracheal segments. These data provide clear evidence for the interaction of mast cells and nerves in the homeostasis of the tracheal epithelium.

In the lung, Leff and colleagues have shown that vagal stimulation causes an enhancement of the secretion of histamine from mast cells after challenge with ascaris antigen in natively allergic dog lung.[101] Other studies have shown changes in permeability following vagal stimulation. Accordingly, we investigated the effect of inhaled antigen in rats sensitized as those in our tracheal studies.[102] Lung epithelium permeability (LEP) was assessed by clearance of an aerosolized radiolabeled molecular probe, 99m technetium diethylenetriamine pentaacetic acid ($^{99M}$Tc-DTPA). The results were expressed as percent clearance from the lungs after 7.5 min (LEP 7.5). Sensitized rats exhibited an approximately threefold increase in LEP 7.5, 20 min after exposure to antigen. Permeability remained significantly elevated 4 h after exposure and returned to normal after 24 h. These changes were not observed in sensitized animals exposed to aerosolized phosphate buffered saline nor in unsensitized, challenged controls. We further confirmed the antigen-specific nature of the increased permeability since no changes were observed when bovine serum albumin was used in place of egg albumin. In addition, capsaicin-treated, sensitized rats showed a smaller change in lung permeability compared with sensitized animals that had received no capsaicin. As in our trachea experiments, mast cell numbers and IgE levels were not significantly different. These data demonstrate a functional consequence of the observations of close association of mast cells and nerves in the lung.

### D. OTHER TISSUE SITES

Additional morphological evidence for the innervation of mast cells has been provided by several groups. Skofitsch et al. demonstrated the close association of substance P-containing nerves and mast cells in the diaphragm and mesentery;[103] Dimlich has shown that some mast cells in the portal tracts in liver are apposed to nerves;[104] and Heine and Forster have shown similar associations in the myocardium.[64,65] Most recently, the intimate membrane/membrane association of type 2 cerebral mast cells with nerves was demonstrated by Dimitriadou et al.[105]

## IV. TISSUE CULTURE STUDIES

Because of the evidence detailed above, we have begun to study whether or not mast cells associate with nerves in culture. In this way we hope to control and determine the conditions under which such interactions occur, and identify the factors involved. In our initial experiments, we have made preparations of superior cervical ganglia from neonatal mice and have cultured them either as explants or as dissociated neuronal cultures after enzymatic digestion. These preparations are grown in coated 35-mm tissue culture dishes in the presence of nerve growth factor (NGF), where rapid neurite outgrowth occurs within 24 h. The majority of our experiments to date have involved the rat basophilic leukemia cell line, RBL-2H3, which has been maintained under standard tissue culture conditions in 10% fetal calf serum. This cell type is considered to be analogous to mucosal mast cells, since they are similar in terms of their proteoglycan and protease enzyme content,[106] and RBL cells have been used widely in studies of the mast cell stimulus-secretion coupling mechanism.[107]

Upon co-culturing RBL cells with nerves, neurite extensions were directed toward the RBL over large intervening distances. These neurites formed contacts with the RBL and branched while maintaining their connections. Time-lapse photography suggests that chemotaxis may be involved in forming these contacts, which are then maintained for periods from 8 to 96 h.[108,109]

An electrophysiological investigation is underway to determine the functional consequences of contact between nerves and RBL cells.[108,110] The membrane potential and input resistances of individual cells have been measured and compared among RBL, either in contact with neurites, or without nerve contact, in cocultures. No differences in membrane potential were detected, with all coculture values not significantly different from RBL grown alone in culture (control value of $-11.2 \pm 0.3$ mV). However, the input resistance of RBL cells in contact with neurites was on average 57% lower than the intrinsic control values ($p < 0.05$) which were the same as the control average values of $62 \pm 3$ MΩ. Therefore, neurite contact increased the conductance of the RBL membrane which can be interpreted as representing increased membrane permeability.

Since we have shown previously that substance P can effect histamine release from mucosal mast cells,[32] and that substance P-containing nerves are associated with mast cells in the rat intestine,[22] we examined the effect of this neuropeptide on membrane potential and input resistance in RBL. The results were analogous to those determined from RBL in contact with neurites; there was no effect on membrane potential, but input resistance was lowered to an equivalent degree. Since neither ACh nor noradrenaline affected the membrane resistance values, this effect may be specific to substance P. This gives support to our hypothesis of a functional mast cell/nerve unit.[110]

Careful ultrastructural examination of the cocultured mast cells and nerves showed that there was intimate membrane-membrane contact between mast cells and nerves.[111] Furthermore, there was some specialization of the cell membranes in the regions of contact, and dense core neurosecretory vesicles were present in the neurites at the points of contact. This strongly resembles our previous findings in the rat nematode model of mast cell hyperplasia.[22]

## V. THE ACTION OF NERVE GROWTH FACTOR (NGF) ON MAST CELLS

In view of the striking microanatomical relationships between mast cells and nerves and the supporting physiological evidence for communication between these two cell types, we wondered if any of the growth factors known to promote or sustain nerve growth might be involved in the regulation of mast cells. A search of the literature revealed that NGF caused both mast cell degranulation *in vitro*[112-114] and increased mast cell numbers *in vivo*.[115] We chose to further investigate these effects and found that in addition to causing phosphatidylserine-dependent histamine release from mast cells, 2.5 S NGF induces enhancement of antigen-mediated histamine release in the absence of exogenous phosphatidylserine.[116] This effect was not dependent upon the continued presence of NGF, since preincubation for 5 min followed by washing for 1 h resulted in a similar enhancement of antigen-mediated histamine release. We also demonstrated that NGF altered the early kinetics of antigen-mediated histamine release and enhanced the actions of concanavalin A, A23187, and 48/80. Thus, NGF could clearly potentiate the responsiveness of mast cells to challenge *in situ*.

Our *in vivo* studies revealed a marked increase in mast cell numbers following the administration of NGF to neonatal animals during the first 2 weeks after birth.[117] In animals given 5 μg of 2.5 S NGF per gram body weight for 14 d, connective tissue mast cell numbers increased approximately 3 times, mucosal mast cell numbers increased approximately 6 times, and those in hematopoietic tissues (spleen and liver) increased up to 63 times. The hyperplastic mast cell populations did not differ phenotypically from those normally found in the tissue sites studied, except for the presence of an increased number of blue staining cells in AB/S stained slides, indicating the presence of a higher precentage of immature cells. In all animals studied, only the connective tissue sites contained mast cells, which stained red in the AB/S sequence, fluoresced with berberine sulfate (indicating the presence of heparin), or exhibited immunoreactivity with anti-RMCPI in an immunoperoxidase technique. Conversely, mucosal mast cells only stained blue, did not fluoresce with berberine sulphate, and only reacted with anti-RMCP-II. In the spleen and liver, the hyperplastic mast cell population contained both RMCP-I and RMCP-II immunoreactive cells, although these were all stained blue in the AB/S sequence and did not react with berberine sulphate. Additional experiments involving the administration of tritiated thymidine at various times during the NGF treatment revealed labeled mast cells by autoradiography. This indicated that NGF caused mast cell proliferation and not just differentiation. Furthermore, athymic animals given NGF also had mast cell hyperplasia, including cells in the intestinal lamina propria, suggesting that immunocompetent T lymphocytes were not necessary at this stage in the animals' development.

In the human, Dr. Matsuda in our laboratory has shown that NGF significantly enhances the growth of colonies in methylcellulose cultures of peripheral blood.[118] This effect was particularly pronounced on colonies of the Eo (eosinophil/basophil) type and was abrogated by depletion of T-cells. There are several other recent reports which suggest that NGF might

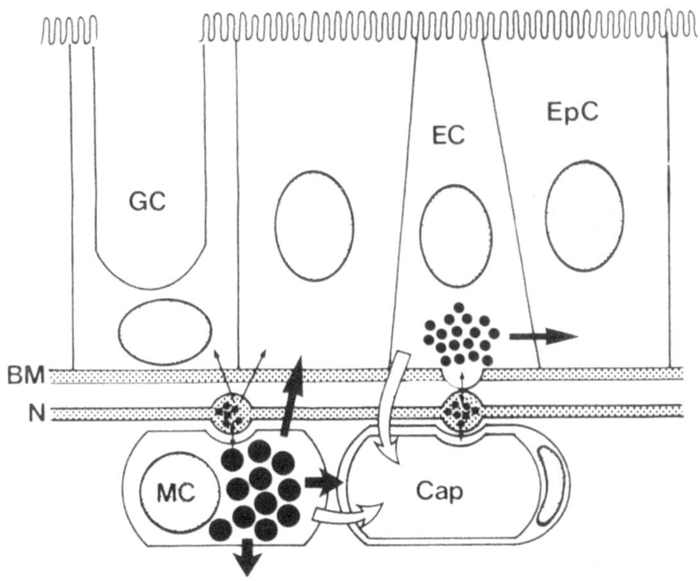


FIGURE 2. Schematic representation of an epithelium, including epithelial cells (EpC), an enteroendocrine cell (EC), a goblet cell (GC), and basement membrane (BM), with subjacent nerve (N), mast cell (MC) and capillary (Cap). The neurocrine (fine arrows), paracrine (bold arrows), and endocrine (open arrows) interactions of the nerve, mast cell, and the enteroendocrine cell are depicted.

have effects on lymphocytes.[119-121] It is not yet evident, however, if the results we obtained in neontal athymic rats, compared to those seen with human peripheral blood cells, reflect species differences, or are attributable to the developmental stage of the animal.

## VI. DISCUSSION

In the introduction to this chapter we stated that we would try to persuade the reader of the substantial evidence favoring the functional innervation of mast cells. We believe that the evidence presented here documents the microanatomical basis for such an interaction as well as providing the physiological support for communication between these two cell types in both directions. Regardless of their involvement in immunological events (and the fact that at least some mast cell types are derived from the bone marrow), we believe that it is imperative not to conceptually restrict this cell type within the immune system. The fact that mast cells exert their effects through paracrine and endocrine interactions, coupled with the neurocrine role documented herein, make it equally acceptable to categorize mast cells within the diffuse neuroendocrine system (depicted schematically in Figure 2). In this respect, it is interesting that most APUD cells are found within epithelia. Are mast cells their connective tissue counterparts?

We have not discussed the fact that regulatory (neuro-) peptides are produced by many APUD cells (some of which have not yet been shown to have the appropriate amine-handling characteristics).[10,14,15] It is of great interest then that mast cells have previously been shown to produce vasoactive intestinal peptide.[122] Other leukocytes also make different "neuropeptides",[123-125] and lymphocytes have the capacity to synthesize proopiomelanocortin-derived peptides.[126] This last observation has led Blalock to suggest that lymphocytes can act as sensory afferent cells of the central nervous system.[127,128] Both the classical "neuropeptides" and the "interleukins" are clearly "regulatory" peptides. Where, then, do our boundaries of cell classification lie?

As already noted, it is not clear if all types of mast cells are closely associated with nerves and are therefore capable of participating in neurocrine-mediated homeostatic events. However, it is clear that both of the current major subsets of mast cells, CTMC and IMMC, are involved in such interactions, since both morphological and physiological evidence supports this in the skin, respiratory tract, and gastrointestinal mucosa, at least in the rat.

With regard to the types of nerves involved in this association, the morphological studies have revealed mainly a peptidergic nature. However, this only reflects the fact that adequate numbers of studies of the association of parasympathetic and sympathetic nerves with mast cells have not been undertaken. The physiological evidence points to an involvement of both cholinergic and peptidergic neurons. This issue is made more cloudy by the heterogenous response of mast cells to different neurotransmitters. In particular, the conflicting data regarding the response of mast cells to ACh have been difficult to interpret. The recent demonstration that mast cells loaded with IgE are more responsive to ACh may be particularly important, especially in terms of augmentation of an immune response.[49] Furthermore, other recent data have shown that the neuropeptide substance P can alter the expression of ACh receptors on nerves.[129] Could this also be true of the mast cell?

We have already mentioned that the different factors responsible for modulating mast cell growth are not well understood and have been mostly characterized in the mouse. We are particularly intrigued by the possibility that NGF produced at the sites of injury and inflammation might be involved in the regulation of mast cell growth and differentiation at these sites. In fact, NGF is well known to be produced by fibroblasts and has been demonstrated in increased amounts in granulation tissue.[130,131] Recent evidence suggests that monocytes/macrophages could also be involved in the production of NGF at the sites of inflammation, either directly or by the paracrine action of IL-1.[132,133] It is also possible that NGF may be important only in the early development of mast cells, which subsequently switch to dependence on other well known interleukins. In this respect, it is interesting that sensory neurones lose their dependence on NGF earlier than sympathetic nerves.[134]

Other factors with nerve-growth promoting actions have been identified and may be involved in either trophic or tropic interactions with mast cells. For example, Coughlin et al. have identified a factor in conditioned medium from heart cells with NGF-like action.[135] Additionally, laminin, an extracellular matrix component, is able to promote neurite outgrowth[136] and this is exaggerated in the presence of heparin sulfate proteoglycan,[137,138] not too dissimilar from the proteoglycans found in some mast cell subsets. Perhaps, then, degranulation of mast cells may also give rise to changes in the local innervation of a tissue. In this respect, increases in the number of nerves and mast cells have been documented in some inflammatory bowel diseases.[139-142] Are these changes related? We have also alluded to changes in nerve function during inflammation. Are these also due to mast cell degranulation?

The data we have presented here regarding changes in epithelial transport and permeability support the interaction of mast cells and nerves and their involvement in homeostasis at these sites. We suggest that this is the "normal" function of mast cells and that only nature's exquisite manipulation has allowed this cell the adaptability to respond specifically to challenge with foreign molecules by the provision of IgE receptors. Thus, the well recognized disease states associated with mast cells, such as allergies and asthma, might reflect the perturbation of normal control mechanisms. If this is true, the strategies for controlling these disorders will need rethinking.

## ACKNOWLEDGMENTS

The authors would like to thank Mrs. J. Butera, Mrs. B. Lahie, and Miss D. Labonte for secretarial assistance; and the Medical Research Council of Canada and the Canadian Foundation for Ileitis and Colitis for financial support.

## REFERENCES


1. **Olsson, Y.,** Mast cells in human peripheral nerve, *Acta Neurol. Scand.*, 47, 357, 1971.
2. **Olsson, Y.,** Mast cells in the nervous system, *Int. Rev. Cytol.*, 29, 27, 1968.
3. **Kiernan, J. A.,** A pharmacological and histological investigation of the involvement of mast cells in cutaneous axon reflex vasodilation, *Q. J. Exp. Physiol.*, 60, 123, 1975.
4. **Lembeck, F. and Holzer, P.,** Substance P as neurogenic mediator of antidromic vasodilation and neurogenic plasma extravasation, *Naunyn-Schmiedeberg's Arch. Pharmacol.*, 310, 175, 1979.
5. **Lee, T. D. G., Swieter, M., Bienenstock, J., and Befus, A. D.,** Heterogeneity in mast cell populations, *Clin. Immunol. Rev.*, 4, 143, 1985.
6. **Befus, A. D., Bienenstock, J., and Denburg, J. A.,** *Mast Cell Differentiation and Heterogeneity,* Raven Press, New York, 1986.
7. **Pease, A. G. E.,** 5-Hydroxytryptophan uptake by dog thyroid "C" cells, and its possible significance in polypeptide hormone production, *Nature (London),* 211, 598, 1966.
8. **Wingren, U., Enerback, L., Ahlman, H., Allenmark, S., and Dahlstrom, A.,** Amines of the mucosal mast cell of the gut in normal and nematode infected rats, *Histochemistry,* 77, 145, 1983.
9. **Pearse, A. G. E. and Polak, J. M.,** Neural crest origin of the endocrine polypeptide (APUD) cells of the gastrointestinal tract and pancreas, *Gut,* 12, 783, 1971.
10. **Pearse, A. G. E. and Takor, T. T.,** Embryology of the diffuse neuroendocrine system and its relationship to the common peptides, *Fed. Proc.*, 38, 2288, 1979.
11. **Cheng, H. and Leblond, C. P.,** Origin, differentiation, and renewal of the four main epithelial cell types in the mouse small intestine. V. Unitarian theory of the origin of the four epithelial cell types, *Am. J. Anat.*, 141, 537, 1974.
12. **Matsuyama, M. and Suzuki, H.,** Differentiation of immature mucous cells into parietal, argyrophil, and chief cells in stomach grafts, *Science,* 169, 385, 1970.
13. **Ponder, B. A. J., Schmidt, G. H., Wilkinson, M. M., and Wood, M. J., Monk, M., and Reid, A.,** Derivation of mouse intestinal crypts from single progenitor cells, *Nature (London),* 313, 689, 1985.
14. **Andrew, A.,** The APUD concept: where has it led us?, *Br. Med. Bull.*, 38, 221, 1982.
15. **Launay, J. M., Tabuteau, F., Haimart, M., Ganem, Y., Villette, J. M., Dreux, C., Debous-Guillemin, M. C., Roseto, A., and Peries, J.,** The diffuse neuroendocrine (APUD) system, *Biomed. Pharmacother.*, 37, 322, 1983.
16. **Crapper, R. M. and Schrader, J. W.,** Frequency of mast-cell precursors in normal tissues determined by an *in vitro* assay: antigen induces parallel increases in the frequency of P-cell precursors and mast cells, *J. Immunol.*, 131, 923, 1983.
17. **Cooke, H. J.,** Neurobiology of the intestinal mucosa, *Gastroenterology,* 90, 1057, 1986.
18. **Lundberg, J. M., Dahlstrom, A., Bylock, A., Ahlman, H., Pettersson, G., Larsson, I., Hansson, H-H., and Kewenter, J.,** Ultrastructural evidence for an innervation of epithelial enterochromaffine cells in the guinea pig duodenum, *Acta Physiol. Scand.*, 104, 3, 1978.
19. **Wade, P. R. and Westfall, J. A.,** Ultrastructure of enterochromaffin cells and associated neural and vascular elements in the mouse duodenum, *Cell Tissue Res.*, 241, 557, 1985.
20. **Newson, B., Ahlman, H., Dahlstrom, A., Das Gupta, T. K., and Nyhus, L. M.,** On the innervation of the ileal mucosa in the rat — a synapse, *Acta Physiol. Scand.*, 105, 387, 1979.
21. **Cheng, H. and Leblond, C. P.,** Origin, differentiation, and renewal of the four main epithelial cell types in the mouse small intestine. III. Enteroendocrine cells, *Am. J. Anat.*, 141, 503, 1974.
22. **Stead, R. H., Tomioka, M., Quinonez, G., Simon, G. T., Felten, S. Y., and Bienenstock, J.,** Intestinal mucosal mast cells in normal and nematode-infected rat intestines are in intimate contact with peptidergic nerves, *Proc. Natl. Acad. Sci. U.S.A.*, 84, 2975, 1987.
23. **Bienenstock, J. and Befus, D.,** Mast cells in host defense, in *Leukocytes and Host Defense,* Alan R. Liss, New York, 1986, 347.
24. **Enerback, L.,** Mast cells in rat gastrointestinal mucosa. I. Effects of fixation, *Acta Path. Microbiol. Scand.*, 66, 289, 1966.
25. **Yurt, R. W., Leid, R. W., Jr., Austen, K. F., and Sibert, J. E.,** Native heparin from rat peritoneal mast cells, *J. Biol. Chem.*, 252, 518, 1977.
26. **Stevens, R. L., Lee, T. D. G., Seldin, D. C., Austen, K. F., Befus, A. D., and Bienenstock, J.,** Intestinal mucosal mast cells from rats infected with *Nippostrongylus brasiliensis* contain protease-resistant chondroitin sulfate di-B proteoglycans, *J. Immunol.*, 137, 291, 1986.
27. **Enerback, L.,** Berberine sulfate binding to mast cell polyanions: a cytofluorometric method for the quantitation of heparin, *Histochemistry,* 42, 302, 1974.
28. **Wingren, U. and Enerback, L.,** Mucosal mast cells of the rat intestine: a re-evaluation of fixation and staining properties, *Histochem. J.*, 15, 571, 1983.
29. **Shanahan, F., Denburg, J. A., Bienenstock, J., and Befus, A. D.,** Mast cell heterogeneity, *Can. J. Physiol. Pharmacol.*, 62, 734, 1984.

30. **Pearce, F. L., Befus, A. D., Gauldie, J., and Bienenstock, J.,** Mucosal mast cells. II. Effects of antiallergic compounds on histamine secretion by isolated intestinal mast cells, *J. Immunol.*, 128, 2481, 1982.
31. **Pearce, F. L., Befus, A. D., and Bienenstock, J.,** Mucosal mast cells. III. Effect of quercetin and other flavonoids on antigen-induced histamine secretion from rat intestinal mast cells, *J. Allergy Clin. Immunol.*, 73, 819, 1984.
32. **Shanahan, F., Denburg, J. A., Fox, J., Bienenstock, J., and Befus, D.,** Mast cell heterogeneity. Effects of neuroenteric peptides on histamine release, *J. Immunol.*, 135, 1331, 1985.
33. **Shanahan, F., Lee, T. D. G., Bienenstock, J., and Befus, A. D.,** The influence of endorphins on peritoneal and mucosal mast cell secretion, *J. Allergy Clin. Immunol.*, 74, 499, 1984.
34. **Johnson, A. R. and Erdos, E. G.,** Release of histamine from mast cells by vasoactive peptides, *Proc. Soc. Exp. Biol. Med.*, 142, 1252, 1972.
35. **Devillier, P., Renoux, M., Giroud, J.-P., and Regoli, D.,** Peptides and histamine released from rat peritoneal mast cells, *Eur. J. Pharmacol.*, 117, 89, 1985.
36. **Goetzl, E. J., Chernov, T., Renold, F., and Payan, D. G.,** Neuropeptide regulation of the expression of immediate hypersensitivity, *J. Immunol.*, 135, 802S, 1985.
37. **Botana, L. M., Espinosa, J., and Eleno, N.,** Adrenergic activity on rat pleural and peritoneal mast cells. Loss of β receptors during the purification procedure, *Gen. Pharmacol.*, 18, 141, 1987.
38. **Koopman, W. J., Orange, R. P., and Austen, K. F.,** Immunochemical and biologic properties of rat IgE. III. Modulation of the IgE-mediated release of slow-reacting substance of anaphylaxis by agents influencing the level of cyclic 3′,5′-Adenosine monophosphate, *J. Immunol.*, 105, 1096, 1970.
39. **Alm, P. E. and Bloom, G. D.,** What — if any — is the role of adrenergic mechanisms in histamine release from mast cells?, *Agents Actions*, 11, 60, 1981.
40. **Alm, P. E. and Bloom, G. D.,** Effects of norepinephrine on transmembrane calcium transport in rat mast cells, *Int. Arch. Allergy Appl. Immunol.*, 66, 427, 1981.
41. **Donlon, M., Hunt, W. A., Catravas, G. N., and Kaliner, M.,** A characterization of beta-adrenergic receptors on cellular and perigranular membranes of rat peritoneal mast cells, *Life Sci.*, 31, 411, 1982.
42. **Phillips, M. J., Barnes, P. J., and Gold, W. M.,** Characterization of purified dog mastocytoma cells. Autonomic membrane receptors and pharmacologic modulation of histamine release, *Am. Rev. Respir. Dis.*, 132, 1019, 1985.
43. **Kaliner, M., Orange, R. P., and Austen, F.,** Immunological release of histamine and slow reacting substance of anaphylaxis from human lung. IV. Enhancement by cholinergic and alpha adrenergic stimulation, *J. Exp. Med.*, 136, 556, 1972.
44. **Fantozzi, R., Moroni, F., Masini, E., Blandina, P., and Mannaioni, P. F.,** Modulation of spontaneous histamine release by adrenergic and cholinergic drugs, *Agents Actions*, 8, 347, 1978.
45. **Masini, E., Blandina, P., Fantozzi, R., Brunelleschi, S., and Mannaioni, P. F.,** Correlation between cholinergic histamine release and quinuclidinyl-benzilate ([$^3$H]-QNB) binding in mast cell membranes, *Agents Actions*, 11, 55, 1981.
46. **Espinosa, J., Botana, L. M., and Fernandez-Otera, M. P.,** Estudio de la actividad secretora de mastocitos er presencia de agonistas colinergicos, *Rev. Esp. Fisiol.*, 39, 441, 1983.
47. **Kazimierczak, W., Adamas, B., and Maslinski, Cz.,** Failure of acetylcholine to release histamine from rat mast cells, *Agents Actions*, 10, 1, 1980.
48. **Fantozzi, R., Masini, E., Blandina, P., Mannaioni, P. F., and Bani-Sacchi, T.,** Release of histamine from rat mast cells by acetylcholine, *Nature (London)*, 273, 473, 1978.
49. **Masini, E., Fantozzi, R., Conti, A., Blandina, P., Brunelleschi, S., and Mannaioni, P. F.,** Mast cell heterogeneity in response to cholinergic stimulation, *Int. Arch. Allergy Appl. Immunol.*, 77, 184, 1985.
50. **Bienenstock, J., Otsuka, H., Dolovich, J., and Denburg, J.,** Growth and differentiation of mast cells and basophils, in *Progress in Immunology VI*, Cinader, B. and Miller, R. G., Eds., Academic Press, Orlando, FL, 1986, 851.
51. **Ihle, J. N., Keller, J., Oroszlan, S., Henderson, L. E., Copeland, T. D., Fitch, F., Prystowsky, M. B., Goldwasser, E., Schrader, J. W., Palaszynski, E., Dy, M., and Lebel, B.,** Biologic properties of homogeneous interleukin-3. I. Demonstration of WEHI-3 growth factor activity, mast cell growth factor activity, P cell-stimulating factor activity, colony-stimulating factor activity and histamine-producing cell-stimulating factor activity, *J. Immunol.*, 131, 282, 1983.
52. **Haig, D. M., McKee, T. A., Jarrett, E. E. E., Woodbury, R., and Miller, H. R. P.,** Generation of mucosal mast cells is stimulated *in vitro* by factors derived from T cells of helminth infected rats, *Nature (London)*, 300, 188, 1982.
53. **Mayrhofer, G. and Bazin, H.,** Nature of the thymus dependency of mucosal mast cells. III. Mucosal mast cells in nude mice and nude rats, in B rats, and in a child with the Di George syndrome, *Int. Arch. Allergy Appl. Immunol.*, 64, 320, 1981.
54. **Hamaguchi, Y., Kanakura, Y., Fujita, J., Takeda, S.-I., Nakano, T., Tarui, S., Honjo, T., and Kitamura, Y.,** Interleukin 4 as an essential factor for *in vitro* clonal growth of murine connective tissue-type mast cells, *J. Exp. Med.*, 165, 268, 1987.

55. **Schmitt, E., Fassbender, B., Beyreuther, K., Spaeth, E., Schwarzkopf, R., and Rude, E.,** Characterization of a T cell-derived lymphokine that acts synergistically with IL3 on the growth of murine mast cells and is identical with IL4, *Immunobiology,* 174, 406, 1987.
56. **Levi-Schaffer, F., Austen, F. K., Gravallese, P. M., and Stevens, R. L.,** Coculture of interleukin-3-dependent mouse mast cells with fibroblasts results in a phenotypic change of the mast cells, *Proc. Natl. Acad. Sci. U.S.A.,* 83, 6485, 1986.
57. **Levi-Schaffer, F., Austen, K. F., Caulfield, J. P., Hein, A., Gravallese, P. M., and Stevens, R. L.,** Co-culture of human lung-derived mast cells with mouse 3T3 fibroblasts: morphology and IgE-mediated release of histamine, prostaglandin $D_2$ and leukotrienes, *J. Immunol.,* 139, 494, 1987.
58. **Isaacson, P.,** Mast cells in benign nerve sheath tumours, *J. Pathol.,* 119, 193, 1976.
59. **Pineda, A.,** Mast cells — their presence and ultrastructural characteristics in peripheral nerve tumors, *Arch. Neurol.,* 13, 372, 1965.
60. **Nennesmo, I., and Reinholt, F.,** Mast cells in nerve end neuromas of mice, *Neurosci. Lett.,* 69, 296, 1986.
61. **Kimura, S.,** Mast cells in solitary glomus tumors: a possible algogenic role, *Acta Dermatovener (Stockholm),* 59, 415, 1979.
62. **Wiesner-Menzel, L., Schulz, B., Vakilzadeh, F., and Czarnetzki, B. M.,** Electron microscopical evidence for a direct contact between nerve fibres and mast cells, *Acta Dermatovener (Stockholm),* 61, 465, 1981.
63. **Kishimoto, S., Nagatani, H., Miyashita, A., and Kobayashi, K.,** Immunohistochemical demonstration of substance P-containing nerve fibres in glomus tumours, *Br. J. Dermatol.,* 113, 213, 1985.
64. **Heine, H. and Forster, F. J.,** Histophysiology of mast cells in skin and other organs, *Arch. Dermatol. Res.,* 253, 225, 1975.
65. **Heine, H. and Forster, F. J.,** Relationships between mast cells and preterminal nerve fibres, *Z. Mikrosk. Anat. Forsch. (Leipzig),* 89, S934, 1975.
66. **Foreman, J. C. and Jordan, C. C.,** Antagonism by neurotensin of the substance P-induced flare in human skin and its relationship to histamine release, *J. Physiol.,* 328, 58, 1982.
67. **Hagermark, O., Hokfelt, T., and Pernow, B.,** Flare and itch induced by substance P in human skin, *J. Invest. Dermatol.,* 71, 233, 1978.
68. **Erjavec, F., Lembeck, F., Florjane-Irman, T., Skofitsch, G., Donnerer, J., Saria, A., and Holzer, P.,** Release of histamine by substance P, *Naunyn-Schmiedeberg's Arch. Pharmacol.,* 317, 67, 1981.
69. **Piotrowski, W. and Foreman, J. C.,** On the actions of substance P somatostatin, and vasoactive intestinal polypeptide on rat peritoneal mast cells and in human skin, *Naunyn-Schmiedeberg's Arch. Pharmacol.,* 331, 364, 1985.
70. **Morton, R. and Chahl, L. A.,** Pharmacology of the neurogenic oedema response. Electrical stimulation of the saphenous nerve in the rat, *Naunyn-Schmiedeberg's Arch. Pharmacol.,* 314, 271, 1980.
71. **Fjellner, B. and Hagermark, O.,** Studies on pruritogenic and histamine-releasing effects on some putative peptide neurotransmitters, *Acta Dermatovener (Stockholm),* 61, 245, 1981.
72. **Foreman, J. C.,** Substance P and calcitonon gene-related peptide: effects on mast cells and in human skin, *Int. Arch. Allergy Appl. Immunol.,* 82, 366, 1987.
73. **Lampe, M. and Kiernan, J. A.,** Degranulation of dermal mast cells: effects of fixation and of antidromic nervous impulses on two histochemically identified cell-types, *Arch. Dermatol. Res.,* 257, 125, 1976.
74. **Kiernan, J. A.,** A study of chemically induced acute inflammation in the skin of the rat, *Q. J. Exp. Physiol.,* 62, 151, 1977.
75. **Chahl, L. A. and Ladd, R. J.,** Local oedema and general excitation of cutaneous sensory receptors produced by electrical stimulation of the saphenous nerve in the rat, *Pain,* 2, 25, 1976.
76. **Stach, W.,** Uber die Nervengeflechte der Duodenalzotten: Licht- und elektronenmikroskopische untersuchungen, *Acta Anat.,* 85, 216, 1973.
77. **Newson, B., Dahlstrom, A., Enerback, L., and Ahlman, H.,** Suggestive evidence for a direct innervation of mucosal mast cells. An electron microscopic study, *Neuroscience,* 10, 565, 1983.
78. **Seelig, L. L., Jr., Schlusselberg, D. S., Smith, W. K., and Woodward, D. J.,** Mucosal nerves and smooth muscle relationships with gastric glands of the opossum: an ultrastructural and three-dimensional reconstruction study, *Am. J. Anat.,* 174, 15, 1985.
79. **Yonei, Y., Oda, M., Nakamura, M., Watanabe, H. H., Tsuskada, N., Komatsu, H. K., Akaiwa, Y., Ichikawa, E., Kaneko, K., Asakura, H., Fujiwara, T., and Tsuchiya, M.,** Evidence for direct interaction between the cholinergic nerve and mast cells in rat colonic mucosa. An electron microscopic cytochemical and autoradiographic study, *J. Clin. Electron Microsc.,* 18, 560, 1985.
80. **Yonei, Y.,** Autonomic nervous alterations and mast cell degranulation in the exacerbation of ulcerative colitis, *Japanese J. Gastroenterol.,* 84, 1045, 1987.
81. **Bani-Sacchi, T., Barattini, M., Bianchi, S., Blandina, P., Brunelleschi, S., Fantozzi, R., Mannaioni, P. F., and Masini, E.,** The release of histamine by parasympathetic stimulation in guinea pig auricle and rat ileum, *J. Physiol.,* 371, 29, 1986.

82. **Ganguly, A. K., Sathiamoorthy, S. S., and Bhatnagar, O. P.,** Effect of subdiaphragmatic vagotomy on gastric mucosal mast cell population in pylorus ligated rats, *Q. J. Exp. Physiol.*, 63, 89, 1978.
83. **Ganguly, A. K. and Gopinath, P.,** Vagus nerves and the gastric tissue histamine concentration in pylorus ligated albino rats, *Q. J. Exp. Physiol.*, 64, 1, 1979.
84. **Perdue, M. H., Chung, M., and Gall, D. G.,** Effect of intestinal anaphylaxis on gut function in the rat, *Gastroenterology*, 86, 391, 1984.
85. **Perdue, M. H. and Gall, D. G.,** Intestinal anaphylaxis in the rat: jejunal response to in vitro antigen exposure, *Am. J. Physiol.*, 250, G427, 1986.
86. **Perdue, M. H. and Gall, D. G.,** Rat jejunal mucosal response to histamine and anti-histamines *in vitro*. Comparison with antigen-induced changes during intestinal anaphylaxis, *Agents Actions*, 19, 5, 1986.
87. **Baird, A. W., Cuthbert, A. W., and Pearce, F. L.,** Immediate hypersensitivity reactions in epithelia from rats infected with *Nippostrongylus brasiliensis*, *Br. J. Pharmacol.*, 85, 787, 1985.
88. **Castro, G. A., Harari, Y., and Russell, D.,** Mediators of anaphylaxis-induced ion transport changes in small intestine, *Am. J. Physiol.*, 253, G540, 1987.
89. **Russell, D. A.,** Mast cells in the regulation of intestinal electrolyte transport, *Am. J. Physiol.*, 251, G253, 1986.
90. **Perdue, M. H. and Gall, D. G.,** Transport abnormalities during intestinal anaphylaxis in the rat: effect of antiallergic agents, *J. Allergy Clin. Immunol.*, 76, 498, 1985.
91. **Perdue, M. H. and Davison, J. S.,** Response of jejunal mucosa to electrical transmural stimulation and two neurotoxins, *Am. J. Physiol.*, 251, G642, 1986.
92. **Perdue, M. H., Galbraith, R., and Davison, J. S.,** Evidence for substance P as a functional neurotransmitter in guinea pig small intestinal mucosa, *Regulatory Peptides*, 18, 63, 1987.
93. **Baird, A. W. and Cuthbert, A. W.,** Neuronal involvement in type 1 hypersensitivity reactions in gut epithelia, *Br. J. Pharmacol.*, 92, 647, 1987.
94. **Perdue, M. H.,** Changes in mast cells, nerves, and the epithelium contribute to ion transport abnormalities during parasite-induced intestinal inflammation in the rat, *Gastroenterology*, 94, A350, 1988.
95. **Perdue, M. H., Marshall, J., and Masson, S.,** Ion transport abnormalities in inflamed rat jejunum. Involvement of mast cells and nerves, *Gastroenterology*, in press.
96. **Kakuta, Y., Stead, R. H., Perdue, M. H., Marshall, J. S., and Bienenstock, J.,** Microanatomical relationships of mast cells and nerves in rat lung and trachea, *Am. Rev. Respir. Disease*, 139, A118, 1989.
97. **Lundberg, J. M., Hokfelt, T., Martling, C.-R., Saria, A., and Cuello, C.,** Substance P-immunoreactive sensory nerves in the lower respiratory tract of various mammals including man, *Cell Tissue Res.*, 235, 251, 1984.
98. **Uddman, R., Luts, A., and Sundler, F.,** Occurrence and distribution of calcitonin gene-related peptide in the mammalian respiratory tract and middle ear, *Cell Tissue Res.*, 241, 551, 1985.
99. **Masini, E., Rucci, L., Cirri-Borghi, M. B., Giannella, E., and Mannaioni, P. F.,** Stimulation and resection of Vidian nerve in patients with chronic hypertrophic nonallergic rhinitis: effect on histamine content in nasal mucosa, *Agents Actions*, 18, 251, 1986.
100. **Bienenstock, J., Perdue, M., Blennerhassett, M., Stead, R., Kakuta, N., Sestini, P., Vancheri, C., and Marshall, J.,** Inflammatory cells and the epithelium: mast cell/nerve interactions in the lung *in vitro* and *in vivo*, *Am. Rev. Respir. Dis.*, 138, 531, 1988.
101. **Leff, A. R., Stimler, N. P., Munoz, N. M., Shioya, T., Tallet, J., and Dame, C.,** Augmentation of respiratory mast-cell secretion of histamine caused by vagus nerve stimulation during antigen challenge, *J. Immunol.*, 136, 1066, 1986.
102. **Sestini, P., Dolovich, M., Vancheri, C., Stead, R. H., Marshall, J. S., Perdue, M., Gauldie, J., and Bienenstock, J.,** Antigen-induced lung solute clearance in rats is dependent on capsaicin-sensitive nerves, *Am. Rev. Resp. Dis.*, 139, 401, 1989.
103. **Skofitsch, G., Savitt, J. M., and Jacobowitz, D. M.,** Suggestive evidence for a functional unit between mast cells and substance P fibers in the rat diaphragm and mesentery, *Histochemistry*, 81, 5, 1985.
104. **Dimlich, R. V. W.,** Electron microscopic examination of mast cells and nerve processes in hepatic portal areas of the rat, *Proc. 42nd Annu. Meet. Electron Microscopy Soc. Am.*, San Francisco Press, San Francisco, 1984, 230.
105. **Dimitriadou, V., Aubineau, P., Taxi, J., and Seylaz, J.,** Ultrastructural evidence for a functional unit between nerve fibers and type II cerebral mast cells in the cerebral vascular wall, *Neuroscience*, 22, 621, 1987.
106. **Seldin, D. C., Adelman, S., Austen, K. F., Stevens, R. L., Hein, A., Caulfield, J. P., and Woodbury, R. G.,** Homology of the rat basophilic leukemia cell and the rat mucosal mast cell, *Proc. Natl. Acad. Sci. U.S.A.*, 82, 3871, 1985.
107. **Fewtrell, C. and Metzger, H.,** Stimulus-secretion coupling in rat basophilic leukaemia cells, in *Biochemistry of the Acute Allergic Reactions*, Becker, E. L., Simon, A. S., Austen, K. F., Eds., Alan R. Liss, New York, 1981, 295.
108. **Blennerhassett, M. G., Stead, R. H., and Bienenstock, J.,** Association and interaction between sympathetic neurons and mast cells in vitro, *Biophys. J.*, 51, 65A, 1987.

109. **Blennerhassett, M. G., Tomioka, M., and Bienenstock, J.,** Interactions of nerves and mast cells. I. Selective formation of contacts *in vitro,* submitted.
110. **Blennerhassett, M. G. and Bienenstock, J.,** Interactions of nerves and mast cells. II. Sympathetic nerve contact alters membrane resistance of rat basophilic leukaemia (RBL-2H3) cell *in vitro,* submitted.
111. **Blennerhassett, M. G. and Bienenstock, J.,** Apparent innervation of rat basophilic leukaemia (RBL-2H3) cells by sympathetic nerves in vitro, submitted.
112. **Bruni, A., Bigon, E., Boarato, E., Mietto, L., Leon, A., and Toffano, G.,** Interaction between nerve growth factor and lysophosphatidylserine on rat peritoneal mast cells, *FEBS Lett.,* 138, 190, 1982.
113. **Sugiyama, K., Suzuki, Y., and Furuta, H.,** Histamine-release induced by 7S nerve growth factor of mouse submandibular salivary glands, *Arch. Oral Biol.,* 30, 93, 1985.
114. **Pearce, F. L. and Thompson, H. L.,** Some characteristics of histamine secretion from rat peritoneal mast cells stimulated with nerve growth factor, *J. Physiol,* 372, 379, 1986.
115. **Aloe, L. and Levi-Montalcini, R.,** Mast cell increase in tissues of neonatal rats injected with the nerve growth factor, *Brain Res.,* 133, 358, 1977.
116. **Tomioka, M., Stead, R. H., Nielsen, L., Coughlin, M. D., and Bienenstock, J.,** Nerve growth factor enhances antigen and other secretagogue-induced histamine release from rat peritoneal mast cells in the absence of phosphatidylserine, *J. Allergy Clin. Immunol.,* 82, 599, 1988.
117. **Tomioka, M., Stead, R. H., Marshall, J., McSharry, C., Nielsen, L., Hamill, R. W., Coughlin, M. D., and Bienenstock, J.,** Nerve growth factor stimulates growth of connective tissue and intestinal mucosal mast cells in neonatal rats, submitted.
118. **Matsuda, H., Coughlin, M. D., Bienenstock, J., and Denburg, J. A.,** Nerve growth factor promotes human hemopoietic colony growth and differentiation, *Proc. Natl. Acad. Sci. U.S.A.,* 85, 6508, 1988.
119. **Thorpe, L. W. and Perez-Polo, J. R.,** The influence of nerve growth factor on the *in vitro* proliferative response of rat spleen lymphocytes, *J. Neurosci. Res.,* 18, 134, 1987.
120. **Thorpe, L. W., Werrbach-Perez, K., and Perez-Polo, J. R.,** Effects of nerve growth factor on the expression of interleukin 2 receptors on cultured human lymphocytes, *Ann. N.Y. Acad. Sci.,* 496, 310, 1987.
121. **Thorpe, L. W., Stach, R. W., Hashim, G. A., Marchetti, D., and Perez-Polo, J. R.,** Receptors for nerve growth factor on rat splenic mononuclear cells, *J. Neurosci. Res.,* 17, 128, 1987.
122. **Cutz, E. W., Chan, W., Track, N. S., Goth, A., and Said, S. I.,** Release of vasoactive intestinal polypeptide in mast cells by histamine liberators, *Nature (London),* 275, 661, 1978.
123. **Goetzl, E. J., Chernov-Rogan, T., Cooke, M. P., Renda, F., and Payan, D. G.,** Endogenous somatostatin-like peptide of rat basophilic leukemia cells, *J. Immunol.,* 135, 2707, 1985.
124. **O'Dorisio, M., O'Dorisio, T., Cotland, S., and Bakerzak, S. P.,** Vasoactive intestinal polypeptide as a biochemical marker for polymorphonuclear leukocytes, *J. Lab. Clin. Med.,* 96, 666, 1980.
125. **Wiedermann, C. J., Goldman, M. E., Plutchok, J. J., Sertl, K., Kaliner, M. A., Johnston-Early, A., Cohen, M. H., Puff, M. R., and Pert, C. B.,** Bombesin in human and guinea pig alveolar macrophages, *J. Immunol.,* 137, 3928, 1986.
126. **Harbour-McMenamin, D., Smith, E. M., and Blalock, J. E.,** Bacterial lipopolysaccharide induction of leukocyte-derived corticotropin and endorphins, *Infect. Immun.,* 48, 813, 1985.
127. **Blalock, J. E. and Smith, E. M.,** A complete regulatory loop between the immune and neuroendocrine systems, *Fed. Proc.,* 44, 108, 1985.
128. **Carr, D. J. J. and Blalock, E.,** A molecular basis for bidirectional communication between the immune and neuroendocrine system, in *Progress in Immunology VI,* Cinader, B., Miller, R. G., Eds., Academic Press, Orlando, FL, 1986, 619.
129. **Weiland, G. A., Durkin, J. A., Henley, J. M., and Simasko, S. M.,** Effects of substance P on the binding of ligands to nicotinic acetylcholine receptors, *Mol. Pharmacol.,* 32, 625, 1987.
130. **Oger, J., Arnason, B. G. W., Pantazis, N., Lehrich, J., and Young, M.,** Synthesis of nerve growth factor by L and 3T3 cells in culture, *Proc. Natl. Acad. Sci. U.S.A.,* 71, 1554, 1974.
131. **Levi-Montalcini, R. and Angeletti, P. N.,** Biological properties of a nerve-growth-promoting protein and its antiserum, in *Regional Neurochemistry;* Proc. 4th Int. Neurochemical Symp., Kety, S. S., Elkes, J., Eds., Pergamon Press, Elmsford, NY, 1961, 362.
132. **Lindholm, D., Heumann, R., Meyer, M., and Thoenen, H.,** Interleukin-1 regulates synthesis of nerve growth factor in nonneuronal cells of rat sciatic nerve, *Nature (London),* 330, 658, 1987.
133. **Otten, U., Weskamp, G., Hardung, M., and Meyer, D. K.,** Synthesis and release of nerve growth factor by rat macrophages, *Soc. Neurosci. Abstr.,* 1987, 184.
134. **Otten, U.,** Nerve growth factor and the peptidergic sensory neurons, *Trends Pharmacol. Sci.,* 5, 307, 1984.
135. **Coughlin, M. D., Bloom, E. M., and Black, I. B.,** Characterization of neuronal growth factor from mouse heart cell-conditioned medium, *Dev. Biol.,* 82, 56, 1981.
136. **Edgar, D., Timpl, R., and Thoenen, H.,** The heparin-binding domain of laminin is responsible for its effects on neurite outgrowth and neuronal survival, *EMBO J.,* 3, 1463, 1984.

137. **Lander, A. D., Fujii, D. K., Gospodarowicz, D., and Reichardt, L. F.,** Characterization of a factor that promotes neurite outgrowth: evidence linking activity to a heparan sulfate proteoglycan, *J. Cell Biol.*, 94, 574, 1982.
138. **Lander, A. D., Fujii, D. K., and Reichardt, L. F.,** Laminin is associated with the ''neurite outgrowth-promoting factors'' found in conditioned media, *Proc. Natl. Acad. Sci. U.S.A.*, 82, 2183, 1985.
139. **Dvorak, A. M., Osage, J. E., Monahan, R. A., and Dickersin, G. R.,** Crohn's disease. III. Target tissues. Proliferation of and injury to smooth muscle and the autonomic nervous system, *Hum. Pathol.*, 11, 620, 1980.
140. **Troster, M. and Grignon, D.,** Enteric nerve fibres, ganglion cells, and Schwann cells in Crohn's disease and ulcerative colitis, *Lab. Invest.*, 54, 64A, 1986.
141. **Lemanske, R. F., Jr., Atkins, F. M., and Metcalfe, D. D.,** Gastrointestinal mast cells in health and disease. II, *Pediatrics*, 103, 343, 1983.
142. **Dvorak, A. M., Monahan, R. A., Osage, J. E., and Dickersin, G. R.,** Crohn's disease: transmission electron microscopic studies. II. Immunologic inflammatory response. Alterations of mast cells. basophils, eosinophils, and the microvasculature, *Hum. Pathol.*, 11, 606, 1980.

Chapter 4

# RECIPROCAL INTERACTIONS BETWEEN MAST CELLS AND THE ENDOCRINE SYSTEM

**Dean Befus**

## TABLE OF CONTENTS

# I. INTRODUCTION

This is the second chapter in this volume on mast cells, and therefore, it is not essential to reproduce a major introduction to mast cell biology. However, in addition to Chapter 3 by Stead et al.,[1] readers interested in a more complete coverage of mast cell biology and function should consult other recent reviews.[2,3] Perhaps most importantly in the context of this chapter, the reader should ponder mast biology in a broad manner, not just with a view to immediate hypersensitivity and immunology.

Mast cells can be activated by a diverse series of stimuli and can release a spectrum of mediators including histamine, serotonin, proteases, lysosomal enzymes, proteoglycans, chemotactic factors, and arachidonic acid metabolites.[2,3] These mediators can act on a number of target cells and induce a variety of metabolic alterations, such as secretion, proliferation, and differentiation. Moreover, it is important to recognize that mast cells do not represent a homogeneous population of cells, but differ in many parameters which encompass mediator content, responsiveness to selected secretagogues or modulatory drugs or endogenous factors, ontogeny, and maturity.[4] The best known example of mast cell heterogeneity is in the rat where intestinal mucosal mast cells differ markedly from peritoneal mast cells.[2-4] This heterogeneity appears to reflect a variety of environmental factors, but most importantly, the tissue site of the mast cell and the tissue physiology or pathophysiology play critical roles in the phenotypic expression of the mast cell genome.[5-7]

The distribution of mast cells has stimulated a potpourri of speculations about their functions. Mast cells occur throughout the body with the exceptions of solid bone and cartilage. They are abundant at mucosal surfaces and presumably play critical roles in host defenses and tissue repair. Their close association with the vascular system has been recognized for many years and the effects of many of their mediators on vessels are well known.[8] More recently, the association of mast cells with peripheral nerves has become carefully scrutinized and this subject is well-reviewed by Stead et al.[1] In this chapter, the relationship between mast cells and the endocrine system will be deliberated. Unfortunately, this relationship has not been a significant thrust in investigations of mast cell biology in recent years, although in the 1950s, a series of reports attempted to address the effects of different endocrine glands and their products on mast cell numbers and metabolic activities, e.g., see Reference 9.

In 1972, Csaba published a tome which described a collection of experiments he had conducted to explore mast cell formation and function and the role of the endocrine system in their regulation.[10] He proposed that, because of their ability to synthesize, store, and secrete extremely potent hormone-like substances, mast cells should be considered unicellular glands widely distributed throughout the body. Moreover, he recognized that mast cells can undergo local hyperplasia and thus provide their mediators in large quantities in local microenvironments. He questioned whether mast cells have identical functions throughout the body and proposed that their functions differ at different times in ontogeny and in different sites. They may release only selected components of their mediator repertoire depending upon the nature of stimulation they are given.

Csaba[10] acknowledged the work of others before him and suggested that "there is not a single endocrine gland whose action on the mast cells has not been repeatedly studied". However, he concluded that the results of these investigations are by no means convincing and are mostly contradictory. Accordingly, in this chapter, literature which has appeared following the review by Csaba will be used as a base, although where deemed appropriate, earlier literature will be reported. The more recent studies have focused upon the thyroid, pituitary, adrenal, and parathyroid glands, and the stomach and reproductive system in the contexts of the effects of mast cell mediators on them, or their effects on mast cell development or secretion.

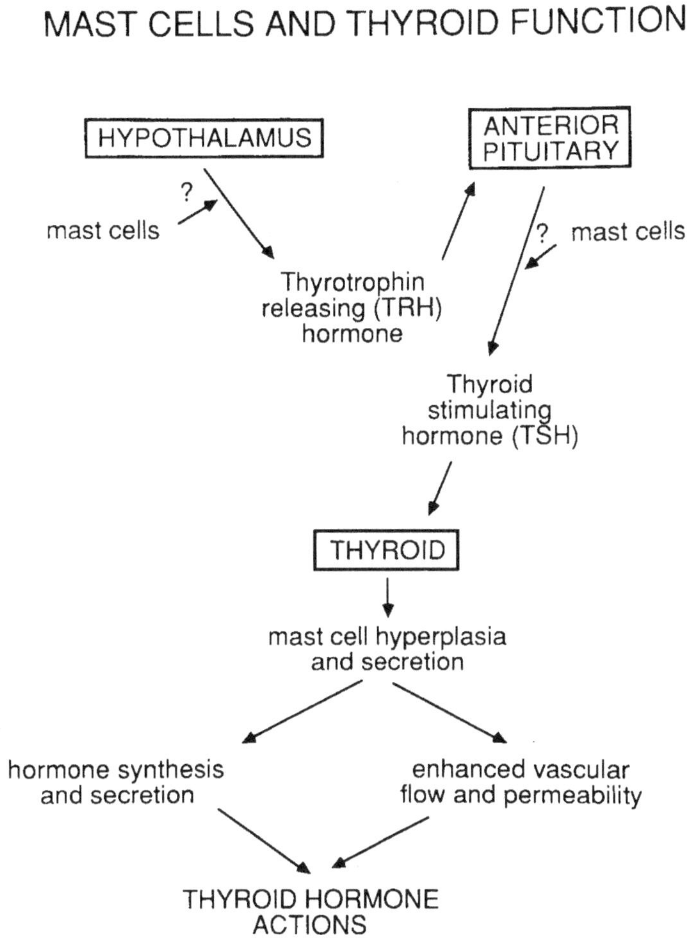


FIGURE 1. The various sites where mast cells may act in the hypothalamic-pituitary-thyroid axis.

## II. MAST CELL-ENDOCRINE INTERACTIONS

### A. THYROID-PITUITARY AXIS

Elucidation of the bidirectional interactions between the endocrine system and mast cells has had a strong basis provided by studies of the thyroid-hypothalamic-pituitary system in rats (Figure 1). The earlier literature in this field received an excellent review by Melander.[11] Mast cells with a histochemical phenotype analogous to peritoneal mast cells are located in both perifollicular and perivascular spaces in the thyroid, where in the rat it has been estimated that there are a total of 200,000 mast cells.[12] Apparently, there have been no studies to establish whether mast cells with a phenotype analogous to intestinal mucosal mast cells are present in the thyroid gland. In contrast to rats, the thyroid glands of mice contain few mast cells.[11] In numerous experimental systems where levels of pituitary thyroid-stimulating hormone (TSH) were increased, thyroid mast cell hyperplasia, but not hyperplasia of mast cells elsewhere, was documented.[11] Treatment with the goitrogen, aminotriazole, which inhibits thyroid hormone synthesis, leads to a fourfold increase in total thyroid mast cells.[12] It may be clinically relevant in hyperthyroidism that the immunoglobulin, "long acting thyroid stimulator" (LATS), present in some patients appears to increase the numbers of thyroid mast cells in mice.[11] Interestingly, associated with lowered thyroid activities in aging rats, thyroid mast cell densities appear to decrease.[13]

In addition to stimulating thyroid mast cell hyperplasia (Figure 1), TSH stimulates serotonin and histamine secretion by thyroid mast cells, but not mast cells outside the thyroid.[11,14] Administration of the mast cell degranulating agent 48/80 to rats caused an 85% reduction of thyroid histamine within 10 min, together with increased thyroid blood flow and capillary permeability and increased synthesis and release of thyroid hormone.[11,14] These 48/80 effects mimicked the effects of TSH on thyroid function, and together these results suggest that a significant portion of the effects of TSH on thyroid function involves mast cell activation, serotonin, and histamine secretion, and in turn their effects on follicular cells and the vasculature (Figure 1).[11] In the rat the effects of histamine appear to be restricted to the vasculature, whereas serotonin influences both hormone synthesis and secretion as well as the vasculature. Interestingly, the effect of serotonin on hormone secretion is inhibited by alpha, but not beta-adrenoreceptor antagonism.[11] Whether alpha- and beta-adrenoreceptors present on mast cells[1] also play a prominent role in cellular network of thyroid function remains to be studied.

Bianco et al.[15] established that disodium cromoglycate, an antiallergic drug with diverse effects (one of which is to inhibit mast-cell secretion) partially inhibits the effects of TSH on the thyroid gland. Moreover, in thyroidectomized rats given L-thyroxine sodium ($T_4$) replacement therapy, 48/80 treatment elevated circulating TSH levels. Disodium cromoglycate completely blocked this elevation of TSH. The authors proposed that histamine and/or serotonin present in mast cells in the median eminence of the hypothalamus must be involved in the secretion of thyrotrophin-releasing hormone (TRH) by the hypothalamus or TSH by the anterior pituitary (Figure 1). How pinealectomy modifies serotonin metabolism in thyroid mast cells is unknown.

The effects of thyroid hormones on mast cells in peripheral tissues has received little study. However, in studies of globule leucocytes in the rat duodenal epithelium, a cell type now recognized to be an activated intraepithelial mast cell,[17] Carriere and Bushke[18] showed that thyroidectomy increased the numbers of globule leucocytes up to tenfold. Moreover, in male rats castration combined with thyroidectomy further enhanced globule leukocyte numbers. Thyroxine, but not testerosterone reduced the globule leukocyte numbers to normal. Castration alone did not alter the numbers of globule leukocytes, but increased the number, size, and acidophilia of the cell granules. It is obvious that additional studies of thyroid effects on mast cells are warranted.

## B. THE ADRENAL-PITUITARY AXIS

Findings such as those of Rasanen[19] in 1960 which established that cortisone and adrenocorticotropic hormone (ACTH) cause a reduction in mast cell numbers, at least in the gastric mucosa (see below), suggested that the pituitary-adrenal system exerts some controls on mast cells.

Unfortunately, early studies which attempted to address this interaction by adrenalectomy, hypophysectomy, or ACTH, adrenal cortical extract, or cortisone administration, were inconclusive. These studies, e.g., see References 20 and 21, are difficult to assess because the histological and quantitative methodologies used are unclear and the magnitude of effects small. However, more recent studies suggest that mast cells stimulate ACTH release, ACTH causes mast-cell activation, and steroids cause mast-cell destruction or inactivation (see below). Thus, it appears clear that mast cells and the pituitary-adrenal system are interactive.

The intraperitoneal administration of histamine to rats stimulates the rise in circulating levels of ACTH.[22] Moreover, both $H_1$ and $H_2$ histamine receptor agonists mimic this response, although only $H_1$ antagonists will inhibit it. The work of Portaleone et al.[23] using hypothalamic membrane preparations suggested that this effect of histamine on ACTH release was directly on the hypothalamus, although others had suggested that the histamine effect might be directly on the pituitary also.[24,25] Interestingly, a number of antidepressant drugs inhibit histamine-induced release of ACTH, including imipramine, doxepin, mianserin, desipramine, and amitriptyline.[22] Perhaps some of the sedative effects of antidepressant drugs relate to their antihistaminic effects. The beta-antagonist, propanolol, does not stimulate ACTH release alone, but markedly enhances the histamine-mediated response.[26] This represents but another example of complexity of interactions among the immune, nervous, and endocrine systems.

In a series of papers over a decade, Stanworth and colleagues reported that the 1 to 24 fragment of ACTH stimulated histamine release from rat peritoneal mast cells and was 15-fold more potent than the parent molecule, ACTH 1 to 39.[27-29] The evidence appears to suggest that this property of stimulating mast cells relates to the charge characteristics of ACTH, and to the distribution of basic amino acid residues in the molecule.[28,30] Many other oligopeptides appear to share these critical properties and activate certain mast cell populations, including substance P, neurotensin, vasoactive intestinal peptide, somatostatin and the opiates, dynorphin, beta-endorphin, and alpha-neoendorphin, as well as the well-known mast cell degranulating compound, 48/80.[2] Perhaps all of these positively charged ligands interact with a single type of receptor on the surface of at least some mast cell populations.

Finally, steroid hormones have major effects on mast cell populations. Csaba[10] reviewed his studies and concluded that glucocorticoids increased the numbers of mast cells in the thymus and lymph nodes, promoted the maturation of mast cells in these sites, and degranulated mast cells in many tissues as well as facilitating formation of new mast cells. It will be interesting to repeat these experiments using some of the methodologies presently available.

Schleimer[31] recently reviewed the ability of corticosteroids to inhibit mast cell secretion. Perhaps of greatest interest is the ability of a single 24 h steroid treatment to markedly reduce the numbers of intestinal mucosal mast cells in the rat without affecting mast cells in other sites in the body.[32] In addition, Lavker and Schechter[33] showed that topical administration of steroid to human skin induced a reduction in skin mast cell numbers. The precise mechanisms of drug action which underlie these various reports remain unknown, although there is sufficient evidence to suggest that adrenal cortical functions play some role in mast cell numbers and activation.

## C. THE PARATHYROID GLAND AND BONE RESORPTION

Exploration of the relationship among mast cells, the parathyroid glands, and bone resorption was stimulated by the original observations that the number of bone marrow mast cells was increased in patients with osteoporosis.[34] It was suggested that mast cells and their heparin play an important role in bone resorption. Subsequently, Rockoff and Armstrong[35] established that administration of parathyroid hormone (PTH) to rats resulted in mast cell hyperplasia in bone. Given that PTH plays such a prominent role in bone resorption and calcium mobilization from bone, the linkages among the parathyroid glands, mast cells, and bone resorption were given some consideration (Figure 2).[36]

The interactions between PTH and mast cells involve not only the stimulation of mast cell hyperplasia, but also the activation of mast cells to release both histamine and serotonin (Figure 2).[37,38] The concentrations of PTH required to stimulate mediator secretion fall in the range found in the plasma of uremic patients (25 to 30 U/ml). In the same studies of mast cell secretion, Wilhelm et al.[37] reported that calcitonin, glucagon, and insulin had no effect on mast cell secretion. The PTH-mast cell relationship may be bidirectional because it has also been reported that the histamine $H_2$ antagonist, cimetidine, inhibits PTH secretion (Figure 2).[39]

The mechanisms involved in bone resorption by mast cells are poorly known, but may incorporate the actions of many different mast cell mediators. For example, heparin, histamine, prostaglandin, and proteases have been implicated in bone remodeling.[36] Thus, in

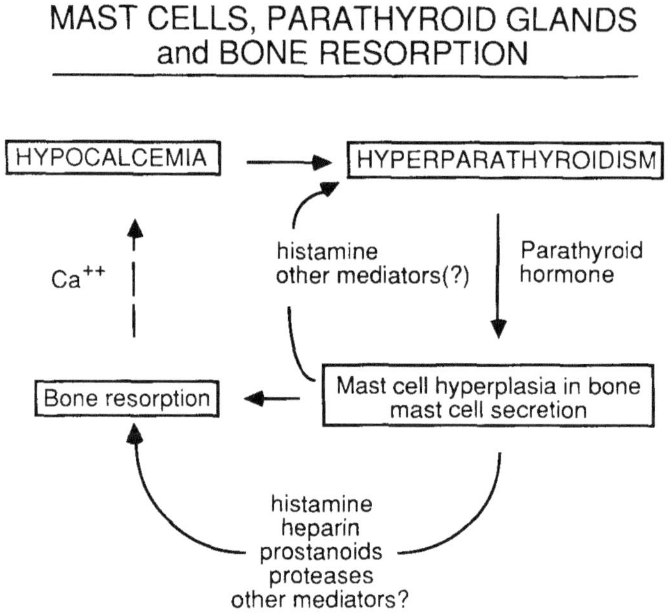


FIGURE 2. Reciprocity of interactions between mast cells and parathyroid gland function.

summary, hyperparathyroidism with its associated elevated production of PTH can induce mast cell hyperplasia and secretion; in turn, mast cell mediators can further stimulate PTH release or initiate and/or perpetuate bone resorption (Figure 2).

## D. REPRODUCTIVE PHYSIOLOGY

There is an interesting literature on the potential relationships between mast cells and their products and urogenital physiology. Unfortunately, this literature is based upon studies in many different species in which there are variable levels of knowledge about mast cell biology and the roles of different mediators. Mast cells have been implicated in physiology changes during the estrous and menstrual cycles, in ovulation, implantation, pregnancy, and parturition (Figure 3). The diversity of mast cell mediators and their roles in inflammation stimulates some interesting thoughts about how they might act in implantation and other aspects of reproductive biology.[40] Some sexual hormones have been studied for their impact upon mast cell biology, although major emphasis has been placed upon estrogens, perhaps largely as a result of the 1955 study by Arvy[41] that showed estrogen injection caused increases in mast cell numbers in numerous tissues in mice.

Mast cells are abundant in the urogenital system, and their numbers and the levels of tissue histamine have been particularly well-studied in the female. Mast cell function in the male reproductive tract has received little investigation, apart from reports of their distribution and abundance,[42,43] and hyperplasia in the testes of patients with idiopathic infertility.[44] Mast cells are not abundant in the ovary, but are found in the vicinity of blood vessels entering and exiting the ovary and in the hilar region.[45] During the surge of preluteinizing hormone (LH) on the day of proestrous in the hamster, there appears to be a marked release of histamine from mast cells in the vicinity of the ovary, and a detectable increase in ovarian histamine with corresponding alterations in blood flow.[45] Thus, the release of histamine and other mast cell mediators may be involved in ovulation. Indeed, cimetidine ($H_2$ antagonist) but not the $H_1$ antagonist, chlorpheniramine, inhibits histamine-induced ovulation in the rabbit.[46] However, neither cimetidine nor chlorpheniramine inhibit human chorionic gonad-

# MAST CELLS and REPRODUCTIVE BIOLOGY

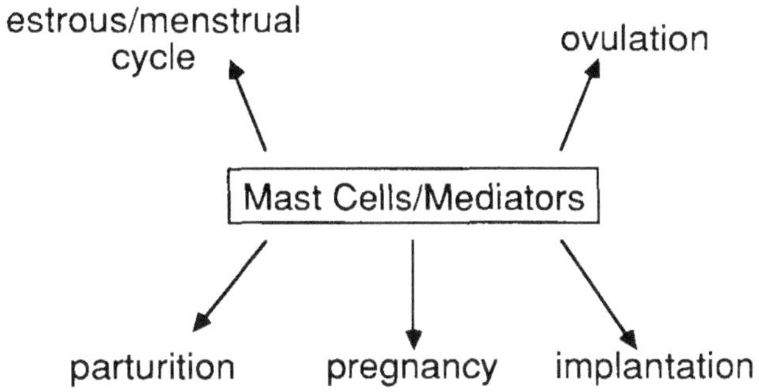


FIGURE 3. Phases in the reproductive cycle that mast cells influence.

otropin (hCG)-induced ovulation. Moreover, in contrast to hCG-induced ovulation, histamine-induced ovulation did not lead to release of mature ova. Thus, histamine may be one component in the mechanism of follicular rupture but does not support the maturation of ova.[46]

In the uterus the myometrium contains the greatest number of mast cells, although they are present in the endometrium as well. Iversen[47] showed that the numbers of mast cells in the human myometrium were significantly elevated in the secretory phase of the menstrual cycle, when compared with the proliferative phase. Subsequently, it was shown in hamsters and rats, that uterine mast cell numbers and histamine levels are highest during estrous, but significantly lower during metestrus and diestrus.[48-50] In humans, uterine mast cell numbers decrease significantly during pregnancy and the postmenopausal period,[47] whereas during the early phase of pregnancy in the rat (up to day 5 following the presence of vaginal sperm or plugs) the numbers of mast cells increase, but then fall dramatically at the time of implantation.[51] In the hamster, senescent (15 to 22 months) uteri have significantly more mast cells than young (3 to 9 months) uteri.[52] The relationship of this observation to changes in sexual hormones in aged animals is unknown.

From studies in the rat, mouse, and rabbit, there is increasing evidence that uterine mast cells, and the histamine which they contain, play a significant role in blastocyst implantation. Shelesnyak[53] showed that 24 h prior to implantation in the rat, there was a significant reduction in uterine histamine levels which could be mimicked by the administration of estrogen. It appears that in the rat there is an estrogen surge approximately 24 h prior to implantation which releases histamine from uterine mast cells,[54,55] in turn, facilitating blastocyst implantation.[56] Gibbons and Chang[57] showed that estrogen treatment caused degranulation of rat uterine mast cells and that pretreatment with progesterone amplified the effects of estrogen on uterine mast cell numbers. The authors proposed that adequate hormonal preparation of the endometrium for implantation involves effects on local mast cell populations that in turn influence the uterine microenvironment.

Brandon and Wallis[58] established that a combination of histamine $H_1$ and $H_2$ receptor antagonists administered to rats at the time of implantation reduced the portion of blastocysts which were implanted. Dey et al.[59] extended these observations to the rabbit by showing that disodium cromoglycate, an inhibitor of mast cell histamine release, reduced the numbers of blastocysts implanted. In addition, in other studies this group established that histamine production by the blastocyst is important in implantation and that the blastocyst expresses $H_1$ receptors, whereas the endometrial cells express $H_2$ receptors for histamine.[60,61] Studies in the mouse also support the hypothesis that histamine is involved in implantation.[62]

Thermal injury to individual uterine horns in the rat induces an irreversible depression in mast cell numbers and histamine levels.[56] Such damaged horns will not support normal blastocyst implantation, presumably due to the depressed levels of mast cells and lower histamine content. Indeed, exogenous histamine administered intralumenally will induce implantation in these rats, but exogenous estrogen will not.

It is interesting that during pregnancy in the mouse and rat, maternal urinary levels of histamine rise dramatically.[63] The origins of this histamine include maternal kidney, stomach, and the fetuses, but it is not clear if the mast cell is the sole source of this histamine. The undoubtedly complex roles that this histamine plays in fetal development and in the maternal-fetal relationship remain to be explored. The hormonal or other factors which activate or modulate mast cell function in the uterus prior to, during, or after pregnancy require careful investigation.

There appears to be a relationship between mast cell numbers and the action of intrauterine contraceptive devices (IUD). The IUD stimulates an increase of mast cells in the uterus and alters their distribution.[64-67] Moreover, the IUD alters the uterine response to exogenous hormones,[64] but to what extent the antifertility effects can be attributed to changes in mast cell function remains to be elucidated.

## E. THE STOMACH AND ACID SECRETION

The success of using histamine $H_2$ receptor antagonists, such as cimetidine and ranitidine, in treating gastric ulcers at least partially resolved the controversy surrounding the importance of histamine in gastric acid secretion. However, it remains unclear precisely how histamine acts and what stimulates its local secretion.[68] This problem is magnified by evidence that the major cell source of histamine in the stomach differs among species. For example, in mammals histamine-immunoreactive endocrine cells are numerous in the rat, but in low numbers in the dog, monkey, and human.[69] By contrast, mucosa of the mouse, rat, and hamster contains few mast cells, but mast cells are numerous in the cat, dog, and human.[69] Moreover, the abundance of histamine-storing endocrine cells can increase markedly in patients with achlorhydria and hypergastrinemia due to gastrinoma.[69] Additionally, the existence of mast cell heterogeneity in different anatomic compartments of the stomach wall has received little study.[70,71] If mast cell subpopulations exist in the gastric wall that differ in histamine content or responsiveness to physiological stimuli, then the elucidation of the role of various histamine-containing cell populations in the gastric wall will represent a most challenging problem.

Potential secretagogues for the release of mast cell-associated histamine in the gastric mucosa are gastrin and pentagastrin. The administration of gastrin to rats causes gastric acid production which is associated with a reduction in gastric mucosal histamine stores.[72] However, this action of gastrin on histamine secretion may not involve mast cell action, but rather the activation of endocrine cells which are the most abundant cells which store histamine in the rat gastric mucosa.[68,69] In humans where gastric mast cells are abundant, pentagastrin infusion over 90 min reduces mucosal histamine content, increases histamine levels in gastric juice and plasma, and stimulates acid secretion.[73] The authors postulated that pentagastrin causes histamine secretion which in turn stimulates parietal cells to secrete $H^+$. Whether histamine secretion is the direct result of pentagastrin acting on mast cells or not has not been determined.

Although direct studies of human gastric mast cells have not been conducted, Tharp et al.[74] employed *in vitro* and *in vivo* studies of human skin mast cells to assess the effects of gastrin and pentagastrin. *In vitro* studies of 200 μm slices of skin established that gastrin

17 and pentagastrin (gastrin 14 to 17 with an additional alanine) stimulated histamine secretion, whereas the N-terminal tridecapeptide of gastrin (gastrin 1 to 13) was not active. Moreover, a similar pattern of reactivity was established for the generation of wheal and flare reactions *in vivo*. These results suggest that at least some human mast cells express gastrin receptors, although the results could be explained by an indirect effect of gastrin rather than a direct effect on mast cells. Furthermore, given evidence of mast cell heterogeneity, it is not clear that skin mast cells should be considered an appropriate representative of mast cells in the gastric mucosa.

In comprehensive studies of $H^+$ secretion in the canine stomach, Soll and his colleagues have isolated gastric mast cells and enriched them by cell fractionation procedures.[68] Isolated canine gastric mast cells do not release histamine in response to gastrin-17 or cholinomimetics, and in ligand binding studies gastrin-17 and muscarinic receptors could not be identified. However, epinephrine inhibited histamine secretion from canine fundic mast cells; an effect blocked by propanolol, the beta-adrenergic receptor antagonist.[68] Thus, in the dog stomach it appears that gastrin does not directly stimulate histamine secretion from mast cells. Accordingly, the action of gastrin on histamine secretion by gastric mast cells of the dog must involve indirect pathways, perhaps incorporating regulatory components of neurocrine and/or paracrine origin. The answer as to whether or not human gastric mast cells are unresponsive to direct gastrin stimulation and lack gastrin receptors, awaits their isolation and appropriate investigation (as in the dog).

As well as the possibility that some hormones may directly activate gastric mast cells, there is evidence that hormonal manipulation will markedly alter the numbers of mast cells in the gastric mucosal.[19] *In vivo* administration of cortisone or ACTH to rats produced a marked decline in mast cells in the gastric mucosa, but interestingly, not in the submucosa (Figure 4) or subcutaneous tissues. Tissues used in these studies were fixed in 4% lead acetate and the mast cells were stained with 1% aqueous toluidine blue (pH not stated in the original publication). Hence, the relationship of the mast cells identified in the gastric mucosa in these studies to the distinct mucosal mast cell of the rat intestine is not clear. Growth hormone (somatotrophin) was also administered to rats by Rasanen[19] and, in contrast to cortisone and ACTH, increased the numbers of gastric mucosal mast cells (Figure 4) and enhanced their granulation. The significance of hormonal regulation of the numbers of gastric mucosal mast cells is unknown, although the association between stress, gastric mast cell degranulation, and gastric erosions[75,76] indicates that important clinical implications exist.

### F. MAST CELLS AND SOMATOSTATIN

Theoharides and Douglas[77] reported that somatostatin stimulates histamine secretion by rat peritoneal mast cells. This result has been confirmed by others[78-80] and extended to human peripheral blood leukocytes.[81] The concentrations of somatostatin effective in histamine secretion were high, namely $5 \times 10^{-7}$ to $10^{-5}$ *M*. Shanahan et al.[80] established that in the rat, somatostatin-induced mediator release was mast cell subtype specific, in that it activated peritoneal mast cells, but was ineffective in the stimulation of intestinal mucosal mast cells.

Using much lower concentrations of somatostatin (i.e., $10^{-15}$ to $10^{-8}$ *M*), Goetzl and Payan[82] established that release of histamine and LTD4 from human basophils and rat leukemic basophils was inhibited. Somatostatin similarly inhibits histamine secretion from mast cells cultured from mouse bone marrow using IL-3,[83] mast cells thought to be analogous to rat intestinal mucosal mast cells. Interestingly, Goetzl et al.[83] also reported that rat basophilic leukemic cells and mouse bone marrow-derived mast cells contain somatostatin. Thus, it appears that mast cells and somatostatin interact in ways that may have considerable clinical relevance in allergic or other inflammatory diseases.

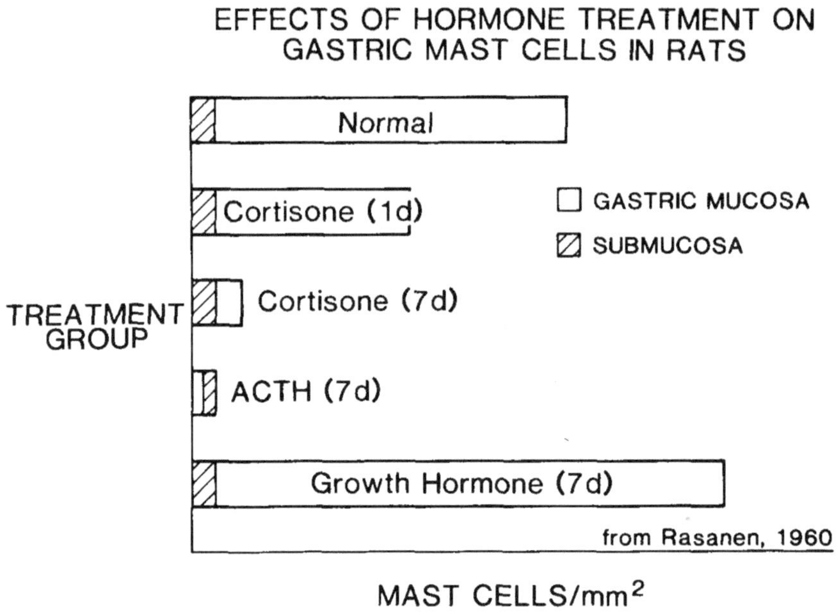


FIGURE 4. Effects of hormone treatments on the numbers of gastric mast cells.

**TABLE 1**
**Hormonal Influences on Mast Cells[a]**

| Hormone | Effect on mast cells: Activation | Inhibition | Proliferation |
|---|---|---|---|
| ACTH | √ | | |
| Cortisone | √ | √ | √ |
| Estrogen | √ | | √ |
| Gastrin/pentagastrin | √ | | |
| Parathyroid hormone | √ | | √ |
| Preluteinizing hormone | √ | | |
| Somatostatin | √ | √ | |
| Somatotropin (growth hormone) | | | √ |
| Thyroid stimulating hormone | √ | | √ |

[a] Effects listed on mast cells from different sites and species. References in text. A √ indicates effect and a blank indicates no effect or not studied.

## III. SUMMARY AND POSTULATIONS

The evidence is overwhelming that there are reciprocal interactions between mast cells and components of the endocrine system. Table 1 summarizes many of the effects of hormones on mast cells, whereas Figures 1 to 3 identify some of the effects of mast cells on the functions of endocrine organs and hormone targets. These reciprocal interactions can involve direct communications employing receptor-ligand linkages or indirect strategies employing other cell types as intermediaries. In the context of the theme of this book and the chapter of Stead et al.,[1] one of the most important collaborating cell types is the nerve. The best example to illustrate this type of interaction may be the vasculature, where the network of nervous, endocrine, and mast cell mediator regulation of flow and permeability is well known. Such networks of cooperation or antagonism must be the normal state of affairs for the regulation of function of organs, tissues, and cells throughout the body, as has been reviewed for the thyroid gland.[11] Unfortunately, the spectrum of neurotransmitters and mast cell mediators which have been studied is limited, especially for mast cells where histamine, and to a lesser extent serotonin, have been almost the only mediators studied in the context of endocrine function.

Future studies must fill some of the gaps which exist in knowledge about mast cells and the endocrine system. As a working model, mast cells should be viewed as heterogeneous populations of cells which are highly adapted to respond appropriately in their specific microenvironments. Possibly the most important component of this heterogeneity is the expression of site-specific surface receptors which make the cell uniquely responsive to factors in the microenvironment. For example, thyroid mast cells may express surface receptors for TSH, whereas mast cells elsewhere do not; or uterine mast cells express receptors for estrogen and mast cells in many other sites do not. With the appropriate molecular probes for such receptors and *in situ* hybridization studies, combined with mast cell histochemistry or electron microscopy, this hypothesis could be addressed. Once activated, mast cells release mediators into the local environment, some of which enter the circulation and may modulate the vasculature to influence hormone delivery. They also act on other cells in the endocrine gland to influence their secretion, differentiation, or proliferation. This hypothesis could be broadly applicable to mast cell function in any organ or tissue and accordingly provides a stimulating basis for future analysis of the reciprocity of mast cell-endocrine communication.

## ACKNOWLEDGMENTS

Ms. Gillian Hardy provided skilled secretarial assistance. The Medical Research Council of Canada has provided financial support for the research activities of the author.

## REFERENCES

1. **Stead, R. H., Perdue, M. H., Blennerhassett, M. G., Katuta, Y., Sestini, P., and Bienenstock, J.,** The innervation of mast cells, in *The Neuroendocrine-Immune Network,* Freier, S., Ed., CRC Press, Boca Raton, FL, 1989.
2. **Befus, D., Lee, T., and Bienenstock, J.,** Immunologic and physiologic features of mast cell function at mucosal surfaces, in *Mucosal Immunity and Infections at Mucosal Surfaces,* Strober, W., Lamm, M. E., McGhee, J., and James, S. P., Eds., Oxford University Press, New York, 1988, 209.
3. **Befus, D., Fujimaki, H., Lee, T. D. G., and Swieter, M.,** Mast cell polymorphisms. Present concepts, future directions, *Dig. Dis. Sci.,* 33, 165, 1988.
4. **Befus, A. D., Bienenstock, J., and Denburg, J. A.,** *Mast Cell Differentiation and Heterogeneity,* Raven Press, New York, 1986, 426.
5. **Galli, S. J.,** New approaches for the analysis of mast cell maturation, heterogeneity and function, *Fed. Proc.,* 46, 1096, 1987.
6. **Otsu, K., Nakano, T., Kanakura, Y., Asai, H., Katz, H. R., Austen, K. F., Stevens, R. L., Galli, S. J., and Kitamura, Y.,** Phenotypic changes of bone marrow-derived mast cells after intraperitoneal transfer into $W/W^v$ mice that are genetically deficient in mast cells, *J. Exp. Med.,* 165, 615, 1987.
7. **Levi-Schaffer, F., Austen, K. F., Gravallese, P. M., and Stevens, R. L.,** Coculture of interleukin 3-dependent mouse mast cells with fibroblasts results in a phenotypic change of the mast cells, *Proc. Natl. Acad. Sci. U.S.A.,* 83, 6485, 1986.
8. **Beaven, M. A.,** Histamine: its role in physiological and pathological processes, *Monogr. Allergy,* 13, 113, 1978.
9. **Padawer, J. and Gordon, A. S.,** Endocrine influences on the mast cells of the rat, *Anat. Rec.,* 124, 659, 1956.
10. **Csaba, G.,** Regulation of mast cell formation, in *Studia Biologica Academiae Scientiarum Hungaricae,* Szentagothai, J., Ed., Publishing House of the Hungarian Academy of Sciences, Budapest, 1972.

11. **Melander, A.,** Aminergic regulation of thyroid activity: importance of the sympathetic innervation and of the mast cells of the thyroid gland, *Acta Med. Scand.*, 201, 257, 1977.
12. **Wynford-Thomas, D. and Stringer, B. M. J.,** Mast cell hyperplasia during goitrogen-induced thyroid growth — a quantitative study, *Acta Endocrinol.*, 101, 365, 1982.
13. **Banerjee, S. N. and Dasgupta, S. R.,** Age-related alteration of thyroidal and mesenteric mast cells of male albino rat, *Exp. Gerontol.*, 16, 475, 1981.
14. **Melander, A., Westgren, U., Sundler, F., and Ericson, L. E.,** Influence of histamine- and 5-hydroxytryptamine-containing thyroid mast cells on thyroid blood flow and permeability in the rat, *Endocrinology*, 97, 1130, 1975.
15. **Bianco, A. C., Nunes, M. T., and Douglas, C. R.,** Influence of mast cells on thyroid function, *Endocrinol. Exp.*, 17, 99, 1983.
16. **Csaba, G. and Barath, P.,** Effect of pinealectomy on the $^3$H-5HT uptake of mast cells in the thyroid gland of the rat, *Acta Anat.*, 89, 442, 1974.
17. **Huntley, J. F., Newlands, G., and Miller, H. R. P.,** The isolation and characterization of globule leucocytes: their derivation from mucosal mast cells in parasitized sheep, *Parasite Immunol.*, 6, 371, 1984.
18. **Carriere, R. and Bushke, M.,** The influence of thyroid and testicular hormones on globule leucocytes in the rat duodenal crypt epithelium, *Anat. Rec.*, 192, 407, 1978.
19. **Rasanen, T.,** Mucosal mast cells of rat stomach; influence of ACTH, cortisone and growth hormone, *Gastroenterology*, 38, 70, 1960.
20. **Smith, D. E. and Lewis, Y. S.,** Influence of hypophysis and adrenal cortex upon tissue mast cell of the rat, *Proc. Soc. Exp. Biol. Med.*, 87, 515, 1954.
21. **Smith, D. E. and Lewis, Y. S.,** Influence of hypophysis and of adrenal cortex upon tissue mast cell of the hamster, *Proc. Soc. Exp. Biol. Med.*, 88, 631, 1955.
22. **Reilly, M. A. and Sigg, E. B.,** Suppression of histamine-induced adrenocorticotropic hormone release by antihistamines and antidepressants, *J. Pharmac. Exp. Therap.*, 222, 583, 1981.
23. **Portaleone, P., DiCarlo, F., and Genazzani, E.,** Hypothalamic $H_1$ and $H_2$ receptors: a tool to investigate the neuroendocrine role of histamine, *Pharmacol. Res. Commun.*, 12, 365, 1980.
24. **Gray, W. D. and Munson, P. L.,** The rapidity of the adrenocorticotropic response of the pituitary to the intravenous administration of histamine, *Endocrinology*, 48, 471, 1951.
25. **Fuche, J. and Kahlson, G.,** Histamine as a stimulant to the anterior pituitary gland as judged by the lymphopenic response in normal and hypophysectiomized rabbits, *Acta Physiol. Scand.*, 39, 327, 1957.
26. **Reilly, M. A.,** Biogenic amine participation in histamine stimulation of ACTH release, *Agents Actions*, 14, 630, 1984.
27. **Jasani, B. and Stanworth, D. R.,** Studies on the mast cell triggering action of certain artificial histamine liberators, *Int. Arch. Allergy*, 45, 74, 1973.
28. **Jasani,B., Kreil, G., Mackler, B. F., and Stanworth, D. R.,** Further studies on the structural requirements for polypeptide-mediated histamine release from rat mast cells, *Biochem. J.*, 181, 623, 1979.
29. **Burt, D. S. and Stanworth, D. R.,** Changes in cellular levels of cyclic amp in rat mast cells during secretion of histamine induced by immunoglobulin decapeptide and ACTH(1-24) peptide: comparison with immunological and ionophore triggers, *Biochim. Biophys. Acta*, 762, 458, 1983.
30. **Stanworth, D. R.,** The role of nonantigen receptors in mast cell signalling process, *Mol. Immunol.*, 21, 1183, 1984.
31. **Schleimer, R. P.,** The mechanisms of antiinflammatory steroid action in allergic diseases, *Annu. Rev. Pharmacol. Toxicol.*, 25, 381, 1985.
32. **King, S. J., Miller, H. R. P., Newlands, G. F. J., and Woodbury, R. G.,** Depletion of mucosal mast cell protease by corticosteroids: effect on intestinal anaphylaxis in the rat, *Proc. Natl. Acad. Sci. U.S.A.*, 82, 1214, 1985.
33. **Lavker, R. M. and Schechter, N. M.,** Cutaneous mast cell depletion results from topical corticosteroid usage, *J. Immunol.*, 135, 2368, 1985.
34. **Frame, B. and Nixon, R. K.,** Bone-marrow mast cells in osteoporosis of aging, *N. Engl. J. Med.*, 279, 626, 1968.
35. **Rockoff, S. D. and Armstrong, J. D., Jr.,** Parathyroid hormone as a stimulus to mast cell accumulation in bone, *Calcif. Tiss. Res.*, 5, 49, 1970.
36. **McKenna, M. J. and Frame, B.,** The mast cell and bone, *Clin. Orthop*, 200, 226, 1985.
37. **Wilhelms, O.-H., Kreusser, W., and Ritz, E.,** Parathyroid hormone elicits histamine release from mast cells, *Mineral Electrolyte Metab.*, 6, 294, 1981.
38. **Tsakalos, N. D., Theoharides, T. C., Kops, S. K., and Askenase, P. W.,** Induction of mast cell secretion by parathormone, *Biochem. Pharmacol.*, 32, 355, 1983.
39. **Sherwood, J., Reinhard, D., and Garcia, M.,** Does cimetidine inhibit parathyroid hormone secretion?, *N. Engl. J. Med.*, 300, 200, 1977.
40. **Finn, C. A.,** Implantation, menstruation, and inflammation, *Biol. Rev.*, 61, 313, 1986.
41. **Arvy, L.,** Effect of injections of estrogen on the mast cells of the white mouse, *Nature (London)*, 175, 506, 1955.

42. **Hermo, L. and Lalli, M.,** Monocytes and mast cells in the limiting membrane of human seminiferous tubules, *Biol. Reprod.*, 19, 92, 1978.
43. **Nistal, M., Santamaria, L., and Paniagua, R.,** Mast cells in the human testis and epididymis from birth to adulthood, *Acta Anat.*, 119, 155, 1984.
44. **Maseki, Y., Miyake, K., Mitsuya, H., Kitamura, H., and Yamada, K.,** Mastocytosis occurring in the testes from patients with idiopathic male infertility, *Fertil. Steril.*, 36, 814, 1981.
45. **Krishna, A. and Terranova, P. F.,** Alterations in mast cell degranulation and ovarian histamine in the proestrous hamster, *Biol. Reprod.*, 32, 1211, 1985.
46. **Kobayashi, T., Wright, K. H., Santulli, R., Kitai, H., and Wallach, E. E.,** Effect of histamine and histamine blockers on the ovulatory process in the *in vitro* perfused rabbit ovary, *Biol. Reprod.*, 28, 385, 1983.
47. **Iversen, O. H.,** Mast cells in the myometrium of the human cervix uteri, and changes caused by androgenic and estrogenic hormones, *Acta Pathol. Microbiol. Scand.*, 49, 337, 1960.
48. **Harvey, E. B.,** Mast cell distribution in the uterus of cycling and pregnant hamsters, *Anat. Red.*, 148, 507, 1964.
49. **Marcus, G. J., Shelesnyak, M. C., and Kraicer, P. F.,** Studies on the mechanism of nidation: X. The oestrogen-surge, histamine-release, and decidual induction in the rat, *Acta Endocrinol.*, 47, 255, 1964.
50. **Brandon, J. M. and Evans, J. E.,** Changes in uterine mast cells during the estrous cycle in the syrian hamster, *Am. J. Anat.*, 167, 241, 1983.
51. **Brandon, J. M. and Bibby, M. C.,** A study of changes in uterine mast cells during early pregnancy in the rat, *Biol. Reprod.*, 20, 977, 1979.
52. **Rahima, A. and Soderwall, A. L.,** Mast cells in uteri of pregnant young and senescent female golden hamsters, *Biol. Reprod.*, 17, 523, 1977.
53. **Shelesnyak, M. C.,** Fall in uterine histamine associated with ovum implantation in pregnant rat, *Soc. Exp. Biol. Med.*, 100, 380, 1959.
54. **Spaziani, E. and Szego, C. M.,** The influence of estradiol and cortisol on uterine histamine of the ovariectomized rat, *Endocrinology*, 63, 669, 1958.
55. **Shelesnyak, M. C.,** Histamine-releasing activity of natural estrogens, *Proc. Soc. Exp. Biol. Med.*, 100, 739, 1959.
56. **Nalbandov, A. V.,** Endocrine control of implantation, in *Biology of the Blastocyst*, Blandau, R. J., Ed., University of Chicago Press, IL, 1971, 383.
57. **Gibbons, A. F. E. and Chang, M. C.,** Number of mast cells in the rat uterus with special reference to its relation to hormonal treatment and decidual response, *Biol. Reprod.*, 6, 193, 1972.
58. **Brandon, J. M. and Wallis, R. M.,** Effect of mepyramine, a histamine $H_1$ – , and burimamide, a histamine $H_2$-receptor antagonist, on ovum implantation in the rat, *J. Reprod. Fertil.*, 50, 251, 1977.
59. **Dey, S. K., Villanueva, C., Chien, S. M., and Crist, R. D.,** The role of histamine in implantation in the rabbit, *J. Reprod. Fertil.*, 53, 23, 1978.
60. **Dey, S. K., Johnson, D. C., and Santos, J. G.,** Is histamine production by the blastocyst required for implantation in the rabbit?, *Biol. Reprod.*, 21, 1169, 1979.
61. **Dey, S. K., Villanueva, C., and Abdou, N. I.,** Histamine receptors on rabbit blastocyst and endometrial cell membranes, *Nature (London)*, 278, 648, 1979.
62. **Saksena, S. K., Lau, I. F., and Chang, M. C.,** Relationship between oestrogen, prostaglandin $F_2$, and histamine in delayed implantation in the mouse, *Acta Endocrinol.*, 81, 801, 1976.
63. **Kahlson, G. and Rosengren, E.,** New approaches to the physiology of histamine, *Physiol. Rev.*, 48, 155, 1968.
64. **Moore, P. J., Velky, P., and Rosenquist, T. H.,** The effect of an intrauterine device on mast cell numbers and distribution in the hamster uterus, *Res. Exp. Med. (Berlin)*, 175, 223, 1979.
65. **Kobayashi, T. K., Okamoto, H., Harami, K., and Yuasa, M.,** The presence of mast cells in IUD smears: a preliminary report, *Acta Cytol.*, 24, 268, 1980.
66. **Tursi, A., Mastrorilli, A., Ribatti, D., Loiudice, L., Contino, R., and Claudatus, J.,** Possible role of mast cells in the mechanism of action of intrauterine contraceptive devices, *Am. J. Obstet. Gynecol.*, 148, 1064, 1984.
67. **Sandvei, R., Wollen, A.-L., Flood, P. R., and Anker, C.,** Mast cells in the tubal wall in women using an intrauterine contraceptive device, *Br. J. Obstet. Gynaecol.*, 93, 758, 1986.
68. **Soll, A. H., Toomey, M., Thomas, L., Shanahan, F., and Beaven, M. A.,** Histamine cells in the canine fundic mucosa, in *Mast Cell Differentiation and Heterogeneity*, Befus, A. D., Bienenstock, J., and Denburg, J., Eds., Raven Press, New York, 1986, 313.
69. **Hakanson, R., Bottcher, G., Ekblad, E., Panula, P., Simonsson, M., Dohlsten, M., Hallberg, T., and Sundler, F.,** Histamine in endocrine cells in the stomach: a survey of several species using a panel of histamine antibodies, *Histochemistry*, 86, 5, 1986.
70. **Heap, B. J. and Kiernan, J. A.,** Histological, histochemical and pharmacological observations on mast cells in the stomach of the rat, *J. Anat.*, 115, 315, 1973.

71. **Saavedra-Delgado, A. M., Turpin, S., and Metcalfe, D. D.,** Typical and atypical mast cells of the rat gastrointestinal system: distribution and correlation with tissue histamine, *Agents Actions,* 14, 1, 1984.
72. **Kahlson, G., Rosengren, E., Scahn, D., and Thunberg, R.,** Mobilization and formation of histamine in the gastric mucosa as related to acid secretion, *J. Physiol.,* 174, 400, 1964.
73. **Man, W. K., Saunders, J. H., Ingoldby, C., and Spencer, J.,** Effect of pentagastrin on histamine output from the stomach in patients with duodenal ulcer, *Gut,* 22, 916, 1981.
74. **Tharp, M. D., Thirlby, R., and Sullivan, T. J.,** Gastrin induces histamine release from human cutaneous mast cells, *J. Allergy Clin. Immunol.,* 74, 159, 1984.
75. **Sathiamoorthy, S. S., Ganguly, A. K., and Bhatnagar, O. P.,** Effect of pylorus ligation on gastric mucosal mast cell population in normal and adrenalectomised albino rats, *Experientia,* 35, 508, 1979.
76. **Diel, F., Borck, H., and Hosenfeld, S.,** Effects of somatostatin on ethanol-induced gastric erosions in the rat: role of mast cells, *Agents Actions,* 18, 273, 1986.
77. **Theoharides, T. C. and Douglas, W. W.,** Somatostatin induces histamine secretion from rat peritoneal mast cells, *Endocrinology,* 102, 1637, 1978.
78. **Baxter, J. H. and Adamik, R.,** Differences in requirements and actions of various histamine-releasing agents, *Biochem. Pharmacol.,* 27, 497, 1978.
79. **Theoharides, T. C., Betchaku, T., and Douglas, W. M.,** Somatostatin-induced histamine secretion in mast cells. Characterization of the effect, *Eur. J. Pharmacol.,* 69, 127, 1981.
80. **Shanahan, F., Denburg, J., Fox, J., Bienenstock, J., and Befus, D.,** Mast cell heterogeneity: effects of neuroenteric peptides on histamine release, *J. Immunol.,* 135, 1331, 1985.
81. **Diel, F., Bethge, N., and Opree, W.,** Histamine secretion in leukocyte incubates of patients with allergic hyperreactivity induced by somatostatin-14 and somatostatin-28, *Agents Actions,* 13, 216, 1983.
82. **Goetzl, E. J. and Payan, D. G.,** Inhibition by somatostatin of the release of mediators from human basophils and rat leukemic basophils, *J. Immunol.,* 133, 3255, 1984.
83. **Goetzl, E. J., Chernov-Rogan, J., Furuichi, K., Goetzl, L. M., Lee, J. Y., and Renold, F.,** Neuromodulation of mast cell and basophil function, in *Mast Cell Differentiation and Heterogeneity,* Befus, A. D., Bienenstock, J., and Denburg, J. A., Eds., Raven Press, New York, 1986, 223.

Chapter 5

# SIGNALING PATHWAYS OF THE NEUROTRANSMITTER-IMMUNE NETWORK

**Thomas L. Roszman and William H. Brooks**

## TABLE OF CONTENTS

# I. INTRODUCTION

Immune function must be considered in the context of the environment in which the immune system resides. Lymphocytes and macrophage responsiveness is susceptible to alteration by a variety of cytokines and growth factors. One of the most provocative of the modulatory influences of immune reactivity is the central nervous system (CNS) and its specific soluble mediators, e.g., neurotransmitters and neurohormones. The interaction of neurohormones with lymphocytes and macrophages has been reviewed previously.[1]

Although a variety of neurotransmitters have been shown to alter lymphocyte function, the present review is confined to catecholamines, acetylcholine, and serotonin. In addition to an overview of the presence (or absence) of receptors for these neurotransmitters on lymphocytes and macrophages, the functional significance of these receptors as a modulatory link between the CNS and immune system is evaluated. Subsequently, these data are collated in a dual receptor regulatory network focusing on the role of neurotransmitters in neuroimmune modulation.

# II. CATECHOLAMINE RECEPTORS

The best characterized as well as most unambiguous of the lymphocyte- and macrophage-associated receptors for catecholamine neurotransmitters is the membrane receptor for β-adrenergic compounds. These receptors appear to be present on lymphoid cells obtained from a number of animal species,[2-6] but the majority of studies have used human peripheral mononuclear cells for definition and functional investigation of the receptor.[7-17] Considerable differences in the number and affinity of β-adrenergic receptors have been reported; affinities ranging from 0.01 to 30 n*M* and numbers varying from 500 to 10,000 binding sites per cell. Although, in part, the range of values can be explained by methodological differences, e.g., the use of broken cell preparations vs. intact cells or choice of radiolabeled ligand, it should be emphasized that peripheral blood mononuclear cell preparations contain a heterogeneous mixture of cells. Thus, B-cells,[8,9,11] T-cells,[8,9,11,15,18] and thymocytes[2,18] all have been demonstrated to have beta-adrenergic receptors. The complexity of the use of unpurified or only partially purified peripheral blood mononuclear cells for these studies is increased further by the three well-described subsets of T-cells; T-helper (Th), T-suppressor (Ts), and T-cytotoxic (Tc). Although the affinity of the receptors on these cells is reasonably similar ($Kd = 2.5 - 5.2 \times 10^{-11}M$) each subset characteristically has a different number of binding sites (Ts = 2900, Tc = 1800 and Th = 750 binding sites/cell).[15] These observations must be kept in mind as one reviews previous reports regarding the definition and function of this receptor on lymphocytes. The potential significance of the heterogeneity of β-adrenergic receptors on the various T-cell subsets will be discussed in the context of the function of these cells.

The evidence for the presence of α-adrenergic receptors on lymphocytes is not as persuasive as it is for the β-adrenergic receptors. It has been demonstrated that membranes prepared from spleen cells obtained from guinea pigs bind $^3H$-clonidine.[19] Both high- and low-affinity sites were detected and the relative potency of agonists and antagonists indicated binding to the $\alpha_2$-type receptor. Additional pharmacologic data also suggest that lymphocytes or macrophages have α-adrenergic receptors.[20-22] Thus, although the presence of lymphoid cell receptors for the α-adrenergics is reported, confirmation and further characterization as demonstrated with β-adrenergic receptors remains to be done. At present, the data must be considered preliminary.

The presence of dopamine receptors on lymphocytes is controversial; a number of investigators have been able to detect them,[23-26] while others cannot.[27,28] Initial studies demonstrated the stereospecific binding of the dopamine antagonist, $^3H$-spiroperidol, to peripheral blood lymphocytes obtained from rats, rabbits, dogs, and human beings.[23] The majority of these studies were performed with rat lymphocytes and demonstrated saturable binding with dissociation constants of 1.9 and 36.2 n*M*. This binding was found to be proportional to cell number and temperature with maximum binding occurring at 37°C. The binding of $^3$H-spiroperidol to rat lymphocytes could be inhibited by (+) butaclamol and other dopaminergic antagonists such as pinozole, haloperidol, chlorpromazine, mezilamine, clozapine, and sulpiride. This same group also has demonstrated $^3$H-spiroperidol binding to B-cells.[24]

Another approach to ascertaining whether neurotransmitter receptors are present on cells is to determine if appropriate agonists activate second messengers. Accordingly, the report that the addition of dopamine to lymphocyte suspension induces significant amounts of cAMP[29] suggests the presence of dopamine receptors of the D1 type which is linked to adenylate cyclase.[30] Thus, addition of $10^{-5}$ to $10^{-6}$ *M* dopamine to rat spleen cells results in a threefold increase in cAMP over basal levels within 10 min. Similar results with rat spleen cell have been obtained by Kouassi et al.[31] who demonstrated $5 \times 10^{-4}$ to $1 \times 10^{-4}$ *M* dopamine was required to elevate cAMP levels approximately 1.3-fold. However, the increase in cAMP was not as dramatic as the 3.5-fold increase noted when $10^{-5}$ *M* isoproterenol was added to these cells. Although these results collectively suggest that the presence of adenylate cyclase linked dopamine receptors (D1) on lymphocytes, they must be viewed with caution because D1 receptors have not been detected by direct-binding in studies employing the radiolabeled dopamine agonist amino-6,7 dihydroxy-1,2,3,4 tetrahydronaphthalene.[32]

The inability of others to demonstrate the presence of either D1 or D2 dopamine receptors on lymphocytes can be attributed to a number of factors. For example, the radiolabeled ligand could be taken up by the cell via either a specific or nonspecific mechanism.[25] Also, it is possible that the labeled ligand is nonspecifically associated with the membrane. A systematic, detailed study of dopamine receptors on lymphocytes taking these factors into consideration is required to resolve these issues. Additionally, it is important to determine whether these receptors are preferentially present on lymphocyte subsets or macrophages, and if they are expressed after cell activation.

## III. CHOLINERGIC RECEPTORS

The presence of muscarinic cholinergic receptors on lymphocytes was first reported by Gordon et al.[33] although there was previous indirect evidence for these receptors based on modulation of lymphocyte function by cholinergic agonists.[34-36] In these studies, using $^3$H-quinuclidinyl benzilate to detect cholinergic receptors on murine splenic lymphocytes, approximately 200 receptors per cell with a Kd of $1 \times 10^9$ *M* were found. This binding was inhibited by atropine and oxotremorine, thereby indicating that the receptors were of the muscarinic type. Subsequently, there have been numerous reports demonstrating the presence of muscarinic cholinergic receptors on T-cells,[37,38] B-cells,[38] monocytes,[39] and thymocytes[40] as determined by $^3$H-quinuclidinyl benzilate binding. However, there are other reports indicating that cholinergic receptors are not detectable on lymphocytes[28,41] or that some of these are of the nicotinic type,[42,43] as determined by pharmacologic studies using cholinergic agonists and antagonists to alter lymphocyte function. Despite these reports, there appears to be general agreement among those laboratories that have found cholinergic receptors regarding the affinity of muscarinic cholinergic receptors on lymphoid cells; the values range between approximately 1 to 65 n*M*.[44-48] However, it should be emphasized that these values are approximately 100-fold lower than the values reported for the same receptors present on brain tissue[49].

Muscarinic receptors predominantly are expressed on T-cells[37] with further evidence to suggest that they are modulated after *in vitro* activation of T-cells.[37] There are no systematic studies of the receptor characteristic on T-cell subsets with the exception of those of Strom et al.,[37] which demonstrated that the binding isotherms for murine T-cells were dissimilar from that of Th-cells. Thus, considerable work remains to be done before the presence and biological relevance of these receptors can be fully established and understood.

## IV. SEROTONIN RECEPTORS

Until recently, there were no reports indicating the presence of serotonin (5-HT) receptors on lymphocytes or macrophages despite the functional data to support the existence of such receptors on these cells (reviewed in next section.) There is now evidence suggesting the presence of 5-HT receptor on murine splenic lymphocytes, but not on thymocytes or peritoneal macrophages.[50] In this study lymphocytes were shown to bind $^3$H-5HT with a Kd of 198 n*M* and a Bmax of 3.53 n*M*. This low-affinity, high-capacity lymphocyte receptor is much different from that detected on brain homogenates where Kd of two to three logs higher are noted.[51] Moreover, the most effective inhibitor of lymphocyte 5-HT binding was imipramine which inhibits a 5-HT uptake system rather than a more classical receptor. More recently, we have examined murine lymphocytes, peritoneal macrophages, and membrane preparations from these cells for the presence of 5-HT receptors using $^3$H-5HT or $^3$H-ketanserin and have been unable to detect high affinity binding sites.[52] Moreover, activation of the lymphocytes with mitogen does not result in the expression of high affinity 5-HT receptors. Parallel assays employing rat cerebral cortex synaptosmal membrane preparations demonstrated high affinity binding with $^3$H-5HT (Kd = 2 n*M*, Bmax 43 fmol/mg tissue) and $^3$H-ketanserin (Kd = 0.8 n*M*, Bmax 24 fmol/mg tissue). Collectively, these results clearly indicate that murine lymphocytes and macrophages do not have high-affinity 5-HT receptors. Nevertheless, in this same study we were able to identify and characterize a high affinity and active 5-HT uptake system in both splenic and thioglycolate elicited macrophages. Splenic macrophages had a Km of 37 n*M* with a Bmax 409 p*M*/3 $\times$ $10^6$ cells/10 min while peritoneal macrophages had a Km of 110 n*M* with a Bmax of 940 p*M*/3 $\times$ $10^6$ cells/10 min. The specificity of this uptake system was demonstrated by employing fluoxetine (a highly selective inhibitor of 5-HT), presynaptic reuptake, and additional 5-HT antagonists. Incubation of the macrophages with $10^{-5}$ *M* fluoxetine reduced $^3$H-5HT uptake to a value comparable to that observed in the presence of excess unlabeled 5-HT. Significant blocking of $^3$H-5HT uptake also was produced by 5-HT receptor antagonists spiperone, ketanserin, and cyproheptadine, yielding $IC_{50}$ values of 0.7, 10, and 100 μ*M*, respectively. These results establish the existence of a specific 5-HT uptake system in macrophages, the significance of which will be discussed later in this review.

## V. MODULATION OF IMMUNE FUNCTION BY CATECHOLAMINES

There are numerous studies demonstrating that lymphocyte function can be modulated by surface adrenergic receptors. Among the earliest is that of Hadden, et al.[53] who demonstrated that high concentrations (2.5 $\times$ $10^{-4}$ *M*) of norepinephrine, epinephrine, and isoproterenol, as well as the α-antagonist phentolamine and β-antagonist propranolol, suppress mitogen responsiveness of human peripheral blood mononuclear cells. However, when more physiologic concentrations ($10^{-9}$ to $10^{-7}$ *M*) of norepinephrine and isoproterenol were employed in these studies, the time and duration of exposure of lymphocytes to these substances became important. Addition of norepinephrine for a brief period at initiation or at the end of the culture period induced increased mitogen responsiveness; similar additions of isoproterenol resulted in depressed lymphocyte responsiveness. These data suggest that α-adrenergic receptor stimulation enhances lymphocyte mitogen responsiveness, while β-receptor stimulation decreases it.

A number of other studies also have demonstrated that mitogen responsiveness of T- and B-cells is suppressed by the addition of β-adrenergic receptor-activating substances.[54-56] While these results agree in general, the interpretation is clouded because β-adrenergic receptors are nonrandomly distributed on B-cells, T-cells, and their subsets.[15] The importance of this becomes apparent in recent studies which demonstrate that Th-cell subsets respond differently to phytohemagglutinin (PHA) in the presence of isoproterenol.[57] In this study, the PHA responsiveness of Leu $3^+8^+$Th-cells was enhanced in the presence of $10^{-5}$ *M* isoproterenol while Leu $3^+8^-$Th-cells were not. These results indicate that modulation of lymphocyte reactivity by catecholamines should be performed by employing highly purified lymphocyte subsets to delineate individualistic, rather than holistic effects.

In an attempt to explore the mechanism(s) responsible for the isoporterenol-induced suppression of mitogen responsiveness, Feldman et al.[58] quantitated effects of isoproterenol on interleukin-2 (IL-2) production and IL-2 receptor (IL-2r) expression on mitogen-stimulated human peripheral blood lymphocytes. While isoproterenol decreased PHA-induced responsiveness and IL-2 production, it had no effect on the expression of high affinity IL-2r. Their results demonstrated that the expression of the low affinity IL-2r (p55 protein) was reduced at 24 h after mitogen-activation, mainly on the T-suppressor/cytotoxic subset. It should be noted that 100 μ*M* of isoproterenol was required to achieve reductions in IL-2 production and IL-2r expression. In a similar study, it was observed that 10 μ*M* isoproterenol inhibited IL-2 production by activated T-cells, but not IL-2r expression.[59]

There are a number of studies which have examined the effects of adrenergic receptor-activating substances on antigen specific *in vitro* immune responses. Among these, are those concerned with the effects of α- and β-adrenergic agonists and antagonists on the *in vitro* primary antibody responses of murine spleen cells (reviewed in Reference 60). Both enhancement[61-64] and suppression[21] of this response has been observed. Overall, these studies suggest that stimulation of $\alpha_1$- and $\beta_2$-receptors on lymphocytes leads to enhancement of the antibody while $\alpha_2$-receptor stimulation is linked to suppression. While the results of these studies pertained more to the initiation and subsequent cellular events which give rise to antibody forming cells, addition of isoproterenol to the plaque-forming cell assay itself also decreases the number of antibody-producing cells detected.[65] In similar studies evaluating cell-mediated immunity, it was observed that isoproterenol depressed the ability of cytotoxic T-cells to kill appropriate target cells (reviewed in Reference 66). Of interest is that if isoproterenol, norepinephrine, or epinephrine were added to the spleen cell cultures at the time of initiation of the mixed lymphocyte reaction, the number of cytotoxic T-cells elicited was enhanced.[67] Thus, while adrenergic receptor activating drugs enhance the generation of either antibody-forming cells and/or cytotoxic T-cells (suggesting their effect to occur at the level of cellular inductive events including proliferation and differentiation), they have the opposite effect on the function of the effector cells elicited, i.e., antibody-forming cells and cytotoxic T-cells. Support for this interpretation comes from the observation that preincubation of lymphocytes with adrenaline ($10^{-7}$ to $10^{-9}$ *M*) for 16 h followed by removal of the drug enhances natural killer (NK)-cell activity to K562 target cells.[68] However, if adrenaline is added directly to the assay system containing the NK cells and targets inhibition of killing is observed. In addition to effects on NK-cell activity, there are data indicating that both epinephrine and norepinphrine can inhibit the gamma interferon activation of murine macrophages to become cytotoxic for herpes simplex virus infected target cells.[69] Because antagonists were not employed in this study, it is not known whether this effect was mediated through α- or β-adrenergic receptors.

The mechanisms responsible for the modulation of lymphocyte and macrophage function by α- and β-receptor agonists are not well defined. It has long been known that activation of β-adrenergic receptors on these cells leads to a rapid increase in cAMP (reviewed in Reference 70), and that increased intracellular levels of this second messenger are generally associated with a decrease in the function of these cells. Thus, activation of the beta-adrenergic receptors on B-cells,[8,71-73] T-cells,[8,71-73] and macrophages[69] results in an early rise in cAMP concentrations. As previously discussed, β-adrenergic receptors are nonrandomly distributed on T-cell subsets[15] which is further reflected in the cAMP response elicited in these subsets when they are activated with isoproterenol.[74] The largest isoproterenol-induced cAMP increase occurs in the Ts-subset followed by Tc-cells and then the Th-cells. The Th-cells can be further subdivided into Leu $3^+8^+$ and Leu $3^+8^-$ with the latter demonstrating a greater accumulation of intracellular cAMP in response to isoproterenol.[54] It is interesting to note that Leu $3^+8^+$ cells exhibit enhanced mitogen responsiveness in the presence of isoproterenol as compared to these cells incubated in the absence of this drug.

Precisely how activation of $\alpha_1$- or $\alpha_2$-adrenergic receptors modulates lymphocyte or macrophage function is not clear. Moreover, it is not yet formally established if lymphocytes or macrophages have α-adrenergic receptors, and thus there is no information on the distribution of these receptors on various lymphocyte subsets. In other cellular systems $\alpha_1$-adrenergic receptor stimulation results in activation of the phosphatidylinositol (PI) cycle and the generation of diacylglycerol (DAG) and inositol-1,4,5 trisphosphate ($IP_3$) while $\alpha_2$-adrenergic receptors are linked to a guanine nucleotide regulatory protein (G protein) in this case, G inhibitory (Gi), which inhibits adenylate cyclase.[75]

Currently there is no unifying hypothesis which can be evoked to account for the varying effects of α- and β-receptor stimulation on lymphocyte and macrophage function. As the activation and progression signals required to bring T-cells,[76] B-cells,[77] and macrophages[78] to a full state of immune readiness are revealed, a better understanding of how α- and β-agonists will influence this will be gained. What is clear is that a lymphocyte receives a number of external signals from antigen and cytokines which initiates a cascade of biochemical events leading to activation. It is in the context of these biochemical events that intracellular signals emanating from stimulation of α- and β-receptors must be interpreted and integrated. In the last section of this review, the implications of this "cross talk" are considered.

As previously discussed, there is controversy over whether lymphocytes or macrophages have dopamine receptors. Moreover, there are no well-delineated studies demonstrating that dopamine can modulate the function of these cells. There are studies showing that dopamine can elevate cAMP levels in lymphocytes, presumably through interaction with the D1 receptor.[30,31] Until experiments are performed relating the presence of dopamine receptors on lymphocytes or macrophages with effects on the function of these cells, the importance of dopamine as an immunomodulator will remain unclear.

## VI. MODULATION OF IMMUNE FUNCTION OF 5-HT

As is the case for dopamine, there is no definitive evidence indicating that "resting" or activated lymphocytes or macrophages have high affinity receptors for 5-HT. Nonetheless, there are data to indicate that 5-HT can modulate the function of these cells. Thus, 5-HT ($10^{-6}$ to $10^{-8}$ *M*) can suppress the gamma interferon (IFN)-induced expression of Ia antigen on murine bone marrow macrophages and the effect is reversed by $5\text{-HT}_2$ serotonergic receptor antagonists.[79] Similarly, 5-HT in the range of $10^{-6}$ to $10^{-7}$ *M* either suppresses or enhances gamma IFN-induced phagocytosis by macrophages.[80] Enhancement of phagocytosis occurs with high concentrations of 5-HT, while suppression is noted with low concentrations. We have obtained data indicating that gamma IFN-induced expression of Ia antigen on thioglycolate-induced peritoneal macrophages is inhibited by 5-HT ($10^{-5}$ to $10^{-7}$*M*) and have further demonstrated that this suppression is the result of elevated levels of prostaglandins ($PGE_2$) produced by the macrophages as a consequence of synergism between the signals delivered by gamma IFN and 5-HT (Jackson, J. and Roszman T., unpublished observations). Moreover, 5-HT ($10^{-4}$ to $10^{-5}M$) can increase the level of cGMP in monocytes as well as increase the chemotactic responsiveness of these cells.[81] It is of interest that prostaglandins can reverse the 5-HT increase in monocyte chemotaxis, suggesting a role for cAMP.

While high affinity 5-HT receptors cannot be detected on macrophages, these cells do have a high affinity (40 n*M*), specific 5-HT uptake system.[52] The relevancy of this uptake system becomes apparent when placed in the context of the concentrations of 5-HT ($10^{-5}$ to $10^{-8}$ *M*) required to modulate macrophage concentration. Although the concentration of 5-HT in platelet free plasma is low (2 to 3 ng/ml), it is released by platelets on aggregation at localized inflammatory sites in concentrations ($10^{-6}$ to $10^{-8}$ *M*) sufficient to interact with the uptake system and modulate macrophage function.

The *in vitro* effects of 5-HT on lymphocyte function also have been studied. Thus, 5-HT inhibits the proliferative response of mitogen[82,83] and alloantigen-stimulated[83,84] lymphocytes. In all cases, $10^{-3}$ *M* 5-HT is required to obtain significant inhibition of proliferation without evidence of a dose-response curve. In one of these studies,[84] it was determined that 5-HT at this concentration was toxic for the lymphocytes. The biologic significance of this 5-HT immunosuppression is questionable although it is possible that 5-HT could achieve a concentration of $10^{-3}$ *M* in a microenvironment.

## VII. MODULATION OF IMMUNE FUNCTION BY CHOLINERGIC AGENTS

The function of lymphocytes is generally enhanced in the presence of muscarinic receptor agonists such as acetylcholine. The incubation of lymphocytes for a period of 24 h in the presence of acetylcholine ($10^{-6}$ *M*) resulted in an increase in RNA and protein synthesis, but not in DNA synthesis.[20] In this study it also was noted that there was augmented PHA-induced RNA, DNA, and protein synthesis by these lymphocytes when acetylcholine, carbachol ($10^{-9}$ *M*), or methylcholine ($10^{-9}$ *M*) were added, and this augmentation could be blocked by the muscarinic antagonist atropine ($10^{-6}$ *M*). Similar results have been obtained employing rat thymocytes.[85,86] Thus, addition of acetylcholine ($5 \times 10^{-9}$ to $1 \times 10^{-7}$ *M*) increased DNA synthesis and proliferation by thymocytes.

Somewhat different results have been obtained with human peripheral blood lymphocytes employing the muscarinic and nicotinic agonist carbamylcholine.[42,43] Addition of various concentrations of carbamylcholine to unstimulated lymphocytes followed by monitoring of proliferation by $^3$H-thymidine incorporation at 72 h of culture revealed a biphasic curve of enhanced proliferation. Carbamylcholine concentration of $10^{-11}$ to $10^{-10}$ *M* and $10^{-6}$ to $10^{-5}$ *M* induced increased proliferation above baseline values, while $10^{-9}$ to $10^{-7}$ *M* did not. Similar results were obtained when lymphocytes were stimulated with mitogen, except that maximum proliferation occurred at 24 h after initiation of the cultures. Using appropriate antagonists, it was concluded that the enhanced proliferation resulted from carbamylcholine binding to muscarinic receptors while the decreased proliferation was blocked by d-tubocurarine and alpha-bungarotoxin, indicating the presence of nicotinic receptors. Thus, lymphocytes or monocytes appear to have three functionally distinct acetylcholine receptors, two of which have muscarinic pharmacologic properties and are immunostimulatory. The third receptor appears to be nicotinic in nature and inhibitory to lymphocyte function in the presence of intermediate carbamylcholine concentrations ($10^{-9}$ *M*), apparently through the induction of Ts-cells.[42]

Enhancement of other lymphocyte functions has been observed after stimulation of the muscarinic receptor. Increased mobility of lymphocytes occurs after the addition of a wide concentration range ($10^{-5}$ to $10^{-10}$ *M*) of either acetylcholine or carbamylcholine which is antagonized by atropine.[87] In a series of experiments, Strom and co-workers[34,66,88,89] have examined the cholinergic influence on *in vivo* generated alloreactive Tc. The Tc were obtained from animals skin grafted with allogeneic tissue and placed in a cell-mediated lympholysis assay employing appropriate target cells. Addition of either carbamylcholine or acetylcholine augmented the activity of the Tc as evidenced by increased cytotoxicity to the target cells. Maximum augmentation of cytotoxicity was observed with $10^{-12}$ *M* carbamylcholine and $10^{-10}$ *M* acetylcholine, and this augmentation was antagonized by atropine, indicating the involvement of muscarinic acetylcholine receptors. Further experiments demonstrated that the effects of the drugs were directed toward the Tc and not the target cells.[89] Moreover, augmentation was not observed if the Tc were preincubated with carbamylcholine for more than 6 min before the introduction of the target cells or if the drug was added after both the Tc and target cells were mixed; indicating a short-lived effect on the Tc and not target cells.

Cholinergic stimulation of lymphocytes increases the intracellular accumulation of cGMP.[35,66,85] In other tissues, the muscarinic acetylcholine receptor mediates various cellular responses including inhibition of adenylate cyclase, breakdown of phosphoinositides, and modulation of potassium channels through the action of proteins.[90,91] Presently, the role of these cellular events in the enhanced lymphocyte activity after stimulation of muscarinic acetylcholine receptors is unknown.

## VIII. NEUROTRANSMITTER-IMMUNE NETWORK OF SIGNALING PATHWAYS

The phenotypic features of the central nervous system and the immune systems include the use of soluble mediators to carry information among the various elements, thereby amplifying other biologically weak signals in efforts to contribute to the well-being of the individual. In addition to architectural similarities, there is growing evidence that the CNS and the mediators of immunity are functionally interrelated.[92] Yet despite the clinical and investigational observations that have led to an operational concept of neuroimmunodulation, the mechanism(s) by which the CNS and mononuclear cells communicate remains elusive. If neurally processed information does indeed pass between the CNS and immune system it must occur via humoral mediators, i.e., neurohormones and/or neurotransmitters. The potential pathways for modulatory signaling between the neuroendocrine and immune system has recently been reviewed.[1] The present review has focused on the other soluble messenger of neural information.

As observed with neurohormone-lymphocyte interactions, the mode of neurotransmitter action on lymphocyte function must be placed in the context of those biochemical and molecular events required for the development of lymphocyte competency and proliferation. Although the differing subpopulations of lymphocytes may vary in their susceptibility to neurotransmitter-induced modulation, as previously discussed, it is unlikely that subtle shifts in neurotransmitter release would induce sufficient quantitative alterations in these subpopulations to affect immune reactivity. Rather, because surface receptors define the cellular response to exogenous agents, e.g., neurotransmitters, antigens, and cytokines, and thus play a critical role in homeostasis and subsequent responsiveness, it may be suggested that simultaneous interactions of neurotransmitters with specific receptors or uptake systems and antigen-binding at the T-cell receptor (TCR) might result in facilitation or down-regulation of lymphocyte responsiveness. Such integrated signaling further suggests the presence of a regulatory network linking the CNS with mononuclear cells. Therefore, it may be hypothesized that it is the events occurring within this network, rather than in suppressor/helper cells and idiotypic/anti-idiotypic networks, that determine the effects of neural modulation of immunity.

The number of surface receptors is thought to be regulated by changes in the rate of receptor synthesis and/or degradation or recycling following receptor internalization. For example, it has been demonstrated that the numbers of receptors for specific neurotransmitters on a cell is influenced by altering concentrations of agonists and antagonists for that particular transmitter.[93] Also, the presence of receptors for IL-2 is influenced by the presence of its target-specific ligand, IL-2,[94] as is the TCR when it is bound by anti-T3 antibody.[95] Although intracellular second messengers induced by neurally derived substances and/or monocyte activation alter expression of various surface receptors, the effects of neurotransmitter and antigen receptor-binding on expression of the specific receptor for the other remains unknown. Modulation of the TCR following neurotransmitter-binding to its receptor is one possible mechanism by which the immune function could be regulated by these neurally linked messengers.

The initial events of receptor-ligand interaction which are amenable to modulation occur in the G proteins. The G proteins which share extensive structural homology are known to couple receptors to a number of different effector molecules, including adenylate cyclase, plasma membrane K channels, and cyclic guanosine 5′-monophosphate phosphodiesterase.[96] These G proteins are known to be involved in signal transduction and have been recently reviewed by Stryer and Bourne.[97] The biochemical cascade of events occurring subsequent to normal TCR-binding and G-protein activation[98,99] include coupling to phospholipase C and progression of phosphoinositide breakdown as well as induction of protein kinases, events that are required for the development of lymphocyte competence and later proliferation. Adrenoreceptor occupancy also induces G-protein activation which stimulates adenylate cyclase, production of cAMP,[100,101] and eventually the induction of a cAMP-dependent protein kinase (PKA).[102] In this manner, cAMP levels modulate the level of receptor phosphorylation, as well as determine the balance between PKC and tyrosine phosphorylation which is refractory to cAMP modulation of the TCR complex.[103] Interestingly, cAMP production also can be modulated by the effects of protein kinase C (PKC) activity on Gi proteins, leading to enhanced adenylate cyclase activity, thereby offering another mechanism of autoregulation.[104] Alternatively, PKC can bypass and/or overcome the inhibitory effects of PKA, thus shifting the balance in favor of enhanced cellular proliferation. Neural-immune interactions, therefore, may occur at the level of G proteins. Accordingly, the modulatory effects of neurotransmitters are determined by a balance of stimulatory and inhibitory G proteins, thereby regulating the earliest steps of transmembrane signaling arising from differential receptor interactions by transmitter and/or antigen. A similar concept of programmable G protein second messengers in modifying hormone-target cell interaction has been proposed by Roddell.[105]

The consequence of TCR-G protein binding is the rapid hydrolysis of membrane phospholipid phosphatidylinositol 4,5 biphosphate by the activation of phospholipase C.[106] This yields two intracellular second messengers, $IP_3$ and DAG. $IP_3$ induces the release of $Ca^{2+}$ from internal stores which results in an increase in cytoplasm from $Ca^{2+}$. In addition, $IP_3$ is further metabolized into $IP_4$ which may have additional signaling functions including regulation of plasma membrane $Ca^{2+}$ channels. This allows influx of extracellular $Ca^{2+}$ into the cell to maintain elevated concentrations of cytoplasmic $Ca^{2+}$ required for continued enzymatic processes leading to transformation. The other breakdown product of the PI cycle, DAG, activates cytosolic PKC which translocates to the surface membrane for phosphorylation of additional important proteins including portions of the TCR.[103]

Induction of G stimulatory (Gs) proteins via neurotransmitter-receptor interaction leads to the activation of adenylate cyclase, while activation of Gi proteins by other neurotransmitters results in inhibition of adenylate cyclase.[107,108] Thus, it is apparent that the balance of Gs to Gi proteins is an important determinant of adenylate cyclase activity and subsequent generation of cAMP and PKA. This balance also is important in governing the generation of cGMP[109] which has been implicated in modulating lymphocyte and monocyte function (reviewed in Reference 70). The consensus appears to be that increased levels of cAMP down-regulate lymphocyte proliferative responsiveness, whereas cGMP facilitates proliferation and monocyte cytokinesis. Therefore, another possible level where neurotransmitters may modify the biological response of appropriately signalled immune cells occurs with variations in the generation of G protein regulated cAMP and cGMP.

In general, protein kinases exist within the cell in an inactive state and are positively regulated by other second messengers binding to regulatory domains or subunits.[110] Regulation of the enzymatic activity of these kinases, therefore, would be expected to set the gain or amplify the signals induced by the specific kinase. Thus, in terms of a neurally modifiable lymphocyte proliferative network, the interrelationships of the various protein kinases are of importance. The cAMP-dependent PKA is capable of modulating PKC activity by limiting PI breakdown and subsequent generation of the specific phosphoinositol-dependent kinase.[103] In this manner. normal TCR-induced lymphocyte activation and proliferation is attenuated by loss of threonine and serine phosphorylation of the receptor, presumably required for full responsiveness. Phosphorylation of the TCR by PI-independent and cAMP-refractory tyrosine kinase may occur in this setting,[111,112] adding further credence to the postulate that neuroimmune modulation is not "all or none", but rather governs or sets the gain of immune responsiveness.

Concurrently, PKC is capable of modifying cAMP production by phosphorylating the catalytic units of adenylate cyclase.[113] The resultant increases in cAMP levels would be expected to increase the gain of potential responsiveness, and thus tend to down-regulate the development of lymphocyte competence and subsequent proliferation. Alternatively, PKC can bypass and/or overcome the inhibitory effects of PKA and hence function to lower the gain and promote activation.[103] Because the normal sequence of events following TCR-activation as well as neurotransmitter interaction can both induce the generation of PKC activity, it becomes evident that a balance in the TCR and transmitter-generated pathways becomes vital in determining whether the amounts and specific activities of this kinase will favor raising or lowering the gain of potential responsiveness. Hence, the balance of these regulatory kinases in the proliferative network is another modulatory interface between the central nervous and immune systems.

In addition to membrane and cytosolic events, nuclear events also are important for lymphocyte responsiveness. Several important growth factors as well as tyrosine-directed protein kinases are encoded by cellular oncogenes and are participants in the biochemical pathways essential for activation and proliferation.[110] Recently, it has been demonstrated that lymphocyte activation results in the early expression of the c-myc proto-oncogene which has been implicated in transcriptional modulation and may be important in the transition of cells from the $G_0$ to $G_1$ phase of the cell cycle.[114] The observation that cAMP represses transcription of c-myc suggests a further point of entry for neuromodulation of lymphocyte function.[115] Thus, neurotransmitter-induced increases in cAMP may shift the balance in favor of attenuated immune reactivity. Alternatively, lowered levels of cAMP or cAMP-dependent kinases would be anticipated to simultaneously promote transcription of these and other proto-oncogenes important for progressive signaling within the lymphocyte.

The immediate results of these early events is the acquisition of receptors for and the secretion of various cytokines required for the proliferative phases of lymphocyte responsiveness. Thus, in addition to regulating the development of lymphocyte competence, another mechanism by which neurotransmitters may govern immune activation is through modulation of the secretion and/or expression of these lymphokine receptors. Support for this hypothesis is gained from the recent report that IL-2 receptor expression is suppressed by β-adrenergic agonists.[58] Alternatively, IL-2 binding to its receptor profoundly inhibits both basal and appropriately stimulated adenylate cyclase activity and cAMP accumulation.[116] Although the mechanism(s) by which this occurs remains to be determined, these findings suggest that the balance of second messengers is equally important in the proliferative phase of lymphocyte activation as that observed in the earlier phases of the development of lymphocyte competence.

In summary, the biological and other biochemical observations involved in the development of lymphocyte competence and subsequent proliferative responsiveness to appropriate receptor-ligand interaction has led to the proposal of a proliferative regulatory network. In the context of neural modulation of immune reactivity, the ability of the lymphocyte or monocyte to appropriately respond to receptor-mediated and transmembrane signaling is differential rather than exclusive and may be determined by the intracellular context in which it is evoked by the dual-receptor regulatory mechanisms as described. Further exploration of this network will reveal the biological and homeostatic importance of the milieu in which the immune system resides and functions.

## ACKNOWLEDGMENT

This chapter was supported in part by an NIH grant, NS 17423.

## REFERENCES

1. **Roszman, T. L. and Brooks, W. H.,** Signaling pathways of the neuroendocrine-immune network, *Prog. Allergy,* 43, 140, 1989.
2. **Singh, V., Millson, D. S., Smith, P. A., and Owen, J. T.,** Identification of beta adrenoceptors during thymocyte ontogeny in mice, *Eur. J. Immunol.,* 9, 31, 1979.
3. **Casale, T. B., Halonen, M., and Kaliner, M.,** Detection of beta-adrenergic receptors on rabbit mononuclear cells isolated free of significant contamination by other cell types, *Life Sci.,* 33, 971, 1983.
4. **Mills, K., Aiweh, S., Olein, G., Arnason, B. G. W., and Chelmicka-Schorr, E.,** Beta adrenergic receptors on splenic lymphocytes from axotomized mice, *Int. J. Immunopharmacol.,* 6, 171, 1984.
5. **Abrass, C. K., O'Connor, S. W., Scarpace, P. J., and Abrass, I. B.,** Characterization of the beta adrenergic receptor of the rat peritoneal macrophage, *J. Immunol.,* 135, 1338, 1985.
6. **Westly, H. J. and Kelley, K. H.,** Down-regulation of glucocorticoid and beta adrenergic receptors on lectin-stimulated splenocytes, *Proc. Soc. Exp. Biol. Med.,* 185, 211, 1987.
7. **Williams, L. T., Snyderman, R., and Lefkowitz, R. J.,** Identification of beta adrenergic receptors in human lymphocyte by (−) $^{3}$H-alprenolol binding, *J. Clin. Invest.,* 57, 149, 1976.
8. **Bishopric, N. H., Cohen, H. J., and Lefkowitz, R. J.,** Beta adrenergic receptors in lymphocyte subpopulations, *J. Allergy Clin. Immunol.,* 65, 29, 1980.
9. **Pochet, R., Delepresse, G., Gausset, P. W., and Collet, H.,** Distribution of beta-adrenergic receptors on human lymphocyte subpopulations, *Clin. Exp. Immunol.,* 38, 578, 1979.
10. **Brodde, O. E., Engel, G., Hoyer, D., Bock, K. D., and Weber, F.,** The beta-adrenergic receptor in human lymphocytes: subclassification by the use of a new radio-ligand, $^{125}$I iodocyanopindolol, *Life Sci.,* 29, 2189, 1981.
11. **Krawietz, W., Werdan, K., Schober, M., Erdmann, E., Rindfleisch, G. E., and Hannig, K.,** Different numbers of beta receptors in human lymphocyte subpopulations, *Biochem. Pharmacol.,* 31, 133, 1982.
12. **Meurs, H., Van Den Bogoard, W., Kauffman, H. F., and Burynzeel, P. L. B.,** Characterization of (−) $^{3}$H-dihydroalprenolol binding to intact and broken cell preparations of human peripheral blood lymphocytes, *Eur. J. Pharmacol.,* 85, 185, 1982.
13. **Halper, J. P., Mann, J. J., Weksler, M. E., Bilezikian, J. P., Sweeney, J. A., Brown, R. P., and Golbourne, T.,** Beta adrenergic receptors and cyclic AMP levels in intact human lymphocytes, *Life Sci.,* 35, 855, 1984.
14. **Brodde, O. E., Kretsch, R. Ikezono, K., Zerkowski, H. R., and Reidemeister, J. C.,** Human beta-adrenoceptors: relation of mycoardial and lymphocyte beta-adrenoceptor density, *Science,* 231, 1584, 1986.
15. **Khan, M. M., Sansoni, P., Silverman, E. D., Engleman, E. G., and Melmon, K. L.,** Beta-adrenergic receptors on human-suppressor, helper and cytotoxic lymphocytes, *Biochem. Pharmacol.,* 35, 1137, 1986.
16. **Thede-Reynolds, K. R., Motulsky, H. J., and Feldman, R. D.,** Temperature-dependent binding of hydrophilic beta adrenergic receptor ligands to intact human lymphocytes, *Life Sci.,* 39, 1325, 1986.

17. **Feldman, R. D. and Lai, C. C.**, Characterization of the human lymphocyte beta-adrenergic receptor by photoaffinity labeling. Alterations with desensitization, *Cir. Res.*, 58, 384, 1986.
18. **Van De Griend, R. J., Astaldi, A., Wijermans, P., Van Doorn. R., and Roos, D.**, Low beta-adrenergic receptor concentration on human thymocytes, *Clin. Exp. Immunol.*, 53, 53, 1983.
19. **McPherson, G. A. and Summers, R. J.**, Characteristics and localization of $^3$H-clonidine binding in membranes prepared from guinea pig spleen, *Clin. Exp. Pharmacol. Physiol.*, 9, 77, 1982.
20. **Hadden, J. W., Hadden, E., and Goldberg, N. D.**, Cyclic GMP and cyclic AMP in lymphocyte metabolism and proliferation, in *Cyclic AMP, Cell Growth, and the Immune Response*, Braun, W., Lichtenstein, L. M., and Parker, C. W., Eds., Springer-Verlag, New York, 1974, 237.
21. **Besedovsky, H. O., Del Ray, A., Sorkin, E., Da Prada, M., and Keller, H. H.**, Immunoregulation mediated by the sympathetic nervous system, *Cell Immunol.*, 48, 346, 1979.
22. **Sanders, V. M. and Munson, A. E.**, Role of alpha adrenoceptor activation in modulating the murine primary antibody response *in vitro, J. Pharmacol. Exp. Ther.*, 232, 395, 1985.
23. **Le Fur, G., Phan, T., and Uzan, A.**, Identification of sterospecific $^3$H-spiroperidol binding sites in mammalian lymphocytes, *Life Sci.*, 26, 1139, 1980.
24. **Le Fur, G., Phan, T., Canton, T., Tur, C., and Uzan, A.**, Evidence for a coupling between dopaminergic receptors and phospholipid methylation in mouse B lymphocytes, *Life Sci.*, 29, 2737, 1981.
25. **Bridart, J. M., Motte, Ph., Assicot, M., Bohuon, C., and Bellet, D.**, Catechol-O-methyltransferase activity and aminergic binding sites distribution in human peripheral blood lymphocyte subpopulations, *Clin. Immunol. Immunopathol.*, 26, 1, 1983.
26. **Ovadia, H. and Abramsky, O.**, Dopamine receptors on isolated membranes of rat lymphocytes, *J. Neurosci. Res.*, 18, 70, 1987.
27. **Fleminger, S., Jenner, P., and Marsden, C. D.**, Are dopamine receptors present on human lymphocytes?, *J. Pharm. Pharmacol.*, 34, 658, 1981.
28. **Maloteaux, J. M., Waterkein, C., and Laduron, P. M.**, Absence of dopamine and muscarinic receptors on human lymphocytes, *Arch. Int. Pharmacodyn.*, 258, 174, 1982.
29. **Stepien, H., Kunert-Radek, J., Karasek, E., and Pawlikowski, M.**, Dopamine increases cyclic AMP concentration in rat spleen lymphocytes *in vitro, Biochim. Biophys. Res. Comm.*, 101, 1057, 1981.
30. **Niznik, H. B.**, Dopamine receptors; molecular structure and function, *Mol. Cell Endocrinol.*, 54, 1, 1987.
31. **Kouassi, E., Li, Y. S., Flacher, M., and Revillard, J. P.**, Cyclic adenosine 3′5′-monophosphate response to dopamine in mouse lymphoid cells and cell lines, *Life Sci.*, 40, 2385, 1987.
32. **Shaskan, E. G., Ballow, M., Lederman, M., Margoles, S. L., and Melchreit, R.**, Spiroperidol-binding sites on mouse lymphoid cells. Effects of ascorbic acid and psychotropic drugs, *J. Neuroimmunol.*, 6, 59, 1984.
33. **Gordan, M. A., Cohen, J. J., and Wilson, I. B.**, Muscarinic cholinergic receptors in murine lymphocytes: demonstration by direct binding, *Proc. Natl. Acad. Sci. U.S.A.*, 75, 2902, 1978.
34. **Strom, T. B., Sytkowski, A. J., Carpenter, C. B., and Merrill, J. P.**, Cholinergic augmentation of lymphocyte-mediated cytotoxicity. A study of the cholinergic receptor of cytotoxic T lymphocytes, *Proc. Natl. Acad. Sci. U.S.A.*, 71, 1330, 1974.
35. **Illiano, G., Tell, G. P. E., Siegel, M. I., and Cuatrecasas, P.**, Guanosine 3′:5′-cyclic monophosphate and the action of insulin and acetylcholine, *Proc. Natl. Acad. Sci. U.S.A.*, 70, 2443, 1973.
36. **Hadden, J. W., Johnson, E. M., Hadden, E., Coffey, E. G., and Johnson, L. D.**, Cyclic GMP and lymphocyte activation, in *Immune Recognition*, Rosenthal, A. S., Ed., Academic Press, New York, 1985, 359.
37. **Strom, T. B., Lane, M. A., and George, K.**, The parallel, time-dependent, bimodal change in lymphocyte cholinergic binding activity and cholinergic influence upon lymphocyte-mediated cytotoxicity after lymphocyte activation, *J. Immunol.*, 127, 705, 1981.
38. **Atweh, S. F., Grayhack, J. J., and Richman, D. P.**, A cholinergic receptor site on murine lymphocytes with novel binding characteristics, *Life Sci.*, 35, 2459, 1984.
39. **Lopker, A., Abood, L. G., Hoss, W., and Lionetti, F. J.**, Stereoselective muscarinic acetylcholine and opiate receptors in human phagocytic leukocytes, *Biochem. Pharmacol.*, 29, 1361, 1979.
40. **Maslinski, W., Grabczewska, E., Laskowska-Bozek, H., and Ryzewski, J.**, Expression of muscarinic cholinergic receptors during T-cell maturation in the thymus, *Eur. J. Immunol.*, 17, 1059, 1987.
41. **Wazer, D.E. and Rotrosen, J.**, Murine lymphocytes lack clearly defined receptors for muscarinic and dopaminergic ligands, *J. Pharm. Pharmacol.*, 36, 853, 1984.
42. **Richman, D. P., Antel, J. P., Burns, J. B., and Arnason, B. G. W.**, Nicotinic acetylcholine receptor on human lymphocytes, *Ann. N.Y. Acad. Sci.*, 377, 427, 1981.
43. **Richman, D. P. and Arnason, B. G. W.**, Nicotinic acetylcholine receptor: evidence for a functionally distinct receptor on human lymphocytes, *Proc. Natl. Acad. Sci. U.S.A.*, 76, 4632, 1979.
44. **Zalcman, S. J., Neckers, L. M., Kaayalp, O., and Wyatt, R. J.**, Muscarinic cholinergic binding sites on intact human lymphocytes, *Life Sci.*, 29, 69, 1981.

45. **Bidart, J. M., Moinageon, Ph., and Bohuon, C.**, Muscarinic cholinergic receptors on circulating human lymphocytes: evidence for a particulate binding site, *Res. Commun. Chem. Pathol. Pharmacol.*, 39, 169, 1983.
46. **Shenkman, L., Rabery, J. M., and Gilad, G. M.**, Cholinergic muscarinic binding by rat lymphocytes: effects on antagonist treatment, strain, and aging, *Brain Res.*, 380, 303, 1986.
47. **Adem, A., Nordberg, A., and Slanina, P.**, A muscarinic receptor type in human lymphocytes: a comparison of $^3$H-ONB binding to intact lymphocytes and lysed lymphocyte membranes, *Life Sci.*, 38, 1359, 1986.
48. **Rabery, J. M., Shenkman, L., and Gilad, G. M.**, Cholinergic muscarinic binding by human lymphocytes: changes with aging, antagonist treatment, and senile dementia of the Alzheimer type, *Ann. Neurol.*, 20, 628, 1986.
49. **Hanley, M. R. and Iversen, L. L.**, Muscarinic cholinergic receptors in rat corpus striatum and regulation of guanosine cyclic 3′5′-monophosphate, *Mol. Pharmacol.*, 14, 246, 1978.
50. **Bonnet, M., Lespinats, G., and Burtin, C.**, Evidence for serotonin (5HT) binding sites on murine lymphocytes, *Int. J. Immunopharmacol.*, 9, 551, 1987.
51. **Peroutka, S. J., and Synder, S. H.**, Multiple serotonin receptors: differential binding of [$^3$H] 5-hydroxytryptamine, [$^3$H] lysergic acid diethylamide and [$^3$H] spiroperidol, *Mol. Pharmacol.*, 16, 687, 1979.
52. **Jackson, J. C., Walker, R. F., Brooks, W. H., and Roszman, T. L.**, Specific uptake of serotonin by murine macrophages, *Life Sci.*, 42, 1641, 1988.
53. **Hadden, J. W., Hadden, E. M., and Middleton, E., Jr.**, Lymphocyte blast transformation. I. Demonstration of adrenergic receptors in human peripheral lymphocytes, *Cell. Immunol.*, 1, 583, 1970.
54. **Smith, J. W., Steiner, A. L., Newberry, W. M., and Parker, C. W.**, Cyclic adenosine 3′,5′-monophosphate in human lymphocytes. Alterations after phytohemagglutinin stimulation, *J. Clin. Invest.*, 50, 432, 1971.
55. **Smith, J. W., Stiner, A. L., Newberry, W. M., and Parker, C. W.**, Human lymphocyte metabolism. Effects of cyclic and noncyclic nucleotides on stimulation by phytohemagglutinin, *J. Clin. Invest.*, 50, 442, 1971.
56. **Johnson, D. L., Ashmore, R. C., and Gordon, M. A.**, Effects of β-adrenergic agents on the murine lymphocyte response to mitogen stimulation, *J. Immunopharmacol.*, 3, 205, 1981.
57. **Khan, M. M., Silverman, E., Engleman, E. G., and Melmon, K. L.**, Responses of subpopulations of human helper T cells to autacoids, *Proc. West. Pharmacol. Soc.*, 28, 225, 1985.
58. **Feldman, R. D., Hunninghake, G. W., and McArdle, W. L.**, β-adrenergic receptor-mediated suppression of interleukin-2 receptors in human lymphocytes, *J. Immunol.*, 139, 3355, 1987.
59. **Chouaib, S., Welte, K., Mertelsmann, R., and Dupont, B.**, Prostaglandin E2 acts at two distinct pathways of T-lymphocyte activation: inhibition of interleukin-2 production and down-regulation of transferrin receptor expression, *J. Immunol.*, 135, 1172, 1985.
60. **Sanders, V. M. and Munson, A. E.**, Norepinephrine and the antibody response, *Pharmacol. Rev.*, 37, 229, 1985.
61. **Burchiel, S. W. and Melmon, K. L.**, Augmentation of the *in vitro* humoral immune response by pharmacologic agents. I. An explanation for the differential enhancement of humoral immunity via agents that elevate cAMP, *Immunopharmacology*, 1, 137, 1979.
62. **Sanders, V. M. and Munson, A. E.**, Beta adrenoceptor mediation of the enhancing effect of norepinephrine on the murine primary antibody response *in vitro, J. Pharmacol. Exp. Ther.*, 230, 183, 1984.
63. **Sanders, V. M. and Munson, A. E.**, Kinetics of the enhancing effect produced by norepinephrine and terbutaline on the murine primary antibody response *in vitro, J. Pharmacol. Exp. Ther.*, 231, 527, 1984.
64. **Sanders, V. M. and Munson, A. E.**, Role of alpha adrenoceptor activation in modulating the murine primary antibody response *in vitro, J. Pharmacol. Exp. Ther.*, 232, 395, 1985.
65. **Melmon, K. L., Bourne, H. R., Weinstein, Y., Shearer, G. M., Kram, J., and Bauminger, S.**, Hemolytic plaque-formation by leukocytes *in vitro, J. Clin. Invest.*, 53, 13, 1974.
66. **Strom, T. B. and Carpenter, C. B.**, Cyclic nucleotides in immunosuppression-neuroendocrine pharmacologic manipulation and *in vivo* immunoregulation of immunity acting via second messenger systems, *Transplant. Proc.*, 12, 304, 1980.
67. **Hatfield, S. M., Peterson, B. H., and Di Micco, J. A.**, Beta adrenoceptor modulation of the generation of murine cytotoxic T lymphocytes *in vitro, J. Pharmacol. Exp. Ther.*, 239, 460, 1986.
68. **Hellstrand, K., Hermodsson, S., and Strannagard, Ö.**, Evidence for a β-adrenoceptor-mediated regulation of human natural killer cells, *J. Immunol.*, 134, 4095, 1985.
69. **Koff, W. C. and Dunegan, M. A.**, Neuroendocrine hormones suppress macrophages-mediated lysis of herpes simplex virus-infected cells, *J. Immunol.*, 136, 705, 1986.
70. **Coffey, R. G. and Hadden, J. W.**, Neurotransmitters, hormones, and cyclic nucleotides in lymphocyte regulation, *Fed. Proc.*, 44, 112, 1985.
71. **Bach, M. A.**, Differences in cyclic AMP changes after stimulation by prostaglandins and isoproterenol in lymphocyte subpopulations, *J. Clin. Invest.*, 55, 1074, 1975.
72. **Niaudet, P., Genevieve, B., and Bach, M. A.**, Differences in effect of isoproterenol stimulation on levels of cyclic AMP in human B and T lymphocytes, *Eur. J. Immunol.*, 6, 834, 1976.

73. **Stachelin, M., Müller, P., Portenier, M., and Harris, A. W.,** β-adrenergic receptors and adenylate cyclase activity in murine lymphoid cell lines, *J. Cyclic Nucleo. Protein. Phosphor. Res.*, 10, 55, 1985.
74. **Khan, M. M., Sansoni, P., Engleman, E. G., and Melmon, K. L.,** Pharmacologic effects of autocoids on subsets of T cells, *J. Clin. Invest.*, 75, 1578, 1985.
75. **Exton, J. H.,** Mechanisms involved in alpha-adrenergic phenomena, *Am. J. Physiol.*, 248, E633, 1985.
76. **MacDonald, H. R.,** T-cell activation, *Annu. Rev. Cell Biol.*, 2, 231, 1986.
77. **Klaus, G. G. B., Bijsterbosch, M. K., O'Garra, A., Harnett, M. M., and Rigley, K. P.,** Receptor signalling and crosstalk in B-lymphocytes, *Immunol. Rev.*, 99, 19, 1987.
78. **Adams, D. O. and Hamilton, T. A.,** Molecular transductional mechanisms by which IFN gamma and other signals regulate macrophage development, *Immunol. Rev.*, 97, 5, 1987.
79. **Sternberg, E. M., Trial, J., and Parker, C. W.,** Effect of serotonin on murine macrophages: suppression of Ia expression by serotonin and its reversal by 5-$HT_2$ serotonergic receptor antagonists, *J. Immunol.*, 137, 276, 1986.
80. **Sternberg, E. M., Wedner, H. J., Leung, M. K., and Parker, C. W.,** Effect of serotonin (5-HT) and other monoamines on murine macrophages: modulation of interferon-gamma induced phagocytosis, *J. Immunol.*, 138, 4360, 1987.
81. **Gallin, J. I., Sandler, J. A., Clyman, R. I., Manganiello, V. C., and Vaughn, M.,** Agents that increase cyclic AMP inhibit accumulation of cGMP and depress human monocyte locomotion, *J. Immunol.*, 120, 492, 1978.
82. **Slauson, D. O., Walker, C., Kristensen, F., Wang, Y., and De Weck, A. L.,** Mechanisms of serotonin-induced lymphocyte proliferation inhibition, *Cell. Immunol.*, 84, 240, 1984.
83. **Bonnet, M., Lespinats, G., and Burtin, C.,** Histamine and serotonin suppression of lymphocyte response to phytohemagglutinin and allogeneic cells, *Cell. Immunol.*, 83, 280, 1984.
84. **Schwartz, A., Askenase, P. W., and Gershon, R. K.,** The effect of locally injected vasoactive amines on the elicitation of delayed-type hypersensitivity, *J. Immunol.* 118, 159, 1977.
85. **Mac Manus, J. P., Boynton, A. L., Whitfield, J. F., Gillan, D. J., and Isaacs, R. J.,** Acetylcholine-induced initiation of thymic lymphoblast DNA synthesis and proliferation, *J. Cell. Physiol.*, 85, 321, 1975.
86. **Whitfield, J. F., Mac Manus, J. P., Rixon, R. H., Boynton, A. L., Youdate, T., and Swierenga, S.,** The positive control of cell proliferation by the interplay of calcium ions and cyclic nucleotides. A review, *In Vitro*, 12, 1, 1976.
87. **Schreiner, G. F. and Unanue, E. R.,** The modulation of spontaneous and anti-Ig-stimulated motility of lymphocytes by cyclic nucleotides and adrenergic and cholinergic agents, *J. Immunol.*, 114, 802, 1975.
88. **Strom, T. B., Deisseroth, A., Morganroth, J., Carpenter, C. B., and Merrill, J. P.,** Alterations of the cytotoxic action of sensitized lymphocytes by cholinergic agents and activators of adenylate cyclase, *Proc. Natl. Acad. Sci. U.S.A.*, 69, 2995, 1972.
89. **Strom, T. B., Carpernter, C. B., Garovoy, M. R., Austin, K. F., Merrill, J. P., and Kaliner, M.,** The modulating influence of cyclic nucleotides upon lymphocyte-mediated cytotoxicity, *J. Exp. Med.*, 138, 381, 1983.
90. **Brown, D.,** Acetylcholine and brains cells, *Nature (London)*, 319, 358, 1986.
91. **Gilman, A. G.,** G proteins and dual control of adenylate cyclase, *Cell*, 36, 577, 1984.
92. **Solomon, G. F.,** Psychoneuroimmunology: interactions between central nervous systems and immune system, *J. Neurosci. Res.*, 18, 1, 1987.
93. **Motulsky, H. J., Cunningham, E. M. S., Deblasi, A., and Insel, P. A.,** Desensitization and redistribution of beta-adrenergic receptors on human mononuclear leukocytes, *Am. J. Physiol.*, 250, E583, 1986.
94. **Smith, K. A.,** Interleukin-2, *Annu. Rev. Immunol.*, 2, 319, 1984.
95. **Reinherz, E. L., Meuer, S., Fitzgerald, K. A., Hussey, R. E., Levine, H., and Schlossman, S. F.,** Antigen recognition by human T-lymphocytes is linked to surface expression of the T3 molecular complex, *Cell*, 30, 735, 1982.
96. **Graziano, M. P. and Gilman, A. G.,** Guanine nucleotide-binding regulatory proteins: mediators of transmembrane signaling, *Trends Pharmacol. Sci.*, 8, 478, 1987.
97. **Stryer, L. and Bourne, H. R.,** G-proteins: a family of signal transducers, *Annu. Rev. Cell Biol.*, 2, 391, 1986.
98. **Mire-Sluis, A. R., Hoffbrand, A. V., and Vickremasinghe, R. G.,** Evidence that guanine-nucleotide binding regulatory proteins couple cell-surface receptors to the breakdown of inositol-containing lipids during T-lymphocyte mitogenesis, *Biochem. Biophys. Res. Commun.*, 148, 1223, 1987.
99. **O'Shea, J. J., Urdahl, K. B., Luong, H. T., Chused, T. M., Samelson, L. E., and Klausner, R. D.,** Aluminum flouride induces phosphatidylinositol turnover, elevation of cytoplasmic free calcium and phosphorylation of the T cell antigen receptor in murine T cells, *J. Immunol.*, 139, 3463, 1987.
100. **Sibley, D. R., Strasser, R. H., Benovic, J. L., Daniel, K., and Lefkowitz, R. J.,** Phosphorylation/dephosphorylation of the beta-adrenergic receptor regulate its functional coupling to adenylate cyclase and subcellular distribution, *Proc. Natl. Acad. Sci. U.S.A.*, 83, 9408, 1986.

101. **Sibley, D. R., Benovic, J. L., Caron, M. G., and Lefkowitz, R. J.**, Regulation of transmembrane signaling by receptor phosphorylation, *Cell*, 48, 913, 1987.
102. **Lefkowitz, R. J. and Caron, M. G.**, Regulation of adrenergic receptor function by phosphorylation, *Curr. Top. Cell. Reg.*, 28, 209, 1986.
103. **Patel, M. D., Samelson, L. E., and Klausner, R. D.**, Multiple kinase and signal transduction. Phosphorylation of the T-cell antigen-receptor complex, *J. Biol. Chem.*, 262, 5831, 1987.
104. **Katada, T., Gilman, A. G., Watanabe, Y., Bauer, S., and Jakobs, K. H.**, Protein kinase C phosphorylates the inhibitory guanine-nucleotide-binding regulatory component and apparently suppresses its function in hormonal inhibition of adenylate cyclase, *Eur. J. Biochem.*, 151, 431, 1985.
105. **Rodbell, M.**, Role of GTP-binding proteins in hormone and drug action: adenylate cyclase and other transduction processes, *Drug Dev. Res.*, 10, 195, 1987.
106. **Isakov, N., Mally, M. I., Scholz, W., and Altman, A.**, T-lymphocyte activation: the role of protein kinase C in the bifurcating inositol phospholipid signal transduction pathway, *Immunol. Rev.*, 95, 89, 1987.
107. **Ashkenazi, A., Winslow, J. W., Peralta, E. G., Peterson, G. L., Schimerlik, M. I., Capon, D. J., and Ramachandran, J.**, An M2 muscarinic receptor subtype coupled to both adenylyl cyclase and phosphoinositide turnover, *Science*, 238, 672, 1987.
108. **Hepler, J. R., Hughes, A. R., and Harden, T. K.**, Evidence that muscarinic cholinergic receptors selectively interact with either cyclic AMP or the inositol phosphate second-messenger response systems, *Biochem. J.*, 247, 793, 1987.
109. **Ho, A. D., Chik, C. L., and Klein, D. C.**, Protein kinase C is involved in adrenergic stimulation of pineal cGMP accumulation, *J. Biol. Chem.*, 262, 10059, 1987.
110. **Hunter, T.**, A thousand and one protein kinases, *Cell*, 50, 823, 1987.
111. **Samelson, L. E., Patel, M. D., Weissman, A. M., Harford, J. B., and Klausner, R. D.**, Antigen activation of murine T cells induces tyrosine phosphoration of a polypeptide associated with the T cell antigen receptor, *Cell*, 46, 1083, 1986.
112. **Klausner, R. D., O'Shea, J. J., Luong, H., Ross, P., Bluestone, J. A., and Samelson, L. E.**, T-cell receptor tyrosine phosphorylation. Variable coupling for different activating ligands, *J. Biol. Chem.*, 262, 12654, 1987.
113. **Yoshimasa, T., Sibley, D. J., Bouvier, M., Lefkowitz, R. J., and Caron, M. G.**, Cross-talk between cellular signaling pathways suggested by phorbol-ester-induced adenylate cyclase phosphorylation, *Nature (London)*, 327, 67, 1987.
114. **Reed, J. C., Alpers, J. D., Nowell, P. C., and Hoover, R. G.**, Sequential expression of protooncogenes during lectin-stimulated mitogenesis of normal human lymphocytes, *Proc. Natl. Acad. Sci. U.S.A.*, 83, 3982, 1986.
115. **Slungaard, A., Confer, D. L., and Schubach, W. H.**, Rapid transcriptional down-regulation of c-myc expression during cyclic adenosine monophosphate-promoted differentiation of leukemic cells, *J. Clin. Invest.*, 79, 1542, 1987.
116. **Farrar, W. L., Cleveland, J. L., Beckner, S. K., Bonvinin, E., and Evans, S. W.**, Biochemical and molecular events associated with interleukin 2 regulation of lymphocyte proliferation, *Immunol. Rev.*, 92, 49, 1986.

Chapter 6

# NEUROENDOCRINE AND IMMUNE FUNCTION IN THE CAPSAICIN-TREATED RAT: EVIDENCE FOR AFFERENT NEURAL MODULATION *IN VIVO*

**Josef Donnerer, Anthony Eglezos, and Robert D. Helme**

## TABLE OF CONTENTS

# I. INTRODUCTION

The immune system is largely considered to be self-contained, being regulated by a variety of intrinsic processes. This regulation is mediated by different subsets of acute inflammatory cells, macrophages, T-cells, antibodies, soluble products, and other immunoregulatory factors including lymphokines. It is this autoregulation that gives the immune system a high degree of autonomy.[1] However, there is rapidly accumulating experimental evidence for the exchange of messages between the immune, endocrine, and nervous systems in a "neuroendocrine immune axis"[2,3] that allows coordinated activation of the specific capacities of each component part.[4] In particular, modulation of the immune response by the endocrine, central, and peripheral nervous systems has been suggested to occur via specific effects of neuropeptides on immunocompetent cells, an effect which is probably receptor-mediated.[5]

However, the experimental results which have been used to support this hypothesis have largely been carried out *in vitro* using pharmacological concentrations of peptides. If the ability of neurotransmitter and neuromodulatory substances to influence the functions of lymphocytes, macrophages, and polymorphonuclear leukocytes is to be proven, supporting evidence must come from *in vivo* studies. For example, in primary afferent nerves, impulses arising from internal (endocrine) and external (nociceptive) stimulation of receptors, must be shown to directly influence the immune responses, and it must be shown that these effects are mediated by presumptive neuromodulator peptides released from those terminals into the microenvironment of the immunocompetent cell.

Of course, the modification of immune function by neuropeptides is dependent upon cellular recognition of the neuropeptide in the milieu in which they are found, and this also must be proven *in vivo*. To date there are very few studies which have taken up this challenge.[6-8] Examples of such studies include those which have shown an inhibition by alpha-endorphin of the plaque forming cell (PFC) response to ovalbumin[8] and inhibition of the PFC response to sheep erythrocytes (SRBC) by alpha-endorphin, the enkephalins,[6] and 6-hydroxy dopamine treatment.[7] Using a PFC assay,[9] we have shown that treatment of neonatal rats with the neurotoxin capsaicin can be used as a tool to explore the *in vivo* modulation of the immune response by primary afferent neurons.

# II. MORPHOLOGICAL AND CHEMICAL EFFECTS OF CAPSAICIN ON THE NERVOUS SYSTEM; LACK OF EVIDENCE THUS FAR FOR EFFECTS ON OTHER SYSTEMS

Capsaicin, 8-methyl-*N*-vanillyl-6-nonenamide, has been very useful as a selective neurotoxin in studying the multiple functions of the afferent nervous system.[10-12] Capsaicin has stimulatory and, depending on the dosage, neurotoxic activity against primary afferents of the C- and Aδ-type.[13]

The importance of the unmyelinated, slow conducting C-fiber afferents is best expressed by the fact that they make up to 70% of all fibers in the dorsal root of rodents, and that their peripheral endings are sensitive to a variety of different stimuli. From electrophysiological experiments it is known that 80% of C-fibers are connected to nociceptors — either mechanical, polymodal (responding to noxious heat, mechanical stimuli, and irritant chemicals) or cold nociceptors; the rest of the C-fiber afferents are connected to low threshold mechanoreceptors or cooling receptors. The closely related Aδ-afferents, thin myelinated nerve fibers, are connected mainly to heat and mechanonociceptors.[14-16] It is also becoming apparent that many other signals from the circulation and interstitial chemical milieu are detected by free nerve endings derived from C- and Aδ-fiber afferents which may lead to modulation of efferent sympathetic nerve activity as well as endocrine and immune functions.

Unmyelinated afferents show branching and sprouting of the nerve fiber at the central and peripheral terminals;[17] a fact that is responsible at the central ending for lack of precision in transmitting messages to spinal cord neurons[18] and at the peripheral endings for spreading of local inflammatory processes. The peripheral phenomenon is based on the ability of C-fiber afferents to transmit excitation in an antidromic direction (experimentally and physiologically) and also to release their transmitters from the peripheral nerve endings (discussed later in Section III).

Passing from the dorsal root, the C-fibers bifurcate and run for some distance rostral and caudal, and branch into the cellular zones of the spinal cord. The central terminals of the C-fibers in lamina I (marginal zone) and lamina II (substantia gelatinosa) show *en passant* enlargements and terminal swellings.[19] Synaptic contacts are made to ascending second-order neurons and interneurons.[20] Nerve enlargements have also been shown at peripheral endings.

A number of peptides and classical transmitters have been localized immunohistochemically in C-fiber afferents,[21] and for some of them good evidence exists for a transmitter role. Capsaicin has been a particularly useful tool in helping to correlate chemical and anatomical findings with functional phenomena (see below).

There are good indications of transmitter roles in C-fiber afferents for substance P (SP),[22] neurokinin A (NKA),[23] calcitonin gene-related peptide (CGRP),[24] and somatostatin (SOM),[25] whereas for several other peptides, such as cholecystokinin (CCK), vasoactive intestinal peptide (VIP),[26] corticotropin-releasing factor (CRF),[27] alpha-neo-endorphin, and dynorphin (DYN),[28] only histochemical or biochemical data exist. Some of these peptides have common precursors and many coexist in sensory neurons.[29-31] The functional consequences are their simultaneous or sequential release upon nerve stimulation,[32] a useful but nevertheless incomplete information for exploring their function.

The stimulatory as well as neurotoxic actions of capsaicin on C- and Aδ-afferents have been investigated following different modes of application, but the neonatal treatment of laboratory rats or mice, with 50 mg/kg subcutaneously, has emerged as the most useful method. Since the treatment of adult rats with capsaicin is always connected with short-lasting nonspecific effects on nerve cells other than C- and Aδ-afferents,[11,33,34] and with longer-lasting secretion of pituitary and adrenal hormones,[35-37] a period of time is necessary to allow for the development of a stable internal environment (several days to weeks). For this reason, and also because of an incomplete toxic effect, the treatment of adult rats with capsaicin is not particularly useful. Despite nonspecific acute effects, the neurotoxicity of neonatal capsaicin is largely restricted to unmyelinated (C-type) primary afferents (72 to 95% of them are destroyed[38,39]), although a small portion of thin myelinated (Aδ-type) fibers (20 to 30%) may also degenerate.[40-42] In spinal ganglia, the corresponding small dark cells are missing (they decrease in number 69 to 75%).

Since the class of damaged nerve fibers is not uniform (chemically and functionally), they are referred to as "capsaicin-sensitive afferents". It is fortunate that the sensor and transmitter systems for C- and Aδ-afferents are closely related. It must be noted that the surviving C-fibers develop normal functional properties.[43]

So far, no changes in amino acid or amine transmitters have been observed following neonatal capsaicin treatment.[44] However, there is evidence that alterations in the pattern of sensory input to the dorsal horn[18,45,46] cause derangement and even secondary degeneration of spinothalamic tract neurons.[47] No changes in other peripheral and central neuron systems[34] and in the unmyelinated postganglionic sympathetic neurons[48] have been observed so far. Why capsaicin has such a selective action on small caliber afferents is not known in detail, but it has to be related to some of the acute effects of the capsaicin application, among them depolarization and increased membrane conductance,[33,49] and changes in membrane permeability,[50] resulting in $Ca^{++}$ accumulation inside the cell.[51] Capsaicin locks the membrane in a depolarized state and disrupts intracellular organelles, blocking retrograde transport of nerve growth factor.[52,53] The lipophilic capsaicin molecule is probably integrated in membranes of sensory neurons and intracellular macromolecules. It seems that developing sensory nerves in the neonatal rat are more susceptible to the neurotoxic action of capsaicin since treatment after the tenth day of life does not produce an irreversible degeneration.[54] Despite the depletion of C-fiber afferents with all the connected functional impairments (as will be described later), the animals grow to adulthood and appear normally active and so provide a useful model for studying the physiology and pharmacology of capsaicin-sensitive afferents. Thus, the way in which the nervous, endocrine, and immune systems compensate for a lifelong loss of such an important source of afferent input can be examined.

## III. NEUROGENIC INFLAMMATION AND ITS RELEVANCE TO THE IMMUNE RESPONSE

The sensory nervous system can exert its influence on the immune system indirectly via the endocrine system or more directly via locally released neurotransmitters. The endocrine system can be activated by orthodromic transmission of various sensory modalities to the central nervous system (CNS).[55] The release of neurotransmitters from the peripheral endings of sensory nerves, as evoked by antidromic activation, induces vasodilatation and plasma extravasation (neurogenic inflammation).[56] The antidromic activation of capsaicin-sensitive afferents has been observed experimentally and seems to be involved in several pathophysiological conditions.[57-60]

Of the proposed transmitters of capsaicin-sensitive afferents, SP and NKA are equally active in causing plasma extravasation in physiological concentrations, and CGRP is very potent in causing vasodilatation[61] and in potentiating the effects of SP and NKA.[62] When injected intradermally in human skin, SP and NKA produce flare and wheal reactions, but CGRP produces an erythema.[63] VIP causes vasodilation in high concentrations,[64] SOM seems to inhibit neurogenic inflammation,[65] and histamine, serotonin, ATP, adenosine, and neurotensin also induce plasma extravasation.[62]

These actions may well enhance the delivery and accumulation of a range of mononuclear leukocytes important for the expression of immune responses.[66,67] In addition, capsaicin-sensitive fibers can be activated by mediators released upon mast-cell degranulation, such as histamine and serotonin,[26] and by factors produced by local tissue damage including $K^+$, bradykinin, prostaglandins, and leukotrienes. As discussed above, the anatomical organization of unmyelinated sensory nerves can rapidly induce a spread of acute inflammatory effects (neurogenic antidromic inflammation) and secondarily facilitate the effector pathways of the immune response.[68] Neurogenic inflammation has been shown to be involved in the flare response in the human skin,[69-71] in reactive hyperemia,[64] heat edema,[72] bronchoconstriction to irritant chemicals and antigen challenge,[73,74] nasal mucosa swelling due to allergic and hyperreactive disorders,[75] ureter inflammation,[76] uveitis,[77] and arthritis.[78]

The sensory fibers involved in neurogenic inflammation are capsaicin-sensitive afferents, because following neonatal capsaicin treatment, plasma extravasation and vasodilatation effects cannot be evoked through antidromic sensory nerve stimulation. In addition, chronic inflammatory processes, such as the adjuvant-induced arthritis model,[78] and specific immune responses, such as antibody production,[9] are impaired after capsaicin treatment.

The anatomical substrate for an involvement of capsaicin-sensitive afferents in local inflammatory processes and immune responses has been demonstrated using immunohistochemistry. Using SP as a marker, nerve fibers are consistently found in the epithelium and in the dermis where they are observed in relation to blood vessels, sweat glands, and hair follicles.[21] In the nasal mucosa, respiratory epithelium, and tracheobronchial tissue (muscular and submucosal connective tissue), in the ureter and urinary bladder, in repro-

**TABLE 1**
**Immunologically Active Cells With Receptors for Sensory Neuropeptides**

| Peptide | Cell source | Cell type | Ref. |
|---|---|---|---|
| Substance P | Human | T Lymphocyte | 5 |
| | | B Lymphocyte | 88 |
| | | Monocytes/macrophages | 88 |
| | | Polymorphs | 88 |
| | | Platelets | 88 |
| | Cell lines | IM9 Lymphoblasts | 5 |
| | | Molt-4 Lymphoblasts | 88 |
| | | Jurkat cells | 88 |
| | | Hut 78 T-cells | 88 |
| | | U-937 Monocytes, differentiated | 88 |
| | | HL-60 cells, differentiated | 88 |
| | Murine | T Lymphocyte | 5 |
| | | B Lymphocyte | 5 |
| | Guinea pig | Macrophages | 92 |
| SOM | Human | Lymphocytes | 5 |
| | | Monocytes | 89 |
| VIP | Human | Lymphocytes | 90 |
| | | Mononuclear leukocytes | 90 |
| | Cell lines | Molt-4 Lymphoblast | 90 |
| | Murine | Lymphocytes | 91 |
| | | Mononuclear leukocytes | 90 |

ductive organs tissue, and other visceral organs, a dense innervation by capsaicin-sensitive afferents is observed.[26,75,79]

It is obvious that the potential exists for neuropeptides to act at a number of different levels of the immune response simply because of the widespread anatomical network of their nerve fibers. In addition to the sympathetic innervation,[80,81] also a fairly dense sensitory peptidergic innervation of lymphoid organs has been demonstrated recently.[81a,81b,81c]

One direct mechanism underlying the interaction of neuropeptides with the immune system is binding of the neuropeptides to specific receptors on immunocompetent cells. Receptors for neuropeptides contained in primary afferent nerves have been demonstrated in the thymus and spleen.[82-84] In the thymus, SP binding appears to be localized in the medulla in association with the thymic vasculature, and using the classification scheme of Lee et al. (1982),[85] the receptor appeared to be of the SP/physalaemin (SP.P) or NK-1 type.[82] In the rat spleen, SP binding sites appear to be localized in areas of nonlymphoid specialized splenic macrophages in the marginal zone.[82,83] A high concentration of CGRP, SP, and VIP binding sites has been shown to be associated with the germinal centers of canine lymph nodes.[81c] Receptors for leu- and met-enkephalin[86] and beta-endorphin[87] have also been shown to exist on lymphocytes.

Neuropeptides in primary afferent nerves such as SP,[88] SOM,[89] and VIP[90,91] have also been shown to bind to specific receptors on the surface of cells important in immune responses (see Table 1). These neuropeptides are potent stimulators of many cells *in vitro*. For example, guinea pig macrophages have been shown to have specific receptors for SP.[92] Binding of SP to this receptor[93] and the receptor on phagocytic cells[94] increases their normal function. In line with these results is a recent finding of SP-induced production of IL-1 by the mouse macrophage cell line P388D1.[94a] Substance P, SOM, and VIP also induce a dose-dependent release of histamine from peritoneal mast cells,[95] and human lymphocytes have been shown to exhibit both high and low affinity receptors for various neuropeptides.[5] The influence of neuropeptides on the proliferative responses of lymphoid cells to mitogenic stimulation varies,[88,96] but it has been shown by many investigators that numerous neuropeptides modulate lymphocyte responses *in vitro*.[5,88,97] Binding of VIP to a specific receptor on lymphocytes inhibits the proliferative responses of lymphocytes to concanavalin A, but not to B-cell mitogens.[90] Payan et al. (1984),[98] have also shown that SP binds preferentially to T-cells, mainly the T helper ($T_h$)-cell subset. Substance P increases the *in vitro* proliferation of rheumatoid synoviocytes[99] and stimulates prostaglandin $E_2$ and collagenase release from synoviocytes.[100] These effects may be responsible for the maintenance of chronic arthritis in which the SP system is hyperactive.[101,102]

Other investigators have also provided evidence for a role of SP in the pathogenesis of immunological disease[78] and in the process of connective tissue repair and wound healing.[103] Since SP receptors have been found on T-, B-, and other cells important in the induction and maintenance of immune responses, it can be presumed that SP-mediated modulation probably occurs at the lymphoid cell level.[103a] For example, SP has previously been shown to stimulate immunoglobulin production from mouse mesenteric lymph nodes *in vitro*.[104] However, there is no study which has reported SP receptors on rat lymphocytes.

The demonstration of specific cell-surface receptors for neuropeptides on lymphocytes and other immunologically active cells, combined with immunohistochemical results showing the presence of neuropeptide binding sites in immune organs have thus provided important evidence to support a direct interaction between the immune and nervous systems. Whether this is an important factor in disease states *in vivo* remains to be fully proven.

## IV. STRESS AND ADRENOCORTICOTROPIN (ACTH) IN CAPSAICIN-TREATED RATS

Adrenocorticotropin (ACTH) release from the pituitary gland and subsequent glucocorticoid secretion from the adrenal cortex represent the uniform hormonal response to a variety of physical and psychological stressors. The predominant trigger for ACTH release is CRF release from hypothalamic neurons. Other mediators involved in this process include vasopressin, oxytocin, angiotensin, and catecholamines.[105] The hypothalamic areas receive signals from circulating humoral factors and from the auditory, visual, olfactory, limbic, and somatosensory system.[55]

Recent investigations have demonstrated that somatosensory forms of stress are dependent on the function of capsaicin-sensitive afferents. Capsaicin-treated rats showed reduced ACTH secretion during cold stress, surgery, formalin stress, and opiate withdrawal (Figure 1).[36,106,107] On the other hand, it could be demonstrated that pituitary function was not impaired in capsaicin-treated rats and also that the stress response to restraint, open field exposure, ether, and hypoglycemia was not affected, suggesting that some "emotional" and humoral factors act independently of capsaicin-sensitive afferents.[106,107] The fact that somatic C-fiber and not A-fiber afferents can induce ACTH secretion was also shown by electrical nerve stimulation using different stimulus parameters.[107] These findings add another important endocrine response to the list of reflexes dependent on afferent messages from peptidergic C-fiber afferents.

So far, it has been demonstrated that capsaicin-sensitive afferents transmit pain sensation to the CNS;[108] induce thermoregulatory mechanisms,[11,109] evoke cardiovascular reflexes,[110] the micturition reflex,[111] tracheobronchial contractions,[112] and visceral reflexes;[48,113] and are responsible for the facilitation of the decidua reaction, and, in part, for lactation in the rat.[114] They also modulate the catecholamine release from the adrenal medulla during hypogly-

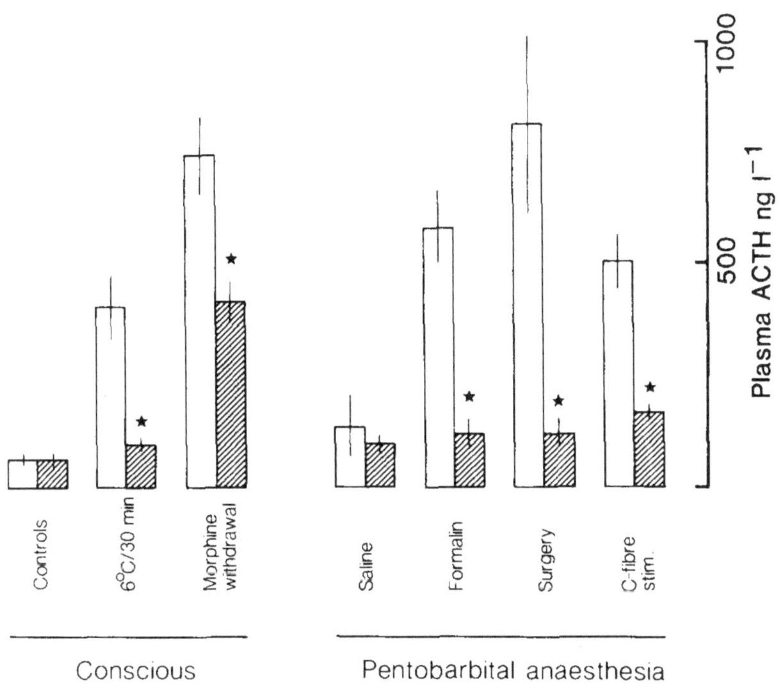


FIGURE 1. ACTH release evoked by somatosensory stimuli in rats pretreated neonatally with capsaicin (hatched columns) or vehicle only (open columns). Capsaicin-treated rats showed an impaired ACTH response to cold exposure (6°C for 30 min), naloxone-precipitated morphine withdrawal, i.p. injection of formalin, neck surgery, and C-fiber stimulation in the sciatic nerve. Significant differences are indicated by asterisks. For details, see References 36, 106, and 107.

cemia.[115,116,116a] It has been demonstrated convincingly that, depending on the type of receptors or sensors stimulated in the periphery or visceral organs and dependent on the anatomical location of the stimulus, one specific or several reflexes are evoked.[12] There might be interactions also between the effector mechanisms. For example, released catecholamines can stimulate ACTH secretion,[117] and corticosteroids also have a complex, multifaceted influence on the immune response manifested predominantly as a suppression of B- and T-cell activity.[118,119] In this way, the somatosensory, and especially the capsaicin-sensitive fibers, may also exert their influence upon the endocrine and immune systems.

## V. HUMORAL IMMUNITY IN CAPSAICIN-TREATED RATS

Substance P induces not only vasodilatation and increased vascular permeability, but has also been shown to bind to the chemotactic receptor of human monocytes[120] and rabbit neutrophils[121] and to stimulate chemotaxis of rat neutrophils.[122] Rats with peripheral nerve lesions accumulate less neutrophils at sites of inflammation.[123] As this is also observed in rats treated as neonates with capsaicin, the effect is most likely due to loss of neuropeptide-containing C- and Aδ-afferent fibers. Substance P is thought to be the likely mediator of this *in vivo* event, but an effect of other neuropeptides has not been excluded.

Neutrophils have also been suggested to be active in the induction of plasma extravasation[67] and the addition of neurally derived SP may increase the accumulation of immunocompetent cells to the site of antigen entry through this mechanism.

Sympathetic modulation of *in vivo* humoral reactivity has also been measured by several investigators using a PFC technique in order to measure immunoglobulin production to a

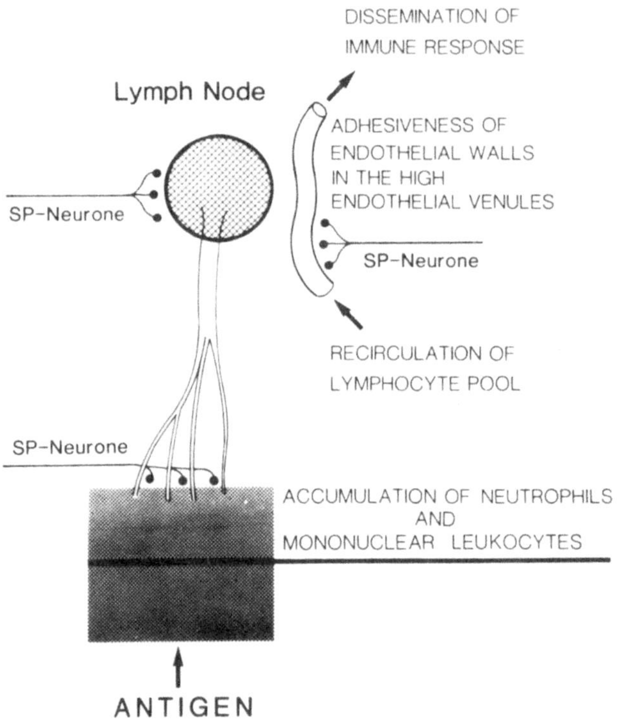


FIGURE 2. Possible sites of action of the primary afferent nerves and associated neuropeptides (such as substance P), on the immune response in the hind limb.

specific antigen.[6-8] Using a modification of the Cunningham PFC assay technique, we have been able to show effects on the humoral immune response to a T-dependent antigen in capsaicin-treated rats. Rats treated neonatally with capsaicin have a markedly decreased PFC response of popliteal lymph node lymphocytes to subcutaneously administered SRBC.[9] However, capsaicin-treated rats given a footpad infusion of SP immediately after, and at the same site as antigen injection, showed a return to normal in the PFC response. The number of PFCs was reversed in the SP-infused animals in an apparent dose-dependent manner, and this effect was demonstrated to be mediated through the C-terminal end of the peptide. Capsaicin pretreatment did not appear to be toxic for the immunocompetent cells as shown by this reversibility of the response in SP-infused rats. We therefore have proposed SP-mediated modulation of the humoral immune response as another important area of communication between the nervous and immune systems. This signal is most likely to occur early following antigen presentation, as reversal of the PFC response occurred with a SP infusion immediately, but not 24 h following immunization.

Neurally derived SP in the footpad may have a modulatory role during an immune response to antigen stimulation for many reasons (see Figure 2). Substance P may mediate plasma extravasation and vasodilatation in the footpad and in the vicinity of the node itself, modulating the rate of lymphocyte transport and recirculation, and as a consequence, the immune response. However, neither an increase in fluid load using normal saline, nor infusion of a protein solution (normal rat plasma at a concentration similar to that found in the neurogenic plasma extravasation response to exogenous SP[124]) at the site of antigen entry reversed the response. Substance P has also been shown to augment lymphocyte traffic and possibly recirculation from the lymph node to the periphery.[125] It is also possible that SP may play a role in controlling the passage of lymphocytes through the high endothelial

**TABLE 2**
**Effects of Capsaicin Treatment on the Induction of T Lymphocytes**

| Antibody | Group | Percentage change[a] |
|---|---|---|
| W3.25 | Normal | NC[b] |
| | Capsaicin | NC[b] |
| MRC OX8 | Normal | −20%[c] |
| | Capsaicin | +23%[d] |

[a] This was measured 6 days after a subcutaneous SRBC injection into both hind footpads.
[b] NC = No change.
[c] Decrease percentage.
[d] Increase percentage.

venules (HEV) of the lymph node. Passage of lymphocytes through HEVs occurs after initial adherence of the cells to the endothelium and then the subsequent passage through the endothelium into the tissue. Immunologic factors such as IL-1 and gamma-interferon[126,127] have been shown to be involved in the development and maintenance of HEVs.[128] These immunologic factors produced at the site of an inflammatory response, may act to attract and direct lymphocytes to the sites of immunologic activity. It has been suggested that both the initial adhesion of cells to the endothelium and migration into and through the tissue may be mediated by the same factor.[129] Chemotactic agents, at high concentrations which suppress/inhibit chemotaxis, lead to an increased adhesiveness of cells to the endothelium.[129,130] Monocyte-derived IL-1 has been shown to stimulate chemotaxis of T and B lymphocytes[131] and also increase binding of these cells to endothelial cells when cultured *in vitro*.[119] Substance P has also been shown to have a modulatory effect on the adhesiveness of human neutrophils,[120] which may be related to its neutrophil chemotactic properties[122] and/or its effects on macrophages.[93]

Immunosuppression by leukotriene $B_4$ has been shown to be due to its selective stimulatory and inhibitory effects on T suppressor ($T_s$)- and $T_h$-cells respectively.[98] As mentioned previously, SP has been shown to bind preferentially to $T_h$-cells. The effects of the SP infusion on the humoral immune response in capsaicin-treated rats may be mediated via its direct effects on $T_h$- and $T_s$-cells.

Flow cytometric methods coupled with the use of specific monoclonal antibodies (MoAbs), have enabled the quantitation of lymphocyte heterogeneity in immunologically active organs. Popliteal lymph nodes from normal and neonatally capsaicin-treated rats (resting and 6 d after antigen (SRBC) stimulation) were quantified using W3.25 and MRC OX8 MoAbs (gift from Dr. H. A. Ward, Monash University). These antibodies have been shown to clearly label mature lymphocytes of the T helper/inducer ($T_{h/i}$) and T cytotoxic/suppressor ($T_{c/s}$) subpopulations, respectively.[132] These two antibodies bind nonoverlapping populations of cells with different functional properties.[133]

Using these MoAbs we have shown that neonatal capsaicin treatment does not have an effect on the cellular profile of the resting rat popliteal lymph node. However, following antigenic stimulation with SRBC, there were certain differences in the lymphocyte heterogeneity of the lymph node (Table 2). After antigenic stimulation, there was a 23% increase and a 20% decrease in the number of MRC OX8 positive cells in the responding popliteal lymph-nodes from neonatally capsaicin-treated, and normal rats, respectively.[133a] There was no change in the percentage of W3.25 positive cells. The differences observed in the percentage of MRC OX8 positive cells after antigenic stimulation suggest that the decreased responsiveness in these rats is due to neurogenic modulation of immunosuppression of immune responses, and not to direct toxicity.

## VI. CONCLUSION

In summary, it is clear that capsaicin-sensitive primary afferent nerve fibers are widely distributed throughout the living organism and that because of their unique ability to respond to polymodal stimuli, they are in a pivotal position to modulate homeostatic control mechanisms. This function encompasses the endocrine, inflammatory, and immune systems, and may well extend to other physiologic and pathologic situations.

This unique neural system, previously referred to as the ''nocifensor system'' in its more restricted role in controlling local blood flow by Lewis[134] and Lembeck,[135] has been available for study only because of the selective neurotoxic action of capsaicin.

It can be safely anticipated that future studies using this *in vivo* model will verify and expand our concepts of neurogenic modulation of humoral and cellular events in the complex decentralized organ systems of the body, namely the immune and endocrine organs. It remains to be seen how this new perspective will find application in current views of psychogenic modulation of pathophysiologic processes.

## ACKNOWLEDGMENTS

Supported by the Austrian Scientific Research Funds (grant Nos. P5616 and P6646), the Austrian National Bank (grant 2216), and the Pain Research Commission of the Austrian Academy of Sciences.

## REFERENCES

1. **Besedovsky, H., del Rey, A. E., and Sorkin, E.,** What do the immune system and brain know about each other?, *Immunol. Today,* 4, 342, 1983.
2. **Blalock, J. E.,** Opinion: the immune system as a sensory organ, *J. Immunol.,* 132, 1067, 1984.
3. **Blalock, J. E. and Smith, E. M.,** The immune system: our mobile brain?, *Immunol. Today,* 6, 115, 1985.
4. **Payan, D. G., McGillis, J. P., and Goetzl, E. J.,** Neuroimmunology, *Adv. Immunol.,* 30, 299, 1986.
5. **Payan, D. G. and Goetzl, E. J.,** Modulation of lymphocyte function by sensory neuropeptides, *J. Immunol.,* 135, 783s, 1985.
6. **Johnson, H. M., Smith, E. M., Torres, B. A., and Blalock, J. E.,** Regulation of the in vitro response using neuroendocrine hormones, *Proc. Natl. Acad. Sci. U.S.A.,* 79, 4171, 1982.
7. **Cross, R. J., Jackson, J. C., Brooks, W. H., Sparks, D. L., Markesbery, W. R., and Roszman, T. L.,** Neuroimmunomodulation: impairment of humoral immune responsiveness by 6-hydroxy dopamine treatment, *Immunology,* 57, 145, 1986.
8. **Heinjen, C. J., Beavers, C., Kavelaars, A., and Ballieux, R. E.,** Effect of alpha-endorphin on antigen induced primary antibody response of human blood cells in vitro, *J. Immunol.,* 136, 213, 1986.
9. **Helme, R. D., Eglezos, A., Dandie, G. W., Andrews, P. V., and Boyd, R. L.,** The effects of substance P on the regional lymph node antibody response to antigenic stimulation in capsaicin pretreated rats, *J. Immunol.,* 139, 3470, 1987.
10. **Nagy, J. I.,** Capsaicin: a chemical probe for sensory neuron mechanisms, in *Handbook of Psychopharmacology,* Vol. 15, Iversen, L. L., Iversen, S. D., and Snyder, S. H., Eds., Plenum Press, New York, 1982, 185.
11. **Szolcsányi, J.,** Capsaicin type pungent agents producing pyrexia, in *Handbook of Experimental Pharmacology,* Vol. 60, Milton, A. S., Ed., Springer-Verlag, Berlin, 1982, 437.
12. **Buck, S. H. and Burks, T. F.,** The neuropharmacology of capsaicin: review of some recent observations, *Pharmacol. Rev.,* 38, 173, 1986.
13. **Winter, J.,** Characterization of capsaicin-sensitive neurones in adult rat dorsal root ganglion cultures, *Neurosci. Lett.,* 80, 134, 1987.
14. **Lynn, B. and Carpenter, S. E.,** Primary afferent units from the hairy skin of the rat hind limb, *Brain Res.,* 238, 29, 1982.
15. **Fleischer, E., Handwerker, H. D., and Joukhadar, S.,** Unmyelinated nociceptive units in two skin areas of the rat, *Brain Res.,* 267, 81, 1983.

16. **Fitzgerald, M. and Gibson, S.,** The postnatal physiological and neurochemical development of peripheral sensory C-fibres, *Neuroscience,* 13, 933, 1984.
17. **Coggeshall, R. E.,** Nonclassical features of dorsal root ganglion cell organization, in *Spinal Afferent Processing,* Yaksh, T. L., Ed., Plenum Press, New York, 1986, 83.
18. **Cook, A. J., Woolf, C. J., Wall, P. D., and McMahon, S. B.,** Dynamic receptive field plasticity in rat spinal cord dorsal horn following C-primary afferent input, *Nature (London),* 325, 151, 1987.
19. **Sugiura, Y., Lee, C. L., and Perl, E. R.,** Central projections of identified, unmyelinated (C) afferent fibers innervating mammalian skin, *Science,* 234, 358, 1986.
20. **Ribeiro-Da-Silva, A. and Coimbra, A.,** Capsaicin causes selective damage to type I synaptic glomeruli in rat substantia gelatinosa, *Brain Res.,* 290, 380, 1984.
21. **Hökfelt, T., Johansson, D., Ljungdahl, A., Lundberg, J. M., and Schultzberg, M.,** Peptidergic neurones, *Nature (London),* 284, 515, 1980.
22. **Pernow, B.,** Substance P, *Pharmacol. Rev.,* 35, 85, 1983.
23. **Nawa, H., Hirose, T., Takashima, H., Inayama, S., and Nakanishi, S.,** Nucleotide sequences of cloned cDNAs for two types of bovine brain substance P precursor, *Nature (London),* 306, 32, 1983.
24. **Goodman, E. C. and Iversen, L. L.,** Calcitonin gene-related peptide: novel neuropeptide, *Life Sci.,* 38, 2169, 1986.
25. **Gamse, R., Lackner, D., Gamse, G., and Leeman, S. E.,** Effect of capsaicin pretreatment on capsaicin-evoked release of immunoreactive somatostatin and substance P from primary sensory neurons, *Naunyn-Schmiedebergs Arch. Pharmacol.,* 316, 38, 1981.
26. **Lundberg, J. M. and Saria, A.,** Polypeptide-containing neurons in airway smooth muscle, *Annu. Rev. Physiol.,* 49, 557, 1987.
27. **Skofitsch, G., Zamir, N., Helke, C. J., Savitt, J. M., and Jacobowitz, D. M.,** Corticotropin releasing factor-like immunoreactivity in sensory ganglia and capsaicin sensitive neurons of the rat central nervous system: colocalization with other neuropeptides, *Peptides,* 6, 307, 1985.
28. **Weihe, E., Hartschik, W., and Weber, E.,** Prodynorphin opioid peptides in small somatosensory primary afferents of guinea pig, *Neurosci. Lett.,* 58, 347, 1985.
29. **Hökfelt, T., Holets, V. R., Staines, W., Meister, B., Melander, T., Schalling, M., Schultzberg, M., Freedman, J., Björklund, H., Olson, L., Lindh, B., Elfvin, L.-G., Lundberg, J. M., Lindgren, J. A., Samuelsson, B., Pernow, B., Terenius, L., Post, C., Everitt, B., and Goldstein, M.,** Coexistence of neuronal messengers — an overview, in *Progress in Brain Research,* Vol. 68, Coexistence of neuronal messengers: a new principle in chemical transmission, Hökfelt, T., Fuxe, K., and Pernow, P., Eds., Elsevier, Amsterdam, 1986, 33.
30. **Gibbins, I. L., Furness, J. B., and Costa, M.,** Pathway-specific patterns of the co-existence of substance P, calcitonin gene-related peptide, cholecystokinin and dynorphin in neurons of the dorsal root ganglia of the guinea-pig, *Cell Tissue Res.,* 248, 417, 1987.
31. **Dalsgaard, C. J., Haegerstrand, A., Theodorsson-Norheim, E., Brodin, E., and Hökfelt, T.,** Neurokinin A-like immunoreactivity in rat primary sensory neurons; coexistence with substance P, *Histochemistry,* 83, 37, 1985.
32. **Saria, A., Gamse, R., Petermann, J., Fischer, J. A., Theodorsson-Norheim, E., and Lundberg, J. M.,** Simultaneous release of several tachykinins and calcitonin gene-related peptide from rat spinal cord slices, *Neurosci. Lett.,* 63, 310, 1986.
33. **Williams, J. T. and Zieglgänsberger, W.,** The acute effects of capsaicin on rat primary afferents and spinal neurons, *Brain Res.,* 253, 125, 1982.
34. **Hajos, M., Svensson, K., Nissbrandt, H., Obal, F., and Carlsson, A.,** Effects of capsaicin on central monoaminergic mechanisms in the rat, *J. Neural Transm.,* 66, 221, 1986.
35. **Meuller, G. P.,** Beta-endorphin immunoreactivity in rat plasma: variations in response to different physical stimuli, *Life Sci.,* 29, 1669, 1981.
36. **Lembeck, F. and Amann, R.,** The influence of capsaicin sensitive neurons on stress-induced release of ACTH, *Brain Res. Bull.,* 16, 541, 1986.
37. **Watanabe, T., Kawada, T., and Iwai, K.,** Enhancement by capsaicin of energy metabolism in rats through secretion of catecholamine from adrenal medulla, *Agric. Biol. Chem.,* 51, 75, 1987.
38. **Nagy, J. I., Hunt, S. P., Iversen, L. L., and Emson, P. C.,** Biochemical and anatomical observations on the degeneration of peptide-containing primary afferent neurons after neonatal capsaicin, *Neuroscience,* 6, 1923, 1981.
39. **Arvidsson, J. and Ygge, J.,** A quantitative study of the effects of neonatal capsaicin treatment and of subsequent peripheral nerve transection in the adult rat, *Brain Res.,* 397, 130, 1986.
40. **Lawson, S. N. and Nickels, S. M.,** The use of morphometric techniques to analyse the effect of neonatal capsaicin treatment on rat dorsal root ganglia and dorsal roots, *J. Physiol.,* 303, 12, 1980.
41. **Holje, L., Hildebrand, C., and Fried, K.,** Proportion of unmyelinated axons in the rat inferior alveolar nerve and mandibular molar pulps after neonatal administration of capsaicin, *Brain Res.,* 266, 133, 1983.

42. **Hiura, A. and Sakamoto, Y.**, Quantitative estimation of the effects of capsaicin on the mouse primary sensory neurons, *Neurosci. Lett.*, 76, 101, 1987.
43. **Welk, E., Fleischer, E., Petsche, U., and Handwerker, H. D.**, Afferent C-fibres in rats after neonatal capsaicin treatment, *Pflügers Arch.*, 400, 66, 1984.
44. **Singer, E. A., Sperk, G., and Schmid, R.**, Capsaicin does not change tissue levels of glutamic acid, its uptake, or release in the rat spinal cord, *J. Neurochem.*, 38, 1383, 1982.
45. **Wall, P. D., Fitzgerald, M., Nussbaumer, J. C., Devor, M., and Van der Loos, H.**, Somatotopic maps are disorganized in adult rodents treated neonatally with capsaicin, *Nature (London)*, 295, 691, 1982.
46. **Cervera, F. and Plenderleith, M. B.**, C-fibre excitation and tonic descending inhibition of dorsal horn neurones in adult rats treated at birth with capsaicin, *J. Physiol.*, 365, 223, 1985.
47. **Saporta, S.**, Loss of spinothalamic tract neurons following neonatal treatment of rats with the neurotoxin capsaicin, *Somatosens Res.*, 4, 153, 1986.
48. **Cervero, F. and McRitchie, H. A.**, Neonatal capsaicin does not affect unmyelinated fibres of the autonomic nervous system, *Brain Res.*, 239, 283, 1982.
49. **Petersen, M., Pierau, F. K., and Weyrich, M.**, The influence of capsaicin on membrane currents in dorsal root ganglion neurones of guinea-pig and chicken, *Pflüegers Arch.*, 409, 403, 1987.
50. **Heyman, I. and Rang, H. P.**, Depolarizing responses to capsaicin in a subpopulation of rat dorsal root ganglion cells, *Neurosci. Lett.*, 56, 69, 1985.
51. **Jancsó, G., Savay, G., and Király, E.**, Appearance of histochemically detectable ionic calcium in degenerating primary sensory neurons, *Acta Histochem.*, 62, 165, 1978.
52. **Miller, M. S., Buck, S. H., Sipes, I. G., Yamamura, H. I., and Burks, T. F.**, Regulation of substance P by nerve growth factor: disruption by capsaicin, *Brain Res.*, 250, 193, 1982.
53. **Otten, U., Lorex, H. P., and Businger, F.**, Nerve growth factor antagonizes the neurotoxic action of capsaicin on primary sensory neurones, *Nature (London)*, 301, 515, 1983.
54. **Gamse, R., Holzer, P., and Lembeck, F.**, Decrease of substance P in primary afferent neurones and impairment of neurogenic plasma extravasation by capsaicin, *Br. J. Pharmacol.*, 68, 207, 1980.
55. **Feldman, S.**, Neural pathways mediating adrenocortical responses, *Fed. Proc.*, 44, 169, 1985.
56. **Lembeck, F. and Holzer, P.**, Substance P as neurogenic mediator of antidromic vasodilation and neurogenic plasma extravasation, *Naunyn-Schmiedebergs Arch. Pharmacol.*, 310, 175, 1979.
57. **Pernow, B.**, Substance P — a putative mediator of antidromic vasodilatation, *Gen. Pharmacol.*, 14, 13, 1983.
58. **Chahl, L. A., Szolcsányi, J., and Lembeck, F.**, in *Antidromic Vasodilatation and Neurogenic Inflammation*, Chahl, L. A., et al., Eds., Akademiai Kiado, Budapest, 1984.
59. **Lembeck, F.**, Substance P and sensory neurones, in *Substance P: Metabolism and Biological Action*, Jordan, C. C. and Oehme, P., Eds., Taylor & Francis, London, 1985, 137.
60. **Holzer, P.**, Local effector functions of capsaicin-sensitive sensory nerve endings: involvement of tachykinins, calcitonin gene-related peptide and other neuropeptides, *Neuroscience*, 24, 739, 1988.
61. **Brain, S. D., Williams, T. J., Tippins, J. R., Morris, H. R., and MacIntyre, I.**, Calcitonin gene-related peptide is a potent vasodilator, *Nature (London)*, 313, 54, 1985.
62. **Gamse, R., Posch, M., Saria, A., and Jancsó, G.**, Several mediators appear to interact in neurogenic inflammation, *Acta Physiol. Hung.*, 69, 343, 1987.
63. **Wallengren, J. and Håkanson, R.**, Effects of substance P, neurokinin A and calcitonin gene-related peptide in human skin and their involvement in sensory nerve-mediated responses, *Eur. J. Pharmacol.*, 143, 267, 1987.
64. **Lembeck, F. and Donnerer, J.**, Postocclusive cutaneous vasodilation mediated by substance P., *Naunyn-Schmiedebergs Arch. Pharmacol.*, 316, 165, 1981.
65. **Lembeck, F., Donnerer, J., and Barthó, I.**, Inhibition of neurogenic vasodilatation and plasma extravasation by substance P antagonists, somatostatin and [D-Met$^2$, Pro$^5$] enkephalinamide, *Eur. J. Pharmacol.*, 85, 171. 1982.
66. **Foreman, J. C. and Jordan, C. C.**, Histamine release and vascular changes induced by neuropeptides, *Agents Actions*, 13, 105, 1983.
67. **Wedmore, C. V. and Williams, T. J.**, Control of vascular permeability by polymorphonuclear leukocytes in inflammation, *Nature (London)*, 289, 646, 1981.
68. **Payan, D. G., Levine, J. D., and Goetzl, E. J.**, Modulation of immunity and hypersensitivity by sensory neuropeptides, *J. Immunol.*, 132, 16014, 1984.
69. **Carpenter, S. E. and Lynn, B.**, Vascular and sensory responses of human skin to mild injury after topical treatment with capsaicin, *Br. J. Pharmacol.*, 73, 755, 1981.
70. **Foreman, J. C., Jordan, C. C., Oehme, P., and Renner, H.**, Structure-activity relationships for some substance P-related peptides that cause wheal and flare reactions in human skin, *J. Physiol.*, 335, 449, 1983.
71. **Lundblad, L., Lundberg, J. M., Änggard, A., and Zetterström, O.**, Capsaicin pretreatment inhibits the flare component of the cutaneous allergic reaction in man, *Eur. J. Pharmacol.*, 113, 461, 1985.

72. **Saria, A.**, Substance P in sensory nerve fibers contributes to the development of oedema in the rat hind paw after thermal injury, *Br. J. Pharmacol.*, 82, 217, 1984.
73. **Saria, A., Lundberg, J. M., Skofitsch, G., and Lembeck, F.**, Vascular protein leakage in various tissues induced by substance P, capsaicin, serotonin, histamine and by antigen challenge, *Naunyn-Schmiedebergs Arch. Pharmacol.*, 324, 212, 1983.
74. **Manzini, S., Maggi, C. A., Gepetti, P., and Bacciarelli, C.**, Capsaicin desensitization protects from antigen-induced bronchospasm in conscious guinea-pigs, *Eur. J. Pharmacol.*, 138, 307, 1987.
75. **Lundblad, L.**, Protective reflexes and vascular effects in the nasal mucosa elicited by activation of capsaicin-sensitive substance P-immunoreactive trigeminal neurons, *Acta Physiol. Scand. Suppl.*, 529, 1, 1984.
76. **Saria, A., Lundberg, J. M., Hua, X., and Lembeck, F.**, Capsaicin-induced substance P release and sensory control of vascular permeability in the guinea-pig ureter, *Neurosci. Lett.*, 41, 167, 1983.
77. **Bynke, G.**, Capsaicin pretreatment prevents disruption of the blood-aqueous barrier in the rabbit eye, *Invest. Ophthalmol. Visual Sci.*, 24, 744, 1982.
78. **Colpaert, F. C., Donnerer, J., and Lembeck, F.**, Effects of capsaicin on inflammation and on the substance P content of nervous tissues in rats with adjuvant arthritis, *Life Sci.*, 32, 1827, 1982.
79. **Traurig, H., Saria, A., and Lembeck, F.**, Substance P in primary afferent neurons of the female rat reproductive system, *Naunyn-Schmiedebergs Arch. Pharmacol.*, 326, 343, 1984.
80. **Giron, L. T., Curtcher, K. A., and Davis, J. N.**, Lymph nodes — a possible site for sympathetic neuronal regulation of immune responses, *Ann. Neurol.*, 8, 520, 1980.
81. **Felten, D. L., Felten, S. Y., Carison, S. L., Olchowska, J. A., and Livnat, S.**, Noradrenergic and peptidergic innervation of lymphoid tissue, *J. Immunol.*, 135, 755s, 1980.
81a. **Fink, T. and Weihe, E.**, Multiple neuropeptides in nerves supplying mammalian lymph nodes: messenger candidates for sensory and autonomic neuroimmunomodulation?, *Neurosci. Lett.*, 90, 39, 1988.
81b. **Weihe, E., Müller, S., Fink, T., and Zentel, H. J.**, Tachykinins, calcitonin gene-related peptide and neuropeptide Y in nerves of the mammalian thymus: interactions with mast cells in autonomic and sensory neuroimmunomodulation?, *Neurosci. Lett.*, 100, 77, 1989.
81c. **Popper, P., Mantyh, C. R., Vigna, S. R., Maggio, J. E., and Mantyh, P. W.**, The localization of sensory nerve fibers and receptor binding sites for sensory neuropeptides in canine mesenteric lymph nodes, *Peptides*, 9, 257, 1988.
82. **Shigematsu, K., Saavedra, J. M., and Kurihara, M.**, Specific substance P binding sites in rat thymus and spleen: in vitro autoradiographic study, *Reg. Peptides*, 16, 147, 1986.
83. **Wiedermann, C. J., Sertl, K., and Pert, C. B.**, Substance P receptors in rat spleen: characterization and autoradiographic distribution, *Blood*, 68, 1398, 1986.
84. **O'Dorisio, M. S.**, Biochemical characterisation of receptors for vasoactive intestinal peptide in nervous, endocrine and immune systems, *Fed. Proc.*, 36, 192, 1987.
85. **Lee, C. M., Iversen, L. L., Hanley, M. R., and Sandberg, B. E. B.**, The possible existence of multiple receptors for substance P, *Naunyn-Schmiedebergs Arch. Pharmacol.*, 318, 281, 1982.
86. **Ausiello, C. M. and Roda, L. G.**, Leu-enkephalin binding to cultured human T-lymphocytes, *Cell Biol. Int. Rep.*, 8, 97, 1984.
87. **Gilman, S. C., Schwartz, J. M., Milner, R. J., Bloom, F. E., and Feldman, J. D.**, Beta endorphin enhances lymphocyte proliferative responses, *Proc. Natl. Acad. Sci. U.S.A.*, 79, 4226, 1982.
88. **Payan, D. G., Brewster, D. R., Missirian-Bastien, A., and Goetzl, E. J.**, Substance P recognition of a subset of human T-lymphocytes, *J. Clin. Invest.*, 74, 1532, 1984.
89. **Bhathena, S. J., Louie, J., Schecter, G. P., Redman, R. S., Wahl, L., and Recant, L.**, Identification of human mononuclear leukocytes bearing receptors for somatostatin and glucagon, *Diabetes*, 30, 127, 1981.
90. **O'Dorisio, M. S., Wood, D. L., and O'Dorisio, T. M.**, Vasoactive intestinal peptide and neuropeptide modulation of the immune response, *J. Immunol.*, 135, 792s, 1985.
91. **Ottaway, C. A. and Greenberg, G. R.**, Interaction of vasoactive intestinal peptide with mouse lymphocytes: specific binding and the modulation of mitogen responses, *J. Immunol.*, 132, 417, 1984.
92. **Hartung, H. P., Wolters, K., and Toyka, K. V.**, Substance P: binding properties and studies on cellular responses in guinea pig macrophages, *J. Immunol.*, 136, 3856, 1986.
93. **Hartung, H. P. and Toyka, K. V.**, Activation of macrophages with substance P: induction of oxidative burst and thromboxane release, *Eur. J. Pharmacol.*, 89, 301, 1983.
94. **Bar-Shavit, Z., Goldman, R., Stabinsky, Y., Gottleib, P., Fridkin, M., Teichberg, V. I., and Blumberg, S.**, Enhancement of phagocytosis — a newly found activity of Substance P residing in the N-terminal tetrapeptide sequence, *Biochem. Biophys. Res. Commun.*, 94, 1445, 1980.
94a. **Kimball, E. S., Persico, F. J., and Vaught, J. L.**, Substance P, neurokinin A, and neurokinin B induce generation of IL-1-like activity in P388D1 cells, *J. Immunol.*, 141, 3564, 1988.
95. **Piotrowski, W. and Foreman, J. C.**, On the actions of substance P, somatostatin and vasoactive intestinal peptide on rat peritoneal mast cells and human skin, *Naunyn-Schmiedebergs Arch. Pharmacol.*, 331, 364, 1985.

96. **Liping, Z., Dan, C., Shuzhen, Z., and Shillian, L.**, Observation of the effect of substance P on human T and B lymphocyte proliferation, *Immunol. Commun.*, 13, 457, 1984.
97. **Norlind, K. and Mutt, V.**, Influence of beta-endorphin, somatostatin, substance P and vasoactive intestinal peptide on the proliferative response of human peripheral blood T lymphocytes to mercuric chloride, *Int. Arch. Allergy Appl. Immunol.*, 80, 326, 1986.
98. **Payan, D. G., Misirian-Bastien, A., and Goetzl, E. J.**, Human T-lymphocyte subset specificity of the regulatory effects of leukotriene-$B_4$, *Proc. Natl. Acad. Sci. U.S.A.*, 81, 3501, 1984.
99. **Cavender, D. E., Haskard, D. O., Yu, C.-L., Iguchi, T., Miossec, P., Oppenheimer-Marks, P., and Ziff, M.**, Pathways to chronic inflammation in rheumatoid synovitis, *Fed. Proc.*, 46, 113, 1987.
100. **Lotz, M., Carson, D. A., and Vaughan, J. H.**, Substance P activation of rheumatoid synoviocytes: neural pathway in pathogenesis of arthritis, *Science*, 235, 893, 1987.
101. **Levine, J. D., Clark, R., Devor, M., Helms, C., Moskowitz, M. A., Basbaum, A. I.**, Intraneuronal substance P contributes to the severity of experimental arthritis, *Science*, 226, 547, 1984.
102. **Oku, R., Satoh, M., and Takagi, H.**, Release of substance P from the spinal dorsal horn is enhanced in polyarthritic rats, *Neurosci. Lett.*, 74, 315, 1987.
103. **Nilsson, J., von Euler, A. M., and Dalsgaard, C. J.**, Stimulation of connective tissue cell growth by substance P and substance K., *Nature (London)*, 315, 61, 1985.
103a. **Mantyh, C. R., Gates, T. S., Zimmerman, R. P., Welton, M. L., Passaro, E. P., Vigna, S. R., Maggio, J. E., Kruger, L., and Mantyh, P. W.**, Receptor binding sites for substance P, but not substance K or neuromedin K, are expressed in high concentrations by arterioles, venules, and lymph nodules in surgical specimens obtained from patients with ulcerative colitis and Crohn disease, *Proc. Natl. Acad. Sci. U.S.A.*, 85, 3235, 1988.
104. **Stanisz, A. M., Befus, D., and Bienenstock, J.**, Differential effects of vasoactive intestinal peptide, substance P and somatostatin on immunoglobulin synthesis and proliferation by lymphocytes from Peyers patches, mesenteric lymph nodes, and spleen, *J. Immunol.*, 136, 152, 1986.
105. **Rivier, C. and Vale, W.**, Effects of corticotropin-releasing factor, neurohypophyseal peptides and catecholamines on pituitary function, *Fed. Proc.*, 44, 189, 1985.
106. **Amann, R. and Lembeck, F.**, Stress induced ACTH release in capsaicin treated rats, *Br. J. Pharmacol.*, 90, 727, 1987.
107. **Donnerer, J. and Lembeck, F.**, Neonatal capsaicin treatment reduces ACTH-secretion in response to peripheral neuronal stimuli but not to centrally acting stressors, *Br. J. Pharmacol.*, 94, 647, 1988.
108. **Gamse, R.**, Capsaicin and nociception in the rat and mouse. Possible role of substance P, *Naunyn-Schmiedebergs Arch. Pharmacol.*, 320, 205, 1982.
109. **Donnerer, J. and Lembeck, F.**, Heat loss reaction to capsaicin through a peripheral site of action, *Br. J. Pharmacol.*, 79, 719, 1981.
110. **Donnerer, J. and Lembeck, F.**, Analysis of the effects of intravenously injected capsaicin in the rat, *Naunyn-Schmiedebergs Arch. Pharmacol.*, 320, 54, 1982.
111. **Holzer-Petsche, U. and Lembeck, F.**, Systemic capsaicin treatment impairs the micturition reflex in the rat, *Br. J. Pharmacol.*, 83, 935, 1984.
112. **Lundberg, J. M. and Saria, A.**, Capsaicin-induced desensitization of airway mucosa to cigarette smoke, mechanical and chemical irritants, *Nature (London)*, 304, 251, 1983.
113. **Lembeck, F. and Skofitsch, G.**, Visceral pain reflex after pretreatment with capsaicin and morphine, *Naunyn-Schmiedebergs Arch. Pharmacol.*, 321, 116, 1982.
114. **Traurig, H., Papka, R., and Rush, M.**, Role of capsaicin-sensitive primary afferent nerves in the neuroendocrine copulation reflex, in *Substance P and Neurokinins*, Henry, J. L., et al., Eds., Springer-Verlag, New York, 1987, 226.
115. **Khalil, Z., Marley, P. D., and Livett, B. G.**, Neonatal capsaicin treatment prevents insulin-stress induced adrenal catecholamine secretion in vivo: possible influence of sensory nerves containing substance P, *Neurosci. Lett.*, 45, 65, 1984.
116. **Amann, R. and Lembeck, F.**, Capsaicin sensitive afferent neurons from peripheral glucose receptors mediate the insulin-induced increase in adrenaline secretion, *Naunyn-Schmiedebergs Arch. Pharmacol.*, 334, 71, 1986.
116a. **Donnerer, J.**, Reflex activation of the adrenal medulla during hypoglycemia and circulatory dysregulations is regulated by capsaicin-sensitive afferents, *Naunyn-Schmiedebergs Arch. Pharmacol.*, 338, 282, 1988.
117. **Axelrod, J. and Reisine, T. D.**, Stress hormones: their interaction and regulation, *Science*, 224, 452, 1984.
118. **Dennis, G. J. and Mond, J. J.**, Corticosteroid-induced suppression of murine B cell immune response antigens, *J. Immunol.*, 136, 1600, 1986.
119. **Dennis, G., June, C. H., Mizuguchi, J., Ohara, J., Witherspoon, K., Finkelman, F. D., McMillan, V., and Mond, J. J.**, Glucocorticoids suppress calcium mobilization and phospholipid hydrolysis in anti-Ig antibody-stimulated B cells, *J. Immunol.*, 139, 2516, 1987.

120. **Collier, D. H., Achten, M. I., Arent, W. P., and Hanson, H. P.**, The neuropeptide substance P causes shape-change and adhesion in human polymorphonuclear leukocytes, *Arthritis Rheum.*, 29, S26, 1986.
121. **Marasco, W. A., Showell, H. J., and Becker, E. L.**, Substance P binds to the formyl-peptide chemotaxis receptor on rabbit neutrophils, *Biochem. Biophys. Res. Commun.*, 99, 1065, 1981.
122. **Helme, R. D., Eglezos, A., and Hosking, C. S.**, Substance P induces chemotaxis of neutrophils in normal and capsaicin-treated rats, *Immunol. Cell Biol.*, 65, 267, 1987.
123. **Helme, R. D. and Andrews, P. V.**, The effect of nerve lesions on the inflammatory response to injury, *J. Neurosci. Res.*, 13, 453, 1985.
124. **Andrews, P. V. and Helme, R. D.**, Tachykinin-induced plasma extravasation in rat skin is mediated through the SP-P receptor, in *Substance P and Neurokinins*, Henry, J. L. et al., Eds., Springer-Verlag, New York, 1987, 203.
125. **Moore, T.**, Modification of lymphocyte traffic by vasoactive neurotransmitter substances, *Immunology*, 52, 511, 1984.
126. **Yu, C. L., Haskard, D. O., Cavender, D., Johnson, A. R., and Ziff, M.**, Human gamma interferon increases the binding of T-lymphocytes to endothelial cells, *Clin. Exp. Immunol.*, 62, 554, 1985.
127. **Cavender, D. E., Haskard, D. D., Joseph, B., and Ziff, M.**, Interleukin-1 increases the binding of human B and T-lymphocytes to endothelial cell monolayers, *J. Immunol.*, 136, 203, 1986.
128. **Hendricks, H. R. and Estermans, I. L.**, Disappearance and reappearance of high endothelial venules and immigrating lymphocytes in lymph nodes deprived of afferent lymphatic vessels: a possible regulatory role of macrophages in lymphocyte migration, *Eur. J. Immunol.*, 13, 663, 1983.
129. **Charo, I. F., Yuen, C., Perez, H. D., and Goldstein, I. M.**, Chemotactic peptides modulate adherence of polymorphonuclear leukocytes to monolayers of cultured endothelial cells, *J. Immunol.*, 136, 3412, 1986.
130. **Hoover, R. L., Briggs, R. T., and Karnovsky, M. J.**, The adhesive interaction between PMN leukocytes and endothelial cells in vitro, *Cell*, 14, 423, 1978.
131. **Miossec, P., Yu, C. L., and Ziff, M.**, Lymphocyte chemotactic activity of interleukin-1, *J. Immunol.*, 133, 2007, 1984.
132. **Mason, D. W., Arthur, R. P., Dallman, M. J., Green, J. R., Spickett, G. P., and Thomas, M. L.**, Functions of rat T-lymphocyte subsets isolated by means of monoclonal antibodies, *Immunol. Rev.*, 74, 57, 1983.
133. **Brideau, R. J., Carter, F. B., McMaster, W. R., Mason, D. W., and Williams, A. F.**, Two subsets of rat T-lymphocytes defined with monoclonal antibodies, *Eur. J. Immunol.*, 10, 609, 1980.
133a. **Eglezos, A. and Helme, R. D.**, Mediation of immunosuppression by capsaicin-sensitive nerves, *Regul. Peptides*, 22, 62, 1988.
134. **Lewis T.**, The nocifensor system of nerves and its reactions, *Br. Med. J.*, 1, 431, 1937.
135. **Lembeck, F.**, Mediators of vasodilation in the skin, *Br. J. Dermatol.*, 109 (Suppl 25), 1, 1983.

Chapter 7

# DIFFERENTIAL SEX-ASSOCIATED DIFFERENCES IN IMMUNE RESPONSES TO AND OSTEONEOGENESIS OF ECTOPICALLY TRANSPLANTED FETAL BONE ALLOGRAFTS: IMPORTANCE OF $H_2$ AND NON-$H_2$ ENCODED ALLOANTIGENS

**Shraga Segal,* Chana Altaratz, Yitzhak S. Engelberg, Boris Tartakovsky, and Laslo Nebel**

## TABLE OF CONTENTS



* S. Segal is the Joseph H. Krupp Professor of Cancer Immunology.

## I. SUMMARY

Fetal bone allografts were transplanted in male and female recipients differing in major histocompatibility (MHC) $H_2$ and non-$H_2$ encoded alloantigens. The aim of this investigation was to test the possibly differential effects of sex-associated inflammatory responses evoked by these alloantigens on osteoneogenesis of the fetal bone allograft and the differentiation of associated cellular elements (i.e., bone marrow). From the results of our present study, it appears that the sex of the recipient and the extent of histoincompatibility at both (MHC) $H_2$ and non-$H_2$ genes of the fetal graft determine the character of the inflammatory processes evoked by the graft, and in turn, the developmental fate in terms of histogenetic ossification processes. In the case of the allograft, the greater the immunogenetic gap the more intense were both the inflammatory responses in the female recipients and the extent of osteoneogenesis, particularly of enchondral bone in the females.

The relevance of these observations to the immunological basis of fetal-maternal relationships and to the higher frequency of autoimmune diseases in females is discussed.

## II. INTRODUCTION

Numerous attempts have been made to elucidate the controversies concerning the immunogenetic basis of fetal-maternal relationships in mammals. Despite these investigations, the immune mechanisms permitting the histoincompatible fetus, which expresses paternal alloantigens *in utero*, to escape immune destruction by the maternal immune system has remained essentially unexplained. Extensive investigations have proved that the mammalian fetus does not escape maternal immune destruction simply as a result of total nondifferential immunosuppression,[1,2] neither do the placenta and its trophoblastic layers provide an absolute immunologically privileged site as the result of its special anatomical architecture or its immunoregulatory function.[3-6] Therefore, one is tempted to assume that nature has equipped the mammalian female with unique immunoregulatory mechanisms, enabling her to respond with specific inflammatory immune processes against fetal tissues. Such differential responses could obviously be expected to be controlled and facilitated by unique immunogenic characteristics of MHC encoded glycoproteins expressed on fetal tissues.[7] In support of this notion is the demonstration that fetal bone allografts (which were used as a model for the study of the development of ectopically grafted fetal tissue to an immunologically nonprivileged site) evoke in female recipients a different humoral response than in males.[8,9] Thus, we observed that female mice produced a high titer of predominantly $IgG_1$ anti-$H_2$ antibodies which were mainly directed against $H_2K$ region controlled alloantigens of the fetal bone graft, and that male recipients produced a lower titer of anti-$H_2$ antibodies which belonged mainly to the $IgG_{2a}$ isotype.[10] These findings were further supported by the results of Chen et al.[11] Such differences may suggest the existence of basic sex-associated differences in the specific immune inflammatory response evoked against the fetal allograft tested. Using the fetal allografts as a model for the study of the immunological basis of fetal-maternal relationships and of sex-associated differences in immune responses, we investigated whether the different inflammatory responses evoked by grafts of fetal tissues in male and female mice would determine the development and differentiation of the allografts. A fetal bone graft was chosen as a model not only due to its fetal characteristics, but mainly because the normal differentiation of its cellular elements is strictly dependent on growth- and differentiation-promoting factors secreted by various inflammatory cells, and especially those dependent on and driven by mature functional T-cells.[12] To assess the role of both major and minor histocompatibility systems, fetal bone allografts were transplanted into male and female hosts of various genetically defined donor-recipient combinations. The transplanted fetal bone grafts were examined histologically to determine the characteristics of the inflammatory infiltrate around and in the allografts and to assess the processes of differentiation and osteoneogenesis.

## III. MATERIALS AND METHODS

### A. MICE AND THEIR $H_2$ HAPLOTYES

All mice were obtained from the Animal Breeding Center of the Weizmann Institute. Female and male mice, 8 to 12 weeks of age, were used as recipients. The strains used were C57BL/6J (bbbbbbbbb), C57BL/10J (bbbbbbbbb), C3H.SW (bbbbbbbbb), and A.TL (skkkkkkkd). Fetuses at 17 to 19 d of gestation were used as donors of fetal bone grafts. These grafts originated in the following strains: C3H.eB (kkkkkkkkk); C3H/eB × C57BL/6J$F_1$ (kkkkkkkkk × bbbbbbbbb); (C3H/DISN × C3H.SW)$F_1$ (kkkkkkkkk × bbbbbbbbb); and A.AL (kkkkkkkkd).

### B. FETAL BONE TRANSPLANATION

Upper and lower limbs were amputated from fetuses. The long bones were cleaned of soft tissue and foot pads were discarded. Bones of four fetal limbs were transplanted intramuscularly in the thigh of each recipient. The grafting procedure was performed under nembutal anesthesia by intraperitoneal injection.

### C. ANTISERA

The recipient mice were bled at 50 d postgrafting and the anti-fetal bone sera were kept at −20°C until use.

### D. STAINING PROCEDURE FOR IMMUNOFLUORESCENT ANALYSIS

A 0.1-ml suspension of $3 \times 10^7$ cells/ml (spleen cells) in phosphate-buffered saline (P.B.S.) was incubated for 30 min at 4°C with dilution of antisera. After two washings with P.B.S. containing 0.02% sodium azide, the cells were further incubated under the same conditions with 20 μl of a fluorescein-isothiocyanate (FITC) conjugate of either rabbit anti-mouse $IgG_1$ or $IgG_2$ (Nordic Laboratories, Tilbert, The Netherlands). After washings, the cells were resuspended in 1 ml of P.B.S. containing 0.02% sodium azide, kept on ice, and immediately analyzed on the FACS II.

### E. FACS II CYTOFLUOROMETRIC ANALYSIS

The frequency and fluorescence profile of the stained cells were determined using the FACS II (Becton-Dickinson). Fluorescein was excited by a 488-nm line at a 400-mW laser light output. The fluorescent signals were detected with an S-11 photomultiplier tube set at 500 V. At least 10,000 viable lymphocytes were analyzed for fluorescence intensity at different fluorescence gains. The percentage of specific fluorescence-positive cells was calculated as described elsewhere,[6] where background staining is always taken into consideration.

### F. HISTOLOGICAL ANALYSIS

At 60 to 75 d posttransplantation, the animals were sacrificed and the grafts were surgically excised and fixed in neutral buffered formalin. Decalcification was obtained by a formic acid-sodium citrate solution. Following decalcification, these grafts were embedded in paraffin and 6-μ-thick serial sections were stained with hemtaoxylin eosin (HE) or periodic acid Schiff reagent plus Alcian blue (PAB) pH 2.5. The stained preparations were examined using a light microscope (×100, ×400 magnification). The histological preparations were examined with respect to the following parameters:

**TABLE 1**
**Donor-Recipient Combinations Used for Fetal Bone Grafting**

| Group no. | Strain of fetal bone graft donor | Strain of recipient | Histocompatibility differences | No. of recipients examined histologically | |
|---|---|---|---|---|---|
| | | | | Males | Females |
| 1 | C3H/eB | C57BL/6J | $H_2$ and non-$H_2$ (fully allogeneic) | 9 | 6 |
| 2 | (C3H/eB × C57BL/6J)$F_1$ | C57BL/10J | $H_2$ and non-$H_2$ (semiallogeneic) | 4 | 6 |
| 3 | (C3H/DISN × C3H.SW)$F_1$ | C3H.SW | $H_2$ | 3 | 6 |
| 4 | A.AL | A.TL | $H_2$K | 8 | — |

1. The state of the grafted fetal cartilage and its ability to be substituted by enchodral bone, namely, bone built by the host under the grafted hypertrophic cartilage or as a result of signals emitted by the grafted hypertrophic cartilage; (this bone, which replaced grafted cartilage, is lamellarly oriented, is generally accompanied by mature bone marrow, and simulates the physiological process of enchodral ossification in fetal long bones).
2. The nature of bone marrow accompanying the enchondral bone, density of cells, and their composition.
3. The process of callus-like formation, namely the induction of *de novo* generation of chondroid by the host which undergoes ossification. This bone is built in the periphery of the graft and is accompanied by osteoblasts and connective tissue cells of host origin. It has the characteristics of immature bone, and is not accompanied by mature bone marrow cell concentrations; (this bone is defined as periosteal bone).
4. The nature of infiltration, namely, the inflammatory cells found around the graft and the intensity of reaction.

## IV. RESULTS

In the first series of experiments aimed at testing the influence of differential sex-associated inflammatory responses on the development of a fetal bone allograft, fetal bones, originating in 17- to 19-d-old fetuses of C3H/eB mice were implanted into adult C57BL/6J male and female mice (Table 1, Group 1). Histological examination of tissue sections from the site of transplantation revealed (Table 2 and Figures 1 to 4) that in this fully allogeneic combination, although principal differences between females and males were not seen, quantitative differences were observed. Both types of bone, i.e., enchondral and periosteal, were found in animals of each sex; in females, enchondral ossification was predominant, while in males the periosteal bone was more developed. The bone marrow accompanying enchondral ossification was dense hypercellular in females, showing almost no fatty cells and containing all types of mature hematopoietic cells, including megakaryocytes. In males, the marrow was less dense and contained more fatty cells. An additional difference between the sexes related to the character of the inflammatory infiltrate around the grafts. Thus, in females the allograft evoked a strong infiltration of numerous granulocytes and mast cells, with great numbers of bone marrow elements in fatty tissue around the grafts ("ectopic" bone marrow formation), while other cells were in the minority. In males, the infiltration was milder, the majority of cells being lymphocytes and fibroblasts, while other cells were in the minority.

Subsequent experiments tested allografts in a donor-recipient combination representing a situation closer to the expected fetal-maternal histocompatibility combinations in nature. In those experiments, fetal bones from (C3H/eB × C57BL/6J)$F_1$ hybrid fetuses were im-

**TABLE 2**
**Histological Observations in a Donor-Recipient Combination Differing in $H_2$ and Non-$H_2$ Alloantigens (Group 1; Fully Allogeneic)**

| | Enchondral bone | | | Infiltration | | |
|---|---|---|---|---|---|---|
| | | | | | Composition | |
| Sex | Intensity | Mature marrow cells | Periosteal bone | Intensity | Predominant cells | Rare cells |
| Female | + +[a] (6/6) | Hypercellular large areas (no fatty cells) | +[a] | Strong | Granulocytes, mature bone marrow cells, mast cells | Lymphocytes, macrophages, plasma cells |
| Male | + + (5/9)<br>+ (2/9)<br>–[a] (2/9) | Less dense areas With fatty cells | + + | Mild | Lymphocytes, fibroblasts | Granulocytes, (more eosinophils), plasma cells, bone marrow cells, mast cells |

[a] Intensity of enchondral and periosteal bone formation: + + = high intensity, + = bone formed, and — = no bone formed.

planted into adult C57BL/10J male and female mice (Table 1, Group 2). The results presented in Table 3 and Figures 5 and 6 demonstrate that in this semiallogeneic combination, the inflammatory reaction and the fate of the fetal bone graft was clearly different in male and female recipients. In this combination, a striking difference was observed in the ossification of the fetal bone grafts. An excellent development of enchondral bone was found in females only, accompanied by a hypercellular "mature" bone marrow (like the fully allogeneic combination), whereas a complete absence of such bone was the rule in the males. No significant differences were observed in the periosteal bone formation. The trabeculae were thicker in both males and females than they were in the fully allogeneic combination, and they contained more osteocytes. The inflammatory response in both sexes was similar to that in the fully allogeneic combination, but milder.

We then examined fetal bones originating in $H_2$ congenic (CSW × C3H/DISN)$F_1$ hybrid fetuses following implantation into CSW adult male and female mice (Table 1, Group 3). Under these conditions, donor and recipient shared the original C3H/DISN background and differed only in MHC ($H_2$)-controlled antigens (Table 4, Figure 7). In contrast to the previous groups, insignificant differences were found between males and females in this group in which suppression of enchondral bone formation was observed concomitantly with "immature" bone marrow rich in mesenchymal cells, osteoblasts, and connective tissue elements. Periosteal bone was developed in both sexes. The overall inflammatory response in this combination was mild. In the female there was a tendency to generate foci of bone marrow cells and clusters of plasma cells, while in the male, one observed foci of lymphocytes. In our previous studies[13] concerning fetal bone allografts in rats, it was demonstrated that the exemption of such grafts from immune destruction was correlated with functional recognition of Ia antigens. It was therefore important to choose a combination in which the congenic donor and recipient shared similar $H_2$I region-encoded specificities, but were incompatible in non-$H_2$I region-controlled transplantation antigens which induce acute rejection of adult allografts. We therefore used a combination in which donor and recipient differed only by alloantigens determined by the $H_2$K region. Thus, fetal bones originating in A.AL fetuses were transplanted into congenic A.TL male mice (Table 1, Group 4; Table 5, and Figure 8). Despite the compatibility in $H_2$I and $H_2$D regions, the fetal bone graft was

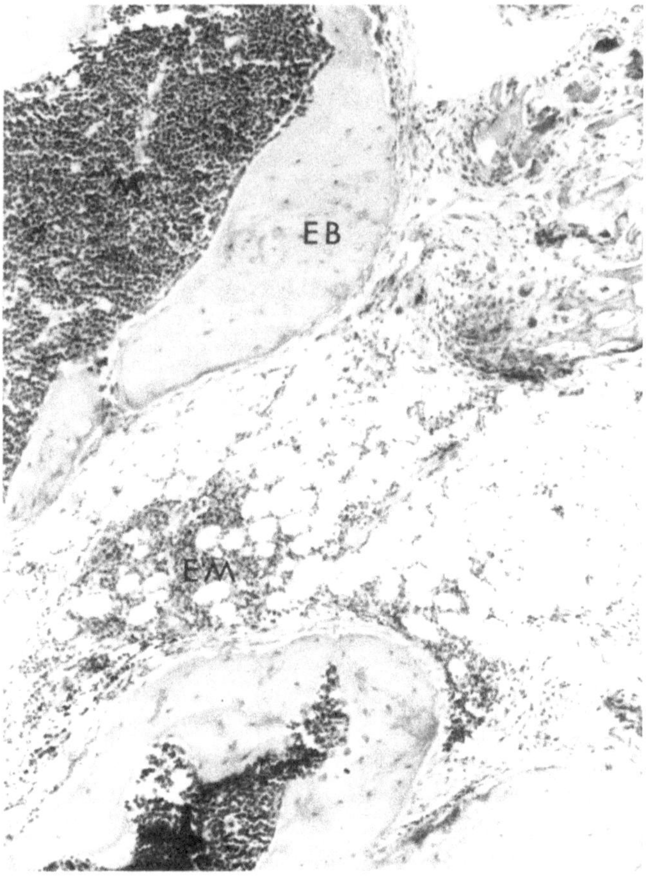


FIGURE 1. Osteoneogenesis in a female host in a fully allogeneic combination. Note new bone of enchondral type (EB), very cellular bone marrow (M), and disperse ectopic bone marrow cells (EM). (HE; magnification × 350.)

infiltrated with inflammatory cells accompanied by focal necrotic processes. Enchondral bone was not observed and the periosteal bone was impaired. Around the graft, a strong inflammatory reaction was seen, including bone marrow cells, especially large lymphocytes, giant cells, and many mast cells. These results show the importance of $H_2I$ and possibly $H_2D$ region-encoded alloantigens for the elicitation of immune reactions involved in the neodifferentiation and ossification of the fetal bone graft. These observations indicate that the fate of the graft can be influenced by the nature of sex-associated, inflammatory immune responses evoked by the implant, which in turn, may be controlled by both MHC and non-MHC transplantation antigens. Different pathways and cellular elements can be involved in the generation of different antibody isotype.[14-16] Therefore, the level of different antibody isotypes directed to MHC-controlled antigens in the various combinations of grafted female and male mice was tested.

The results shown in Table 6 denote that the observed differences in the inflammatory response and fetal bone development are correlated with differences in the subclasses of antibodies generated in the grafted mice. Thus, in the fully allogeneic combination and in the congenic semiallogeneic combination, only a low level of $IgG_1$ anti-$H_2$ alloantigens was induced in both sexes. The major response was observed in the $IgG_2$ antibody subclass. The most striking difference was observed in the noncongenic semiallogeneic combination where an $IgG_1$ predominant response was observed in female recipients only.

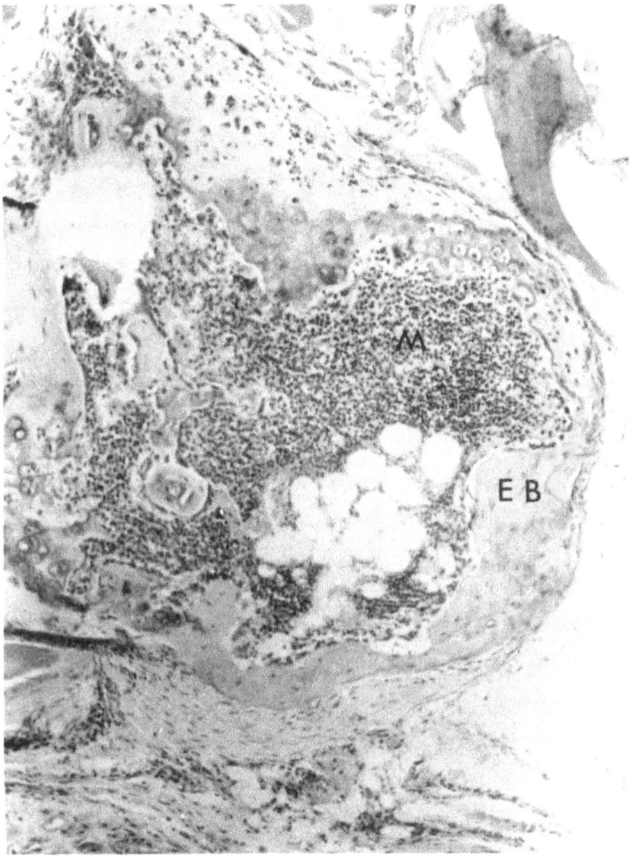


FIGURE 2. Osteoneogenesis in a male host in a fully allogeneic combination. Note less developed enchondreal bone (EB) and less cellular bone marrow (M) with fatty aggregation. (HE; magnification × 350.)

## V. DISCUSSION

One could have expected that to exempt the mammalian histoincompatible fetus from destruction by the maternal immune system the female might have developed, during evolution, unique immunoregulatory strategies which would enable her to react differently from the male to transplantation alloantigens, especially to MHC encoded alloantigens expressed on fetal tissues.

Our previous studies have indeed demonstrated that (a) principle sex-associated differences exist in the humoral immune response of male and female mice to $H_2$ determined alloantigens expressed on adult and fetal tissue;[8-20] and (b) in accordance with these findings, differences exist in the MHC-associated and especially $H_2I$ region-dependent antigen-presenting processes between male and female mice.[17] These differences which are seemingly under the control of both steroid and nonsteroid hormones,[18,19] suggested the existence of principal differences in the immune-regulated inflammatory processes evoked by allogenic tissues in male and female mice. This led us to further investigate whether and to what extent such differences may influence the normal development of an allogeneic fetal graft in males and females. As a basic model system, we decided to examine the inflammatory and osteogenic processes which have developed in the sites of implantation of a fetal bone graft transplanted ectopically to male and female recipients, histoincompatible in different combinations of $H_2$- and non-$H_2$-controlled alloantigens (Table 1). In previous investigations

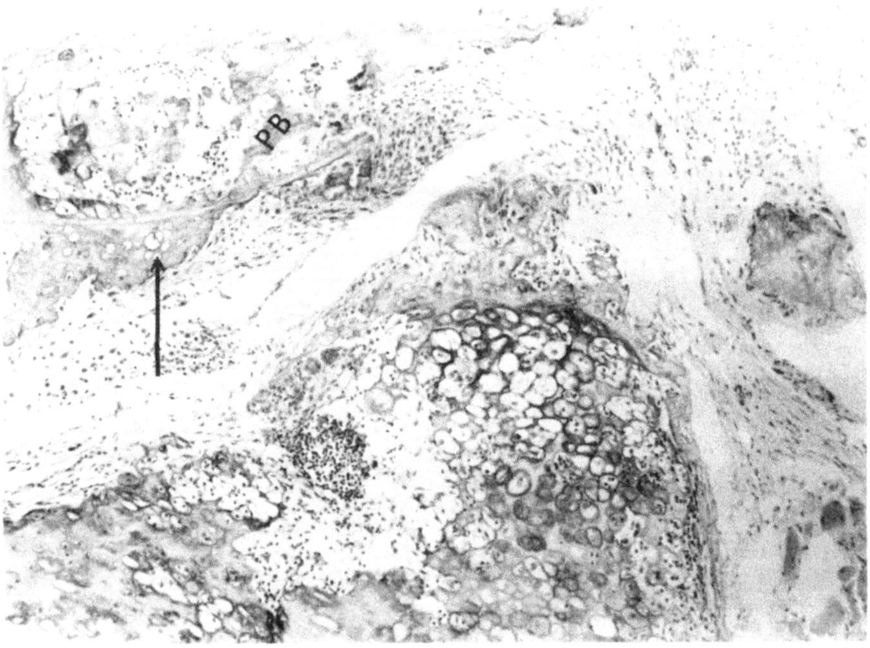


FIGURE 3. Periosteal bone formation in a female host in a fully allogeneic combination. Note small chondroid (arrow), very slight bone formation (PB). (HE; magnification × 350.)

performed in rats,[13] we demonstrated that fully allogeneic fetal bone grafts were well tolerated by immunocompetent recipients and underwent ossification processes in spite of strong allogeneic barriers, thus attesting the unique immunogenic characteristics of fetal tissues, as well as its possible enhanced development by some products of inflammatory responses.

In the present study, a histological examination of serial tissue sections from the site of the ectopically transplanted fetal bone graft revealed the following:

1. When a fetal bone graft was transplanted to an adult congenic male recipient differing in the $H_2K$ region (Table 5), we observed an inflammatory infiltrate inside and around the graft, followed by immature bone marrow cellular elements. The cartilagenous fetal tissue was partly degenerated. Periosteal bone elements were observed. No formation of enchondral bone was found.
2. When fetal bone grafts originating from $F_1$ $H_2$ congenic mice were transplanted to paternal male and female recipients (i.e., (C3H/DISN × CSW)$F_1$ transplanted to C3H.SW mice), we again observed an infiltration of inflammatory elements as described in Table 4. In this combination, however, we observed a well-developed periosteal bone, but a poor formation of enchondral bone accompanied by poorly differentiated, immature bone marrow elements. No significant sex-associated differences were observed in fetal bone development between the male and female recipients.
3. When a fetal bone graft was implanted into completely allogeneic recipients differing in both $H_2$ and non-$H_2$ alloantgiens, we observed an intense inflammatory infiltrate surrounding the allograft. In contrast to the aforementioned observations, we found in addition to the development of periosteal bone, which was more developed in the male, the development, particularly in the female, of massive enchondral bone with hypercellular bone marrow containing the characteristic elements of fully differentiated adult marrow (Table 2).

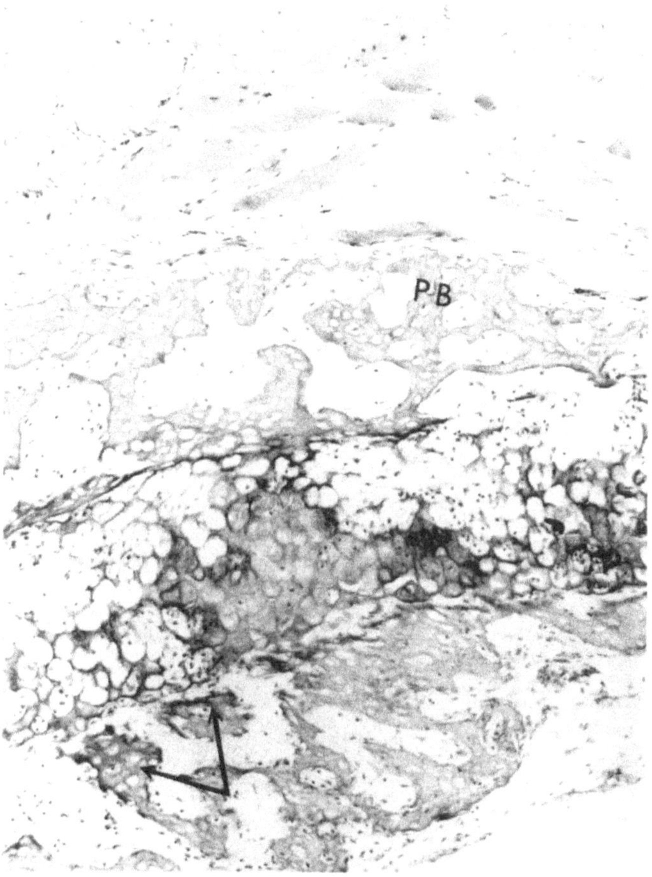


FIGURE 4. Periosteal bone formation in a male host in a fully allogeneic combination. Note more developed chondroid (arrow) in a callus-like ossification process (PB), no bone marrow formation. (HE; magnification × 350.)

4. When $F_1$-originated fetal bone was grafted in a non-$H_2$ congenic semiallogeneic combination, namely into paternal male and female mice, which differed in both $H_2$ and non-$H_2$ alloantigens, (yet still facilitate alloantigen dependent cell cell interaction by semiallogeneic compatibility with the implanted host), marked differences in fetal bone development was observed. Thus, in both males and females, we found a heterogeneous inflammatory infiltrate composed of various cells and periosteal bone formation. Yet, while massive enchondral bone was formed in the female accompanied by the concomitant presence of large foci of fully differentiated bone marrow cells, neither enchondral bone nor detectable bone marrow formation was detected in the male recipient (Table 3).

At this point it should be mentioned that when syngeneic fetal bone grafts were transplanted ectopically to adult female and male mice (Segal, S., et al., unpublished data), no differences were observed in males and females as to the fate of the graft. The cartillagenous bones matured and ossified normally. The new bone was lamellar and vital and was accompanied by mature bone marrow containing few fatty cells. No inflammatory infiltrate was seen around the grafts.

The results described above may be the consequence of differential immune processes evoked by allogeneic fetal graft in females as compared to males. This assumption is

## TABLE 3
## Histological Observations in a Donor-Recipient Combination Differing in $H_2$ and Non-$H_2$ Alloantigens (Group 2; Semiallogeneic)

| | Enchondral bone | | | Infiltration | | |
|---|---|---|---|---|---|---|
| | | | | | Composition | |
| Sex | Intensity | Mature marrow cells | Periosteal bone | Intensity | Predominant cells | Rare cells |
| Female | + +[a] (4/6)<br>–[a] (2/6) | Hypercellular<br>Large areas | + + | Strong | Granulocytes, bone marrow cells, mast cells, fibroblasts | Lymphocytes, plasma cells, macrophages |
| Male | – (4/4) | – | + + | Mild | Lymphocytes, plasma cells, mast cells, fibroblasts | |

[a] Intensity of enchondral and periosteal bone formation: + + = high intensity and — = no bone formed.

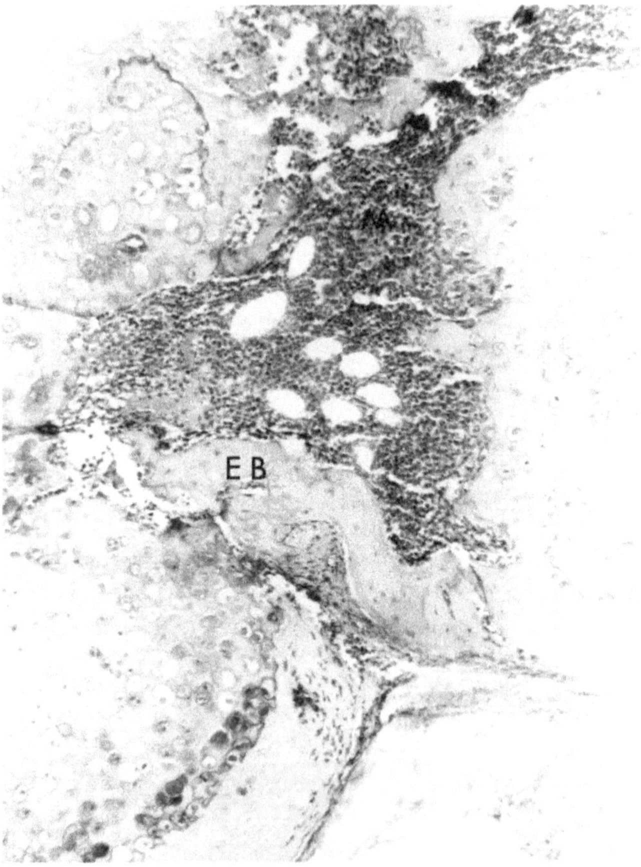


FIGURE 5. Osteoneogenesis in a female host in a semiallogeneic combination. Note well-developed enchondral bone (EB) with very cellular marrow (M). (HE; magnification × 350.)

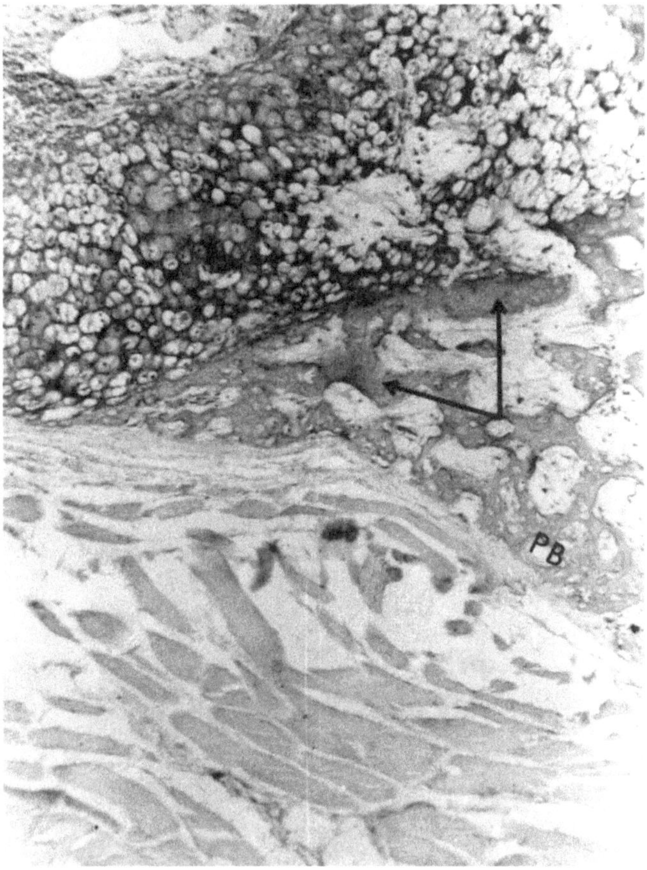


FIGURE 6. Periosteal bone formation in a male host in semiallogeneic combination. Note chondroid formation (arrow) of periosteal type, scarce bone formation (PB). No marrow cells. (HE; magnification × 350.)

**TABLE 4**
**Histological Observations in a Donor-Recipient Combination Differing in $H_2$ Only (Group 3)**

| | Enchondral bone | | Periosteal | Infiltration | |
|---|---|---|---|---|---|
| Sex | Intensity | Mature marrow cells | bone | Intensity | Composition |
| Female | ±[a] | ± (Immature cells) | + +[a] | Mild | Plasma cells (foci), bone marrow cells, mast cells, lymphocytes, granulocytes |
| Male | ± | ± (Immature cells) | + + | Mild | Lymphocytes (foci), mast cells, plasma cells |

[a] Intensity of enchondral and periosteal bone formation: + + = high intensity, + = bone formed, and — = no bone formed.

supported by the fact that in testing the nature of the various antibody isotypes elicited against the alloantigens by the fetal graft, a predominant induction of $IgG_1$ in the presence of $IgG_2$ antibodies was observed in the female recipient in the fully semiallogeneic combination. The fact that the antibodies elicited under the present experimental conditions interact (as attested by FACS analysis) only with a fraction of spleen cells (Table 6), may

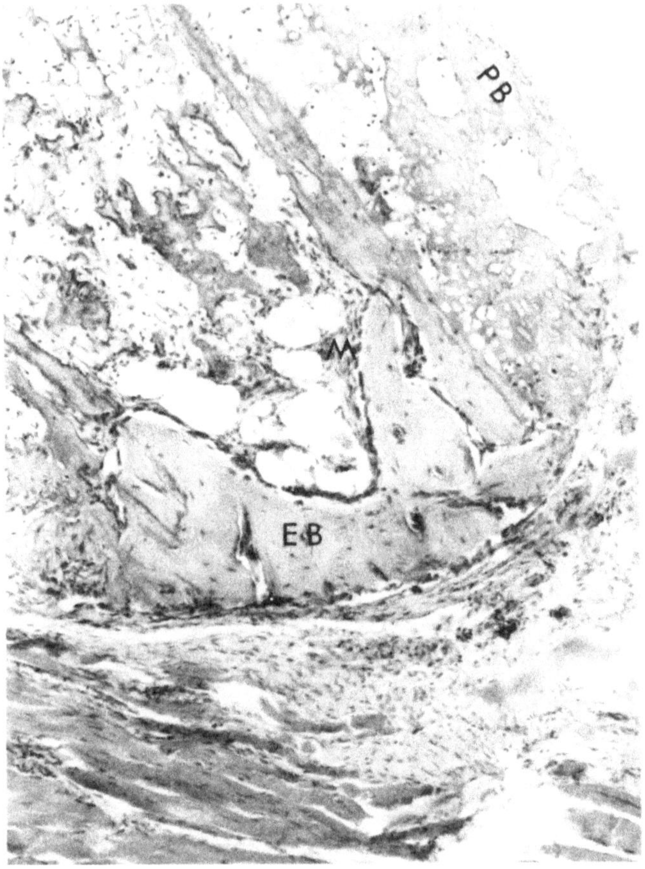


FIGURE 7. Osteogenesis in a female host in a congenic combination differing in $H_2$. Note focal enchondral bone formation (EB) with immature marrow cells (M). Definite periosteal bone formation (PB). (Similar findings in the male). (HE; magnification × 350.)

**TABLE 5**
**Histological Observations of a Donor-Recipient Combination Differing in $H_2K$ Region Only (Group 4)**

| | Enchondral bone | | Periosteal | Infiltration | |
|---|---|---|---|---|---|
| **Sex** | **Intensity** | **Mature marrow cells** | **bone** | **Intensity** | **Composition** |
| Male | —[a] | — | +[a] | Strong | Various bone marrow elements, macrophages, mast cells, fibroblasts |

[a] Intensity of enchondral and periosteal bone formation: + = bone formed and — = no bone formed.

indicate that the immunization with fetal bone grafts may induce humoral responses to MHC-associated epitopes which are differentially expressed only on some, but not all, splenic cells. Since the induction of $IgG_1$ and $IgG_2$ seems to be controlled by different regulatory pathways;[14-16] these results support the assumption that principal basic differences exist in the regulatory mechanisms of the immune response between males and females against MHC-encoded transplantation antigens of fetal tissues, especially in cases of incompatibility

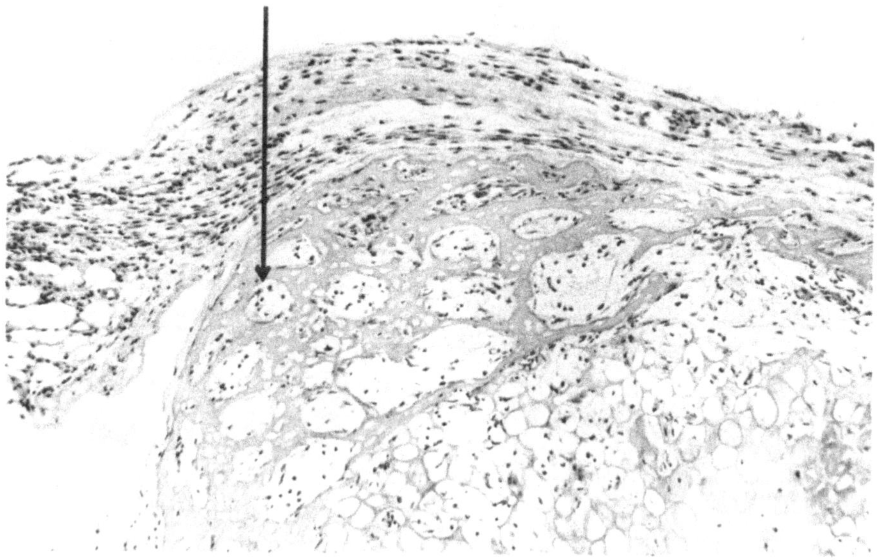

FIGURE 8. Periosteal bone formation in a congenic male host differing in $H_2K$ region. Note developed chondroid (arrow) with scarce bone formation. (HE; magnification × 350.)

## TABLE 6
## Different Anti-$H_2$ Alloantigen Isotypes Produced by Females and Males Following Transplantation with Genetically Different Fetal Bones[a]

| | $H_2$-specific IgG isotypes | | | |
|---|---|---|---|---|
| **Sex** | **$IgG_1$** | **$IgG_2$** | **Genetic combination** | **Group no.** |
| Female | 21[b] | 53 | Fully allogeneic | 1 |
| Male | 25 | 58 | | |
| Female | 28 | 59 | $H_2$ Congenic $F_1$ | 3 |
| Male | 27 | 14 | | |
| Female | 6.5 | 45 | $H_2$ Congenic $F_1$ | 3 |
| Male | 6 | 26 | | |
| Female | 50.5 | 21.3 | Semiallogeneic | 2 |
| Male | 5.2 | 18.1 | | |

[a] Female and male mice were transplanted with fetal bones. The antisera thus obtained 50 d after grafting were analyzed for the presence of $H_2$-specific IgG isotypes by means of immunofluorescence. The donor spleen cells were stained with the different antisera and subsequently developed for fluorescence, using FITC-labeled rabbit anti-mouse $IgG_1$ or goat antimouse $IgG_2$. The percentage of the fluorescent cells was scored on the FACS II at fluorescence gain 4.

[b] Percent fluorescent cells.

at both the MHC and the non-MHC antigens. At this point, it is of interest to note that $IgG_1$ antibodies were found as a major antibody subclass directed against fetal antigens which were extracted from the placenta of pregnant female mice mated with allogeneic males. These, unlike similarly isolated $IgG_2$ antibodies, were found to enhance the growth of tumors of paternal origin.[20-23] Since $IgG_1$ antibodies, similar to IgE, were found to regulate the degranulation processes of mast cells in different mammalian species[24] and since mast cells have been found to influence enchondral bone formation[25,26] and remodelling, the function and production of $IgG_1$ by the fetal bone graft may be of particular importance to the presently observed differences in the generation of enchondral bone in the various genetic combinations.

Steroid sex hormones have been shown to influence the differentiation of osteogenic tissues.[27] Yet, the phenomena we observed cannot be attributed solely to a hormonal effect. Had this been the case, we should have found similar differences in fetal bone in males and females in the $H_2$ congenic combination, yet in this combination such differences were not found.

Assuming that periosteal bone is generated by ossification of fibroblastoid cell-derived osteoblasts and enchondral bone is derived from bone marrow originating osteoblasts, we tend to suggest the following cascade of processes to explain the observed differences in the development of the fetal bone graft. In general, when a fetal bone graft is transplanted in a congenic combination differing in the $H_2K$ region only, an inflammatory process is evoked which is sufficient to activate some pathways of the inflammatory reaction restricted to and regulated by the $H_2K$ region. This reaction will lead to partial degeneration of the implanted fetal cartilage, yet it is sufficient to support the development of some possibly host-derived chondroid, providing signals which will lead to its calcification and periosteal bone formation, possibly originating from surrounding connective tissue, host-derived osteogenic elements (i.e., tissue-derived and pericytic fibroblasts). Such signals can be induced by chemotactic factors derived from the hypertrophic new chondroid undergoing ossification.[28-30]

Although in this combination some precursors of bone marrow are attracted to the inflammatory products, signals permitting the development of mature bone marrow cells, in particular those leading to the generation of osteogenic bone marrow elements, are not generated. When $F_1$ $H_2$ congenic fetal bones are transplanted to parent recipients, allogeneic $H_2I$ region-encoded antigen-induced factors will be secreted. These factors (i.e., lymphokines, monokines, and growth factors) may support (1) a better periosteal bone formation of the implanted bone graft and enhanced proliferation and differentiation of mesenchymal elements, supported by a proper microenvironment conditioning; and (2) migration and partial development of bone marrow elements which are attracted to the site of implantation. Yet this combination is insufficient with respect to the requirements for massive enchondral bone formation which might be dependent on more mature stage of cartilage[30,31] and especially on bone marrow and its associated osteoblast differentiation.[32] These requirements seemingly are met by inflammatory processes and secretion of appropriate growth factors evoked by non-$H_2$-controlled transplantation antigens. Qa and Tla region encoded class I MHC glycoproteins, the genes of which are mapped outside the classical $H_2$ region and which were found to be expressed on bone marrow cellular elements and other cells, may be excellent candidates for such interactions.[33] The host immune response against such antigens may lead to secretion of factors and generation of signals which will result in a better development and maturation of the grafted fetal cartilage. Thus, transplantation of fully allogeneic or fully semiallogeneic fetal bone grafts to females is followed by massive osteogenesis and enchondral bone formation. Although the present experiments were performed in nonpregnant females and therefore are not under the influence of hormone imbalances associated with pregnancy, one may assume that the presently observed sex-associated differences may even be more profound in pregnant females. Experiments performed in our laboratory have demonstrated that castrations of both males and female mice abolish the above described differential humoral response (IgG, VS $IgG_2$) elicited against allogeneic fetal bone grafts (Segal, S., et al., unpublished results), indicating the importance of sex hormones in the regulation of these responses.

The importance of non-$H_2$ genes in the regulation of maternal immune responses against allogeneic embryonic and fetal tissues, where $H_2$ differences between mother and conceptus exist, is further emphasized by the findings of Chaouat et al.,[34] who demonstrated that the frequency of spontaneous abortions occurring in mating of CBA/J females ($H_2K$) with DBA/2J males ($H_2{}^d$) can be reduced by preimmunizing the CBA females with spleen cells of BALB/c ($H_2{}^d$) males, but not of CBA or DBA males. Although the mechanism by which nonimmunoglobulin non-MHC ($H_2$) encoding genes may influence such immune responses is still obscure, one should emphasize in this respect that genes outside chromosome 17 were recently found to control the expression of ($H_2IJ$) molecules which control the function of regulatory suppressor T lymphocytes.[56]

Furthermore, Wernet et al.[35-37] observed that non-$H_2$ genes or genes found outside the classicial $H_2$ region may regulate the induction of T-cell dependent IgG antibodies against $H_2D$ region determined antigens.

The main notion that emerges from our present investigation is that both $H_2$ and non-$H_2$ transplantation antigens expressed on embryonic or fetal tissues may evoke a differential inflammatory response in males and females. These may not only result (especially in the female) in the failure to reject the fetal graft, but the inflammatory factors secreted by the activated immunocytes may interact with the embryonic cells and facilitate their enhanced growth and directions of differentiation. $H_2I$ region-evoked immune processes in males and females may be of the utmost importance for the promotion and enhancement of allogeneic fetal tissue.

These assumptions are supported by the following facts:

1. Activated lymphocytes and macrophages were found to secrete factors which promote the regulation of proliferation and differentiation of cells from both mesenchymal and nonmesenchymal origin, i.e., Interleukin 1, 2, 3, etc.,[38-42] interferon,[44] colony-stimulating factor (CSF),[41,43] epidermal growth factor (EGF),[45] fibroblast growth factor (FGF),[38] angiogenic factor,[41] B-cell growth factor (BCGF),[43] fibronectin,[46] and prostaglandins.[47,48]
2. Factors secreted by activated immunocytes regulate hemopoiesis.[41,49]
3. Inflammatory processes increase the release of platelet-derived growth factor (PDGF).[45]
4. Sex hormones regulate antigen presentation processes in mice.[12]
5. $IgG_1$ antibodies, which are predominant in the females responses in the semiallogeneic combination (Table 6) in addition to IgE, are potent activators of basophils and mast cells[19] which surround the implanted fetal bone. These cells are known to inhibit some T-cell functions due to release of histamine and other biogenic amines.[43]
6. Inappropriate mixed lymphocyte reaction (MLR) responses are associated in humans with a high frequency of premature abortions.[44,45]
7. Brown Norwegian (BN) rats which are defective in mounting a normal delayed type hypersensitivity (DTH) response (unlike immune competent Lewis rats) were unable to support the normal development of an allogeneic fetal bone graft. On the contrary, they rejected this graft promptly.[8]
8. Thymectomized mice, unlike their sham-operated immunocompetent control counterparts, rejected fetal bone allografts (Segal, S., et al., unpublished results). These results therefore are in line with our previous assumptions and the results and suggestions of Billingham et al.[1] and other investigators[1] and tend to indicate that strong

alloantigenic differences between mother and conceptus occurring in a heterogeneous randomly mating population, may not only fail to endanger pregnancy, but on the contrary, a qualitatively selective controlled immune response under conditions which prevent the rejection of the embryo is supportive and possibly advantageous to facilitate the enhanced normal growth of the fetus, thus encouraging the evolutionary advantageous heterogenization of the species.

These findings therefore, if further pursued, may shed new light on (a) fetal maternal relationship; (b) differences in occurrence growth and progression of various malignant tumors in males and females; (c) the influences of inflammatory processes on bone development and resorption; (d) the relation and association between primary and secondary immune deficiencies, skeletal development, bone malformations (osteopetrosis[48]), tissue development, and normal homeostasis; and (e) differences in both the morbidity and mortality from infections caused by various infectious agents, especially the intracellular ones, (i.e., tuberculosis).

Furthermore, the present findings taken together with our previous findings strongly suggest the possible existence of principal, basic sex-associated differences in the regulation of immune responses in males and females, expecially those responses restricted by the constituents of the MHC system which govern self, nonself, and modified self immune recognition. If indeed this is the case, our findings may also shed new light as to the predominant higher frequency and the relative refractory nature of most of the autoimmune diseases in females. The elucidation of mechanisms responsible for these mainly qualitative sex-associated differences may lead to new and more rational treatments for various autoimmune disorders.

## REFERENCES

1. **Billingham, R. E. and Head, J. R.,** The riddle of allogeneic conceptus, *Transplant. Proc.*, 4, 877, 1983.
2. **Beer, A. E.,** *The Immunobiology of Mammalian Reproduction*, Prentice-Hall, Englewood Cliffs, NJ, 1976.
3. **Chaouat, G., Kolb, J. P., and Wegmann, T. G.,** The murine placenta as an immunological barrier between the mother and the fetus, *Immunol. Rev.*, 75, 31, 1983.
4. **Imakawa, K., Anthony, R. V., Kazemi, M., Marotti, K. R., Polites, H. R., and Roberts, R. M.,** Interferon-like sequence of ovine trophoblast protein secreted by embryonic trophectoderm, *Nature (London)*, 330, 377, 1987.
5. **Pollard, J. W., Bartocci, A., Areci, R., Drlofsky, A., Ladner, M. B., and Stanley, R. E.,** Apparent role of the macrophage growth factor, CSF-1, in placental development, *Nature (London)*, 330, 484, 1987.
6. **Chaouat, G.,** Placental immunoregulatory factors, *J. Reprod. Immunol.*, 10, 179, 1987.
7. **Wegmann, T. G.,** The presence of class I MHC antigens at the maternal-fetal interface and hypotheses concerning the survival of the murine fetal allograft, *J. Reprod. Immunol.*, 3, 267, 1981.
8. **Segal, S., Tartakovsky, B., Katzav, S., and De Baetselier, P.,** The immunologic basis of the fetal-maternal relationship, *Transplant Proc.*, 7, 582, 1980.
9. **Tartakovsky, B., De Baetselier, P., Feldman, M., and Segal, S.,** Sex-associated differences in the immune response against fetal major histocompatibility antigens, *Transplantation*, 32, 395, 1981.
10. **Tartakovsky, B., De Baetselier, P., Feldman, M., and Segal, S.,** Sex-associated differences in the immune response against fetal major histocompatibility antigens. II. Association of the $IgG_1$ with H-2K-encoded alloantigens, *Transplantation*, 36, 191, 1983.
11. **Chen, R. E. W. and Loke, Y. W.,** Antibody isotype produced by pregnant mice immunized with fetal allogeneic lymphocytes is mainly IgG1, *J. Reprod. Immunol.*, 10, 189, 1987.
12. **Askalanov, A. A., Gordienko, S. M., Avoyunicheva, O. E., Bondarenko, A. V., and Voronkov, S. F.,** The role of T-system immunity in repatory regeneration of the bone tissue in animals, *J. Hyg. Epidemiol. Microbiol. Immunol.*, 31, 210, 1987.
13. **Segal, S., Siegal, T., Altaratz, H., Lev-El, A., Nevo, Z., Nebel, L., Katznelson, A., and Feldman, M.,** Fetal bone grafts do not elicit allograft rejection because of protecting anti Ia alloantibodies. Implications to the immune survival of fetuses in allogeneic mothers, *Transplantation*, 28, 88, 1979.

14. **Elson, C. O., Heck, J. A., and Strober, W.**, T-cell regulation of murine IgA synthesis, *J. Exp. Med.*, 149, 632, 1979.
15. **Torrigiani, G.**, Quantitative estimation of antibody in the immunoglobulin classes of the mouse. II. Thymic dependence of the different classes, *J. Immunol.*, 108, 161, 1972.
16. **Bosma, C. G., Owen, J., Eaton, G., Marshall, G., Dewitt, C., and Bosma, M. J.**, Concentration of $IgG_1$ and IgG2a allotypes in the serum of nude and normal allotype cogenic mice, *J. Immunol.*, 124, 879, 1980.
17. **Weinstein, Y., Ran, S., and Segal S.**, Sex-associated differences in the regulation of immune responses controlled by the MHC of the mouse, *J. Immunol.*, 137, 656, 1984.
18. **Stiles, D. P. and Siiteri, P. K.**, Steroids as immunosuppressants in pregnancy, *Immunol. Rev.*, 75, 117, 1983.
19. **Siiteri, P. K. and Stiles, D. P.**, Immunologic and endocrine interrelationships in pregnancy, *Biol. Reprod.*, 26, 1, 1983.
20. **Bell, S. C. and Billington, D.**, Antifetal alloantibody in the pregnant females, *Immunol. Rev.*, 75, 5, 1983.
21. **Voisin, G. A. and Chaouat, G.**, Demonstration, nature and properties of maternal antibodies fixed on placenta and directed against paternal antigens, *J. Reprod. Fertil.*, 21, 89, 1974.
22. **Bell, S. C. and Billington, W. D.**, Humoral immune responses in murine pregnancy. III. Relationship between anti-paternal alloantibody levels in maternal serum, placenta, and fetus, *J. Reprod. Immunol.*, 5, 299, 1983.
23. **Bell, S. C. and Billington, W. D.**, Major anti-paternal alloantibody induced by murine pregnancy is non complement fixing $IgG_1$, *Nature (London)*, 288, 387, 1980.
24. **Voisin, G. A.**, Role of antibody classes in the regulatory facilitation reaction, *Immunol. Rev.*, 49, 3, 1980.
25. **Severson, A. R.**, Mast cells in areas of experimental bone resorption and remodelling, *Br. J. Exp. Pathol.*, 50, 17, 1969.
26. **Lindholm, R., Lindholm, S., and Liukko, P.**, Fracture healing and mast cells. I. The periosteal callus in rats, *Acta Orthop. Scand.*, 38, 115, 1967.
27. **Cornell, C., Burnett, M. F., and Reddi, A. H.**, Influence of estrogen and progesterone on matrix-induced enchondral bone formation, *Calcif. Tissue Int.*, 35, 609, 1983.
28. **Somerman, M., Hewitt, A. T., Varner, H. H., and Schiffmann, E.**, Identification of bone matrix derived chemotactic factor, *Calcif. Tissue Int.*, 35, 481, 1983.
29. **Urist, M. P., Lietze, A., Mizutani, H., Takagi, K., Trifitt, J. T., Amstutz, J., De Lange, R., Termine, J., and Finerman, G. A. M.**, A bovine low molecular weight bone morphogenetic protein (BMP) fraction, *Clin. Orthop. Relat. Res.*, 162, 329, 1982.
30. **Shimomura, Y., Yoneda, T., and Suzuki, F.**, Osteogenesis by chondrocytes from growth cartilage of rat rib, *Calcif. Tissue Res.*, 19, 179, 1975.
31. **Hanaoka, H.**, The fate of hypertrophic chondrocytes of the epiphyseal plate, *J. Bone Jt. Surg.*, 58A, 226, 1976.
32. **Friedenstein, A. J., Petrakova, K. W., Kurolegova, A. I., and Frolova, G. P.**, Heterotrophic transplants of bone marrow. Analysis of precursor cells for osteogenic and haematopoietic tissues, *Transplantation*, 6, 230, 1968.
33. **Margulies, D. H., Evans, G. A., Flaherty, L., and Seidman, J. G.**, H-2-like genes in the T1a region of mouse chromosome 17, *Nature (London)*, 295, 168, 1982.
34. **Chaouat, G., Kiger, N., and Wegmann, T. G.**, Vaccination against spontaneous abortions in mice, *J. Reprod. Immunol.*, 5, 4, 1983.
35. **Wernet, D. and Lilly, F.**, Genetic regulation of the antibody response to $H\text{-}2D^b$ alloantigens in mice. II. Tolerance to non-H-2 determinants abolishes the antibody response in BI0.A(5R) mice, *J. Exp. Med.*, 144, 266, 1976.
36. **Wernet, D., Shafran, H., and Lilly, F.**, Genetic regulation of the antibody response to H-2D alloantigens in mice. III. Inhibiton of IgG response to noncongenic cells by preimmunization with congenic cells, *J. Exp. Med.*, 144, 654, 1976.
37. **Wernet, D. and Lilly, F.**, Genetic regulation of the antibody response to $H\text{-}2D^b$ alloantigens in mice. I. Differences in activation of helper T cells in C57BL/10J and BALB/c congenic strains, *J. Exp. Med.*, 141, 573, 1975.
38. **Schmidt, J. A. et al.**, Interleukin-1, a potential regulator of fibroblast proliferation, *J. Immunol.*, 128, 2177, 1982.
39. **Watson, J., Frank, M. B., Mochizuki, D., and Gillis, S.**, The biochemistry and biology of IL-2, *Lymphokines*, 6, 95, 1982.
40. **Ihle, J. N., Rebar, L., Keller, J., Lee, J. C., and Hapel, A. J.**, Interleukein-3: possible roles in the regulation of lymphocyte differentiation and growth, *Immunol. Rev.*, 63, 5, 1982.
41. **Stanley, E. R.**, Colony stimulating factor, in *The Lymphokines*, Hadden, J. W. and Stewart, W. E., Eds., Humana Press, Clifton, NJ, 1981, 101.

42. **Davis, P. and Bonney, R. J.,** Secretory products of mononuclear phagocytes: a brief review, *J. Reticuloendothel. Soc.*, 26, 37, 1979.
43. **Stewart, C. C.,** Factors which influence the proliferation and differentation of mononuclear phagocytes, in *Heterogeneity of Mononuclear Phagocytes,* Forster, O. and Laudy, C., Eds., Academic Press, New York, 1980, 377.
44. **Smith, K. A.,** Lymphokine regulation of T cell and B cell function, in *Fundamental Immunology,* Paul, W. E., Ed., Raven Press, New York, 1984, 559.
45. **Seppa, H. E. J., Grotendorst, G., Seppa, S. T., Schiffmann, E., and Martin, G. R.,** Platelet derived growth factor is chemotactic for fibroblast, *J. Cell Biol.*, 92, 584, 1981.
46. **Tsukamoto, Y., Helsel, W. E., and Wahl, S. M.,** Macrophage production of fibronectin, a chemoattractant for fibroblasts, *J. Immunol.*, 127, 673, 1981.
47. **Parker, C. W.,** Mediators: release and function, in *Fundamental Immunology,* Paul, W. E., Ed., Raven Press, New York, 1984, 731.
48. **Barry, M. A.,** Prostaglandins: fine tuning the immune system?, *Immunol. Today,* 65, 69, 1980.
49. **Segal, S., Melmon, K. L., and Weinstein, Y.,** Isolation of suppressor T Lymphocytes adherent to histamine conjugated albumin and their role in suppression of IgG response to SRBC, *Cell. Immunol.*, 59, 171, 1981.
50. **Dinarello, C. A.,** Biology of interleukin 1, *Fed. Am. Soc. Exp. Biol. J.*, 2, 108, 1988.
51. **Beer, A. E., Avebeeman, J. F., Ayers, J. W. T., and Haines, R. F.,** MHC antigens, maternal and paternal responses and chronic habitual abortions in humans, *Am. J. Obstet. Gynecol.*, 141, 987, 1981.
52. **Lauritsen, J. G., Kirstensen, T., and Grunnet, N.,** Depressed, mixed lymphocyte culture reactivity in mothers with recurrent spontaneous abortion, *Am. J. Obstet. Gynecol.*, 125, 35, 1976.
53. **Rocklin, R. E., Kitzmiller, J. L., Capenter, C. B., Garvoy, M., and David, J. R.,** Maternal fetal relation. Absence of an immunologic blocking factor from the serum of women with chronic abortions, *N. Engl. J. Med.*, 295, 1209, 1986.
54. **Beer, A. E., Scott, J. R., and Billingham, R. E.,** Histocompatibility and maternal immunological status as determinants of fetoplacental weight and litter size in rodents, *J. Exp. Med.*, 142, 180, 1975.
55. **Milhaud, G. and Labat, M. L.,** Thymus and osteoperosis, *Clin. Ortho. Relat. Res.*, 135, 260, 1978.
56. **Hayes, C. E., Lyczek, K. K., Krum, D. P., Whitcomb, R. M., Hullet, D. A., and Cantor, H.,** Chromosome 4 Jt gene controls murine T cell surface I-J expression, *Science,* 223, 559, 1984.

Chapter 8

# AUGMENTATION OF NATURAL IMMUNITY BY CONDITIONING AND POSSIBLE MECHANISMS OF ENHANCEMENT

**Vithal K. Ghanta, H. Brent Solvason, and Raymond N. Hiramoto**

## TABLE OF CONTENTS

# I. INTRODUCTION

A large resource of older literature exists which documents many aspects of conditioned "immunobiologic" responses.[1] These conditioning experiments involved both the specific and nonspecific immune responses. Specific antibody responses were measured by agglutinin titers, and the nonspecific responses ranged from peritoneal exudates to complement levels and activity of lysozyme. These studies demonstrated many instances of enhanced responses due to conditioning.[1] The difficulty replicating these early studies has been attributed to the use of different species, varied environmental conditions, the schedule of conditioned stimulus (CS) and unconditioned stimulus (US) pairing, and the interval after the CS for measuring the conditioned response (CR).[1]

Recently, several articles have been published which report a conditioned suppression of antibody response and the graft vs. host response.[2,3] These studies have exploited a conditioned taste aversion paradigm which uses cyclophosphamide as the US. This model was first used by Ader and Cohen[2] to show a slight but consistent and statistically significant decrease in antibody titers in conditioned animals. Other investigators have shown that within a classical conditioning-type paradigm, modulation of allograft transplantation, tumor transplantation, and NK cell activity can be seen.[4-7]

It has been demonstrated that antibody responses can be modulated by discrete central nervous system (CNS) lesions,[8-11] by stimulation of specific sites in the CNS,[12-14] and by classical (Pavlovian) conditioning.[15,16] There are a variety of physiologic means by which the CNS could modulate the level of the immune response to produce the different effects observed. One could envisage changes resulting from the interactions between neuropeptides or neuroendocrines and the immune system.[17] To support this contention, there are now numerous reports indicating receptors for neurotransmitters on lymphocytes,[18] the stimulation of which can lead to altered cellular activity.[19] Gorczynski et al.[5] have demonstrated the modulation of growth of a transplanted syngeneic plasmacytoma tumor in BALB/c mice which were conditioned by a classic conditioning taste aversion paradigm with the initial use of drug-induced immunosuppression. They have also shown that this conditioned effect was abolished by exposure to cimetidine, a histamine type II-receptor antagonist.

Our studies have investigated the potential for conditioned enhancement of two linked nonspecific responses — interferon levels and natural killer (NK) cell activity. The reported studies describe conditioned enhancement of NK cell activity, the functional relevance, and the possible mechanisms of conditioned enhancement.

# II. CONDITIONING OF THE NK RESPONSE WITH CAMPHOR ODOR

A classical conditioning paradigm using the odor of camphor as the CS and an injection of Poly I:C as the US was used to show a conditioned increase in an immune-type response (NK cell activity) without restimulation by its inducer. The conditioned response (CR) in this conditioning protocol was the subsequent elevation in NK cell activity.

Female BALB/c mice of 6 weeks of age were maintained 12:12 h light:dark cycle in our laboratories for 2 weeks prior to utilization in the experiments. Appropriate groups of mice were given 20 μg/mouse poly I:C in 0.1 ml i.p. just prior to exposure to the odor of camphor. Depending on the experiment, mice were exposed to odor of camphor for either 1 or 4 h in their own cages in an enclosed area. They were transferred to new cages with clean bedding prior to returning them to the facility.

The association trial sessions were repeated once every 72 h, 9 times. The test for conditioning was the tenth session and occurred 72 h after the ninth session (Table 1). Animals were sacrificed 24 h after the conditioning test and their spleen cells were assayed

**TABLE 1**
**Schedule of Camphor and Poly I:C Association Trials**

| Groups | Treatment: Association trials sessions 1 to 9 (every 72 h) | Treatment: Conditioning test session 10 (72 h after session 9) |
|---|---|---|
| 1 Saline control | Saline only<br>No CS | Saline only<br>No CS |
| 2 Saline + CS control | Saline + CS | Saline + CS |
| 3 Poly I:C control | Poly I:C only<br>No CS | Poly I:C only<br>No CS |
| 4 Poly I:C, positive control | Poly I:C + CS | Poly I:C + CS |
| 5 Poly I:C, negative control | Poly I:C + CS | No treatment, no CS; animals not disturbed |
| 6 Conditioned | Poly I:C + CS | Saline + CS |

*Note:* CS = Odor of camphor; Poly I:C = 20 μg Poly I:C i.p.

**TABLE 2**
**Enhanced NK Activity in Response to a Conditioned Stimulus**

| Groups | % $^{51}$Cr released at effector:target cell ratios: 200:1 | 100:1 | 50:1 |
|---|---|---|---|
| 1 Saline only | 1:39 ± 0.23[a] | 0.80 ± 2.05 | 0.87 ± 0.61 |
| 2 Saline + camphor | 0.55 ± 0.80 | 2.19 ± 1.32 | 0.17 ± 0.92 |
| 3 Poly I:C only | 15.90 ± 2.69 | 16.41 ± 1.87 | 11.43 ± 2.49 |
| 4 Poly I:C + camphor (positive control) | 20.02 ± 3.15 | 14.97 ± 1.66 | 14.06 ± 2.51 |
| 5 Poly I:C + camphor (negative control) | 4.67 ± 1.87 | 6.62 ± 0.84 | 2.02 ± 0.88 |
| 6 Conditioned response (camphor only) | 11.23[b] ± 1.24 | 9.63[b] ± 0.92 | 6.66[b] ± 1.21 |

[a] Values are mean ± SEM.
[b] Statistically different from all five control groups ($p$ from <0.02 to <0.0001).

for NK cell activity. The percent $^{51}$Cr released from lysed tumor-targets indicating the NK cell cytotoxic activity for each group is given in Table 2. There was no effect of camphor inhalation on NK cell activity in either the saline control groups 1 and 2, or in the groups receiving Poly I:C injections. Group 6, the CR group, had a significantly elevated release of $^{51}$Cr above the Poly I:C negative control group 5 ($p < 0.012$) at the 200:1 ratio. The NK activity in Group 5, the Poly I:C negative control, at 72 h past its last Poly I:C injection was approaching the unstimulated control groups 1 and 2 (baseline values), through it was slightly elevated at this time. At all the ratios tested, the Poly I:C positive control group 4 had a significantly elevated percent release above the CR group (Table 2). These results indicate that group 6, the CR group, did show a conditioned enhancement of NK activity. A statistically significant increase in NK activity was obtained for the conditioned group (group 6) when compared with the negative control group which was conditioned nine times, but not given camphor smell on the test date (group 5). The conditioning of the NK response with camphor has been repeated many times.

## III. CONDITIONED ENHANCEMENT OF NK ACTIVITY, WITH SACCHARIN AND LiCl AS THE CONDITIONING STIMULUS

Jenkins[20] attempted to condition an enhancement of the immune response to sheep red blood cells in rats. They assumed that classical conditioning will be most efficient when a CS and US that are likely to be correlated in the natural environment are paired. They chose

**TABLE 3**
**Procedure for Conditioned Enhancement of the NK Response**

| Groups | Treatment days 0 | 7 | 9 | 10 |
|---|---|---|---|---|
| CND | S, L, P | S, L | S, L | NK-cell activity assay |
| CNDo | S, L, P | $H_2O$ | $H_2O$ | NK-cell activity assay |
| NC | $H_2O$, P | S, L | S, L | NK-cell activity assay |
| US | S, L, P | $H_2O$, P | $H_2O$ | NK-cell activity assay |

*Note:* S = saccharin, L = LiCl, and P = poly I:C

**TABLE 4**
**Enhancement of the NK Response by Conditioning to Poly I:C and Saccharin and LiCl**

| Groups | n | % $^{51}$Cr released, effector:target cell ratios 200:1 | 100:1 | 50:1 |
|---|---|---|---|---|
| CND | 6 | 23.99 ± 1.83[a] | 17.98 ± 1.39 | 13.42 ± 1.23 |
| CNDo | 5 | 16.83 ± 2.09 | 11.11 ± 1.43 | 9.09 ± 1.67 |
| NC | 5 | 15.13 ± 0.98 | 11.83 ± 0.77 | 9.23 ± 1.35 |
| US | 5 | 33.85 ± 2.36 | 28.36 ± 2.05 | 21.98 ± 2.27 |

*Note:* Comparisons were statistically significant by the Wilcoxon Rank Test as follows:
CND vs. CNDo 200:1 $p < 0.015$; 100:1 $p < 0.004$; 50:1 $p < 0.04$.
CND vs. NC 200:1 $p < 0.004$; 100:1 $p < 0.004$; 50:1 $p < 0.04$.

[a] Values are mean ± SEM.

a combination of a novel sweet taste (saccharin) and an injection of LiCl which induces gastric upset as a combination CS. The US was inoculated with sheep red blood cells. This was based on their idea that ingestion of pathogens in the natural environment would be associated with a specific (novel) taste and a gastrointestinal response (illness). We adopted this strategy, using poly I:C as the US, and a combination CS identical to that used by Jenkins et al.

Mice were housed five to six per cage and first trained to consume a daily ration of water within a 1-h period. On day 0, CND, CNDo, and US groups were presented with 0.1% saccharin between 10 and 11 a.m., and injected i.p. with 125 mg/kg of LiCl (CS). The animals were then given 20 μg poly I:C (P) i.p. All animals were kept throughout the experiment on the schedule for water. On days 7 and 9, the CND and NC groups were given saccharin (S) followed by LiCl(L). CNDo received tap water only and the US group was given tap water and injected with poly I:C on day 7. All animals were tested for NK cell activity on day 10 (Table 3).

If an association was made between the CS and the response to poly I:C, an elevation of the NK cell function would take place in the conditioned (CND) group when the animals were reexposed to the CS only (Table 4). The CNDo group is a control group which was exposed to CS/US association signal, but not reexposed to CS. The NC (nonconditioned) group was exposed to CS on days 7 and 9 and was never given CS/US together. The NK response in the CNDo and NC groups were virtually identical. It is clear that at the three effector to target cell (E:T) ratios tested, the CS alone was able to recall the NK cell response in the CND group, but not in the CNDo or NC groups. The response in the CND group

**TABLE 5**
**Median Survival, Longest Survival, and Disease-Free Survivors of Mice Exposed to Camphor and Treated with Poly I:C, Poly I:C and Control[a]**

| Group | Conditioning treatment | Test treatment | Median survival (days) | Longest survival (days) | % Disease-free survivors[b] |
|---|---|---|---|---|---|
| I | Poly I:C + camphor | Camphor | 43 | >100[c] | 20 (2/10) |
| II | Poly I:C + camphor | None | 34 | 48 | 0 |
| III | Poly I:C | Poly I:C + camphor | 38 | 61 | 0 |
| IV | No treatment | None | 37 | 46 | 0 |

[a] Mice in groups I and II were given 0.1 ml (20 μg/mouse) of poly I:C i.p. and exposed to camphor smell for 4 h. Group III received poly I:C only. Following transplantation with 1 × $10^4$ MOPC 104E cells i.v., mice were reexposed to camphor or other treatment as indicated under test treatment.

[b] The two surviving mice had returned to baseline IgM levels at 97 d and survived > 120 d (Figure 2).

[c] The overall survival curves of the four groups were compared by log rank test. The $p$ value is 0.08 when group I was compared against group IV. The $p$ value is 0.06 by chi-square by comparing the proportions at day 66 (70 and 100% mortality in group I and group IV, respectively).

was lower than that of the US group given poly I:C on the day 7. In this experiment, at the three E:T ratios, the CS alone recalled in the CND animals on the average, 71, 63, and 61% of the maximum NK cell response seen in the US group. The recalled response was about 1.5 times greater than that of the CNDo or NC control groups.

## IV. CONDITIONING OF THE NK CELL RESPONSE AND RESISTANCE TO MOPC 104E MYELOMA

A BALB/c myeloma which secretes anti-dextran IgM was used for the tumor transplantation studies. One-tenth milliliter (1 × $10^4$ cells) was given to each mouse. The mice were bled from the tail vein (~80 μl of blood) every 4 or 5 d to monitor for the tumor-specific anti-dextran IgM from day 10 post-tumor transplantation. First, it was established that exposure of mice to camphor odor prior to and post tumor transplantation had no effect on the growth of MOPC 104E myeloma. Following this, four groups of BALB/c mice (10 mice/groups) were used in this experiment. Groups I and II were given poly I:C (20 μg/mouse, i.p.) prior to exposure to camphor for 4 h. The poly I:C treatment and exposure to camphor was repeated every third day for a total of ten times. Group III was given poly I:C only, and group IV received no treatment and served as a control. One day after final exposure to camphor and treatment with poly I:C of the appropriate groups, all four groups were injected with 1 × $10^4$ MOPC 104E cells intravenously. Following tumor transplantation, group I was reexposed to camphor only from day 2 following tumor transplantation, Group II was left untreated, group III received camphor and poly I:C, and group IV served as a control for tumor growth due to handling only. The mice were followed for survival every day. The median survival of mice in group I, which were given poly I:C and exposed to camphor prior to tumor transplantation and subsequently reexposed to camphor, was 43 d compared to 34, 38, and 37 d of groups II, III, and IV, respectively (Table 5). Figure 1 shows the survival curves for each of the groups. All of the mice in groups II, III, and IV were dead by day 48, 61, and 46, respectively. There were no disease-free survivors in the three groups. The conditioned group (group I) had 2 mice living over 100 d without any evidence of clinical disease. The two mice from the conditioned group had low levels of IgM compared to the IgM levels of the mice in the control group. Seven mice of the

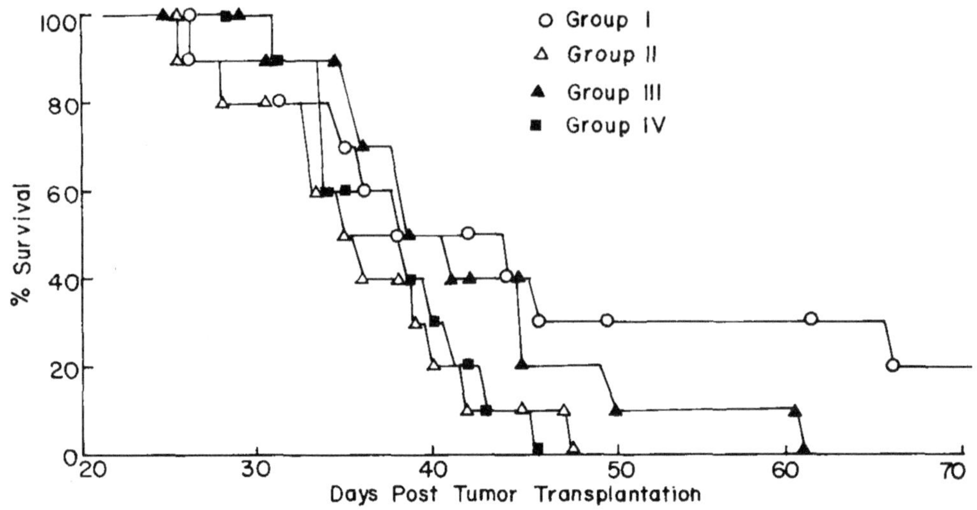


FIGURE 1. Plot of survival curves of Group I (○), conditioned and reexposed to camphor; Group II (△), conditioned, not reexposed to camphor; Group III, (▲), poly I:C treatment followed by poly I:C and exposure to smell of camphor; and Group IV, (■), no treatment, control mice. In Group I, two mice are alive at >100 d with no evidence of disease.

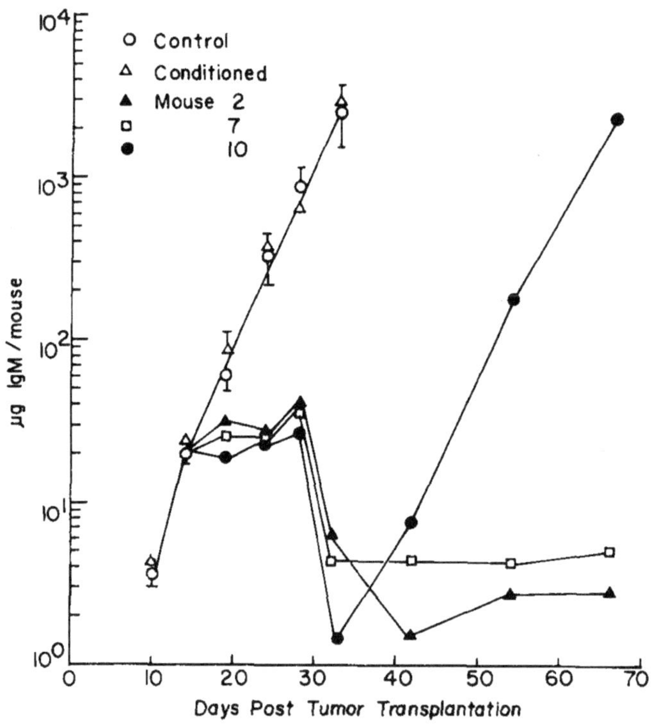


FIGURE 2. Plot of circulating MOPC 104E IgM in μg/mouse vs. days post-tumor implantation. The IgM (average ± SE) of the 10 control mice (○) showed logarithmic increase with time. In the conditioned group, three animals, mouse 2, (▲); 7, (□) and 10 (●), showed deviation in IgM pattern from the control group. The average IgM of the seven remaining animals of the conditioned group (△) is also plotted. Mouse 10 showed regression of its myeloma, but relapsed after day 32 and died with tumor.

**TABLE 6**
**Comparison of Poly I:C, α-IFN, and β-IFN, to Serve as a US Signal with S-LiCl as the Conditioned Stimulus[a]**

| Groups | % Specific $^{51}$Cr released at E:T Ratios 200:1 | 100:1 | 50:1 |
|---|---|---|---|
| Experiment 1: Poly I:C (20 μg) | | | |
| CND | 15.85 ± 3.19[b] | 14.18 ± 1.77 | 11.46 ± 1.78 |
| NC | 9.11 ± 1.75* | 9.14 ± 2.05* | 6.57 ± 2.17* |
| CND vs. NC | *$p$ = 0.001 | *$p$ = 0.002 | *$p$ = 0.004 |
| Experiment 2: β-IFN ($10^4$ IU) | | | |
| CND | 10.95 ± 5.19 | 7.94 ± 3.75 | 4.32 ± 2.30 |
| NC | 3.70 ± 1.41* | 2.72 ± 0.68* | 1.45 ± 0.50* |
| CNDo | 5.31 ± 3.12$^+$ | 4.76 ± 2.76$^+$ | 2.82 ± 2.40$^+$ |
| US | 33.50 ± 14.61 | 24.69 ± 11.45 | 11.47 ± 8.98 |
| CND vs. NC | *$p$ = 0.002 | *$p$ = 0.002 | *$p$ = 0.002 |
| CND vs. CNDo | +$p$ = 0.030 | +$p$ = 0.009 | +$p$ = 0.123 |
| Experiment 3: α–IFN ($10^4$ IU) | | | |
| CND | 4.50 ± 1.18 | 3.96 ± 0.72 | 2.73 ± 0.69 |
| NC | 4.28 ± 0.65 | 3.47 ± 0.91 | 2.76 ± 1.25 |
| CNDo | 5.29 ± 1.26 | 3.93 ± 0.98 | 2.48 ± 0.83 |
| US | 19.57 ± 5.59 | 13.93 ± 4.50 | 9.15 ± 2.90 |
| CND vs. NC and CND vs. CNDo not significant at any E:T ratio tested | | | |

[a] This is the 6-d schedule (CS on days 3 and 5). These were all separate experiments.
[b] NK-cell activity is % $^{51}$Cr release from labeled YAC-1 tumor targets in a 4-h assay. All values are mean ± standard deviation, $p$ values were calculated by the Wilcoxon Rank Test Method for the unpaired case.

conditioned group died from progressive growth of the myeloma. One mouse regressed its tumor initially followed by tumor relapse and died at day 66 due to tumor growth. We did not observe any complete regression in group III (given poly I:C repeatedly).

## V. CONDITIONING WITH INTERFERON

A combination of α- and β-interferon (IFN) is produced by leukocytes in response to poly I:C by mice. To assess whether IFN is the molecule elicited by poly I:C that is important in the conditioned association, we tested purified α- and β-IFN for their potency to form a conditioned association with the S-LiCl conditioned stimulus. Titers of IFN (5 × $10^3$ IU in plasma) in response to a 20 μg poly I:C injection had been observed at 6 h in preliminary experiment. Based on this observation a, 1 × $10^4$ IU dosage was selected to test α- and β-IFN as the US. Purified murine α- and β-interferons were obtained from Lee Biomolecular Research Laboratories (San Diego, CA), dissolved in sterile distilled water, and kept at 4°C. Saccharin-LiCl combination was used as the CS. A 0.1% solution of saccharin was given for 1 h following 24 h of water deprivation. Immediately following 1-h exposure to the sweet tasting saccharin water, LiCl at 125 mg/kg was injected i.p. At 15 to 20 min later interferon was injected i.v. at 1 × $10^4$ IU/mouse.

The studies show that α-IFN does not act as a potent US at this dose, whereas β-IFN does act as a potent US. The conditionally augmented NK cell activity observed with β-IFN as US is similar to that elicited by the poly I:C injection (Table 6). The results with α-IFN have been tested three times, and on one occasion, a slight conditioned effect was noted. This study has been repeated using camphor as the CS.

## VI. DISCUSSION

Immune responses can be modulated by a number of physiologic means, such as through the CNS, via endocrine, and/or direct neural pathways. One way of demonstrating the interaction between the CNS and immune system (IS) is by conditioning studies. There is a revival of interest in this field to demonstrate the link between these physiological functions. A paradigm frequently used in recent studies of conditioned immune activity, regardless of the outcome variable measured, is the saccharin-cyclophosphamide, illness-induced, taste-aversion model described by Ader and Cohen.[2] In this model the taste of saccharin, the CS, is paired with an i.p. injection of cyclophosphamide, the US. The animal is then immunized with an antigen and the response to the antigen is assessed at some later time. A number of different types of responses were measured with this protocol. Both T-dependent response to sheep red blood cells and T-independent response to trinitrophenylated lipopolysaccharide antigen were conditioned using this protocol.[21,22] Conditioned suppression of graft-vs.-host response using saccharin-cyclophosphamide protocol was demonstrated by Bovbjerg et al.[23] Another approach to conditioning the T-cell response paired saccharin-flavored water with antilymphocyte serum (ALS).[24] In this study an *in vitro* mixed lymphocyte response (MLR) was measured.

Gorczynski et al.[4] used a slightly different protocol and found that the host-vs.-graft response to a skin allograft could be conditionally enhanced. They showed that conditioned enhancement of precytotoxic T lymphocytes (pCTL) could be elicited by pairing the grafting procedure itself as the CS and the allograft as US.

In an elegant study reported by Gorczynski et al.,[5] using the saccharin-cyclophosphamide paradigm, a significant decrease in host resistance to a mouse plasmacytoma in conditioned animals was blocked by an $H_2$ receptor antagonist, cimetidine. The $H_2$ receptor is found on T-suppressor cells and the data suggest at least one possible mechanism of immunosuppression. This tumor system provides a model for a more detailed analysis of the conditioned immunosuppression phenomena.

A recurring concern in taste-aversion conditioning model is whether conditioning is really due to the "stress" associated with a conflict in avoiding sweet tasting water in the conditioning paradigm. This was addressed by Kelly et al.,[25,26] who showed that LiCl, a nonimmunosuppressive drug, could induce taste aversion and also cause a conditioned decrease in delayed-type hypersensitivity response to sheep red cells. Ader and Cohen[2] have suggested that corticosteroids, per se, are unlikely to be the mediators of conditioned suppression, because a dissociation exists between the extinction of the conditioned taste aversion and the conditioned immunosuppression. It has been assumed that conditioned immunosuppression was due to an association within the CNS of some US-derived signal elicited by the immune system in response to cyclophosphamide.

Of the recent conditioned immunobiologic responses, a study by Russell et al. which showed a conditioned histamine release was impressive because plasma histamine values (the conditioned response) nearly equaled the effect of the US itself.[27] The above mentioned studies are some examples of recently reported research in the area of conditioned immunobiologic responses.

Natural killer (NK) cells are found in normal animals which have the capacity to kill tumor cells without prior sensitization and regulate immune responses and bone-marrow graft survival.[28] The activity of the NK cells is modulated by humoral factors produced by other lymphocytes and fibroblasts, as well as neuropeptides and neuroendocrines. Thus, NK-cell activity seems to be a good model to study control by humoral substances derived from the CNS. Second, NK-cell activity could have profound effects on the immunocompetence of the conditioned host.

The unidirectional, nonspecific modulation of the immune system by the CNS via neuroendocrine and autonomic pathways has been reported by many investigators.[29-31] Classical conditioning of NK-cell activity seem to be well suited as a model for bidirectional communication between the CNS and IS. The critical issue in assigning bidirectionality to this interaction is whether an immune derived signal is required for the conditioned association. Though this is assumed to be the case in many published studies of conditioned immunobiologic responses, it has not been established. Our data showing conditioning with purified IFN support the concept of bidirectional signals.

We report the data of conditioned augmentation of NK-cell activity with camphor and saccharin-LiCl as the conditioned stimuli. Poly I:C is a synthetic double stranded RNA which mimicks viral infection by eliciting an interferon response. Natural killer cells respond to interferons with increased cytotoxic activity. An augmentation of NK activity observed in spleen cells of mice was the result of an association formed between the CS (camphor odor or saccharin-LiCl) and an injection of poly I:C. The association of camphor odor with poly I:C was made every third day for nine times, and the test for the conditioned response was made with a single exposure to camphor odor. NK-cell activity of spleen cells was assessed 24 h later. Every animal in the conditioned group had increased NK-cell activity over the controls. This protocol was shortened significantly to a single CS/US association and two reexposures to CS on days 3 and 5, or days 5 and 7, or days 7 and 9. Any of the three reexposure combinations are effective in producing the augmented NK-cell activity observed.

The question of IFN's role in this conditioned association was explored by direct i.v. administration of either α- or β-IFN. α-IFN was unable to consistently induce a conditioned association at the same dose as β-IFN. β-IFN was equally capable of forming an association with either S-LiCl or the odor of camphor as CS. This same dose augmented NK activity equally using either α- or β-IFN. It should be noted, however, that these two species of IFN have different specific activities.

Intracerebral injections of human leukocyte IFN caused potent endorphin-like opioid effects and human leukocyte IFN, but not fibroblast-derived IFN bound to opiate receptors *in vitro*.[32] These effects were blocked by naloxone. Further, human leukocyte IFN, both purified and recombinant, could significantly increase the discharge rate of spontaneously firing cells in the hippocampus and somatosensory cortex, but γ-IFN preparations did not.[33] In addition, β-endorphin has been found to stimulate NK cells directly via an opiate binding site[34,35] and also augment IFN production by NK cells after stimulation by poly I:C or phytohemagglutinin.[19] All these studies suggest a link between opiate effector function and IFN functions. The basis for this interaction is unknown, but may involve some cross-reactivity between receptors.

The physiological mechanism(s) of conditioned augmentation of NK-cell activity and conditioned immune response in general remain to be determined. The NK-cell activity model system is well suited for these studies. One of the possible mediators of the afferent arm (peripheral to CNS) could be IFN, which is induced in response to poly I:C. We have clearly shown that β-IFN can serve as a potent unconditioned stimulus in eliciting a conditioned response. This is not a direct proof to indicate that IFN released upon injection with poly I:C serves as the US. Elimination of the conditioned association between IFN produced in response to poly I:C and CS by treatment *in vivo* with anti-α and anti-β-interferon antisera may clarify the role of β-IFN in forming the conditioned association. We have been unable to show an elevation of interferon in the serum of conditioned mice reexposed to the CS. This may indicate that IFN is not the effector molecule in the CR.

At the CNS (efferent pathway) and the endocrine level, it is necessary to define the neuropeptides and/or the neuroendocrines involved in the conditioned immunobiologic responses. Some of the possible mediators of this system are the opioids (endorphins and enkephalins) and the catecholamines.

## VII. SUMMARY

Natural killer cell system does provide a good model for studies of conditioned immunobiologic responses. The NK-cell activity can be conditioned using both camphor odor (a neutral CS) and saccharin-LiCl paradigm. Second, as demonstrated by us, β-IFN can serve as a potent US. Both β- and α-interferons are produced by the host in response to poly I:C. Third, NK cells play a critical role in the regulation of tumor growth in addition to other functions, such a modulation of bone marrow graft survival and regulation of immune responses. The functional role of the conditionally enhanced NK-cell activity has been demonstrated in the MOPC 104E model *in vivo*. Using the neutral stimulus, odor of camphor and pure β-IFN we are getting closer to the establishment of the mechanism at least for one conditioned response.

## VIII. CONCLUSIONS

These findings suggest the bidirectionality of the conditioned immune responses. Using IFN as US and NK-cell activity as a conditioned response, we are constantly getting closer in establishing the mechanism for this sytem. Additional research is needed to establish whether IFN is one of the afferent mediators. Studies are also needed to identify mediators that are involved in the efferent pathway.

## ACKNOWLEDGMENTS

The work discussed here has been supported by United States Public Health Grant CA 37570 and a grant from the American Cancer Society (ACS), IM-509.

## REFERENCES

1. **Ader, R.,** A historical account of conditioned immunobiologic responses, in *Psychoneuroimmunology*, Ader, R., Ed., Academic Press, New York, 1981, 321.
2. **Ader, R. and Cohen, N.,** Conditioned immunopharmacologic responses, in *Psychoneuroimmunology*, Ader, R., Ed., Academic Press, New York, 1981, 281.
3. **Bovbjerg, D., Ader, R., and Cohen, N.,** Acquisition and extinction of conditioned suppression of a graft-vs-host response in the rat, *J. Immunol.*, 132, 111, 1984.
4. **Gorczynski, R. M., Macrae, S., and Kennedy, M.,** Conditioned immune response associated with allogeneic skin grafts in mice, *J. Immunol.*, 129, 704, 1982.
5. **Gorczynski, R. M., Kennedy, M., and Ciampi, A.,** Cimetidine reverses tumor growth enhancement of plasmacytoma tumors in mice demonstrating conditioned immunosuppression, *J. Immunol.*, 134, 4261, 1985.
6. **Ghanta, V., Hiramoto, R., Solvason, B., and Spector, H.,** Influence of conditioned natural immunity on tumor growth, *Ann. N.Y. Acad. Sci.*, 496, 637, 1987.
7. **Ghanta, V., Hiramoto, R., Solvason, B., and Spector, H.,** Neural and environmental influences on neoplasia and conditioning of NK activity, *J. Immunol.*, 135, 848S, 1985.
8. **Stein, M., Schiavi, R. C., and Camerino, M. S.,** Influence of brain and behavior on the immune system, *Science*, 191, 435, 1976.
9. **Macris, N., Schiavi, R. C., and Camerino, M. S.,** Effect of hypothalamic lesions on immune processes in the guinea pig, *Am. J. Physiol*, 219, 1205, 1970.
10. **Korneva, E. A. and Khai, L. M.,** Effect of destruction of hypothalamic areas on immunogenesis, *Fed. Proc. Trans. Suppl.*, 23, 88, 1964.
11. **Brooks, W. H., Cross, R. J., Roszman, T. L., and Markesbery, W. R.,** Neuroimmunomodulation: neural anatomical basis for impairment and facilitation, *Ann. Neurol.*, 12, 56, 1982.

12. **Spector, N. H.**, in *Immunoregulation*, Fabris, N., Garaci, E., Hadden, J., and Mitchison, N. A., Eds., Plenum Press, New York, 1983.
13. **Spector, N. H.**, The central state of the hypothalamus in health and disease: old and new concepts, in *Physiology of the Hypothalamus*, Morgane, P. and Panksepp, J., Ed., Marcel Dekker, New York, 1980, chap. 6.
14. **Roszman, T. L., Cross, R. J., Brooks, W. H., and Markesbery, W. R.**, in *Experimental and Clinical Interventions in Aging*, Walker, R. F. and Cooper, R. L., Eds., Marcel Dekker, New York, 1983, 163.
15. **Luk'ianenko, V. I.**, The problem of conditioned reflex regulation of immunobiologic reactions, *Usp. Sovrem. Biol. (Moskva)*, 51, 170, 1961.
16. **Dolin, A. O. and Dolina, S. A.**, *Pathology of Higher Nervous Functions*, Vysshaya Shkola, Moscow, 1972, 332.
17. **Ahlqvist, J.**, Hormonal influences on immunologic and related phenomena, in *Psychoneuroimmunology*, Ader, R., Ed., Academic Press, New York, 1981, 355.
18. **Blalock, J. E., Harbour-McMenamin, D., and Smith, E. M.**, Peptide hormones shared by the neuroendocrine and immunologic systems, *J. Immunol.*, 135, 858S, 1985.
19. **Mandler, R., Biddison, W., Mandler, R., and Serrate, S.**, Beta-endorphin augments the cytolytic activity and interferon production of natural killer cells, *J. Immunol.*, 136, 934, 1986.
20. **Jenkins, P. E., Chadwick, R. A., and Navin, J. A.**, Classically conditioned enhancement of antibody production, *Bull. Psychon. Soc.*, 21, 485, 1983.
21. **Wayner, E. A., Flannery, G. R., and Singer, G.**, The effects of taste aversion conditioning on the primary antibody response to sheep red blood cells and *Brucella abortus* in the albino rat, *Physiol. Behav.*, 21, 995, 1978.
22. **Cohen, N., Ader, R., Green, N., and Bovbjerg, D.**, Conditioned suppression of a thymus independent antibody response, *Psychosom. Med.*, 41, 487, 1979.
23. **Bovbjerg, D., Ader, R., and Cohen, N.**, Behaviorally conditioned suppression of a graft-vs-host response, *Proc. Nat. Acad. Sci. U.S.A.*, 79, 583, 1982.
24. **Kusnecov, A. W., Sivyer, M., King, M. G., Husband, A. J., Cripps, A. W., and Clancy, R. L.**, Behaviorally conditioned suppression of the immune response by anti-lymphocyte serum, *J. Immunol.*, 130, 2117, 1983.
25. **Kelley, K. W., Dantzer, R., Mormede, P., Salmon, H., and Aynaud, J. M.**, Conditioned taste aversion suppresses induction of delayed-type hypersensitivity immune reactions, *Physiol. Behav.*, 34, 189, 1985.
26. **Kelley, K. W., Dantzer, R., Mormede, P., Salmon, H., and Aynaud, J. M.**, Conditioned taste aversion induces immunosuppression in the absence of an immunosuppressive drug, *C. R. Acad. Sci. (Paris)* (3rd Series), 299, 123, 1984.
27. **Russell, M., Dark, K. A., Cummings, R. W., Ellman, G., Callaway, E., and Peeke, H. V. S.**, Learned histamine release, *Science*, 225, 733, 1984.
28. **Hansson, M., Kiessling, R., and Anderson, B.**, Human fetal thymus and bone marrow contain target cells for natural killer cells, *Eur. J. Immunol.*, 11, 8, 1981.
29. **Monjan, A. A.**, Stress and immunologic competence: studies in animals, in *Psychoneuroimmunology*, Ader, R., Ed., Academic Press, New York, 1981, 185.
30. **Kasahara, K., Tanaka, S., Ito, T., and Hamashima, U.**, Suppression of the primary immune response by chemical sympathectomy, *Res. Commun. Chem. Pathol. Pharmacol.*, 16, 687, 1977.
31. **Besedovsky, H. O., del Rey, A., Sorkin, E., Daprada, M., and Keller, H. H.**, Immunoregulation mediated by the sympathetic nervous system, *Cell. Immunol.*, 48, 346, 1979.
32. **Blalock, J. E. and Smith, E. M.**, Human leukocyte interferon (HuIFN-alpha): potent endorphin-like opioid activity, *Biochem. Biophys. Res. Commun.*, 101, 472, 1981.
33. **Reyez-Vasquez, C. and Prieto-Gomez, B.**, Alpha and gamma interferon's effects on cortical and hippocampal neurons: microiontophoretic application and single cell recording, *Int. J. Neurosci.*, 25, 113, 1984.
34. **Mathews, P. M., Froelich, C. J., Sibbitt, W. L., and Bankhurst, A. D.**, Enhancement of natural cytotoxicity by beta-endorphin, *J. Immunol.*, 130, 1658, 1983.
35. **Faith, R. E., Huey, H. J., Murgo, A. J., and Plotnikoff, N. P.**, Neuroimmunomodulation with enkephalin's enhancement of human natural killer cell activity *in vitro*, *Clin. Immunol. Immunopathol.*, 31, 412, 1984.

Chapter 9

# HYPOTHALAMIC-IMMUNE INTERACTIONS: MODULATION OF IMMUNE FUNCTION BY ELECTROLYTIC AND CHEMICAL LESIONS IN THE CENTRAL NERVOUS SYSTEM

**Richard J. Cross**

## TABLE OF CONTENTS

# I. INTRODUCTION

The immune system is composed of a variety of interacting mononuclear cells and cytokines which act in concert to generate immune responses to soluble, particulate, and cell bound antigens. Within this system, counter-balancing positive and negative signals from suppressor cells,[1] idiotypic networks[2,3] and genetic restriction[4,5] modulate immune responsiveness while inhibiting the development of autoimmune diseases. Until recently, it was believed that the immune system was autonomous and self-regulating with minimal influence from other systems. However, new lines of evidence have rapidly accrued which indicate that factors regulated by or emanating from the central nervous system (CNS) are instrumental in modulating immune responsiveness.

# II. CENTRAL NERVOUS SYSTEM (CNS)-IMMUNE SYSTEM INTERACTIONS DURING DEVELOPMENT

Interactions between the CNS and immune system appear to begin during development. These observations were made using the Snell-Bagg hypopituitary dwarf mouse and athymic "nude" mouse models. The hypopituitary dwarf mouse expresses deficiencies in growth hormone (GH), thyroid stimulating hormone (TSH), and thyroxine.[6] In addition to obvious endocrine abnormalities, these animals also display abnormal immunologic development. Their immunological deficiency is characterized by an underdeveloped spleen, thymus, and lymph nodes with an accompanying peripheral lymphopenia. Functionally, they do not produce normal levels of antibody to soluble antigens or respond to alloantigens by inducing cytotoxic T lymphocyte activity. The mice can be restored to a state of normal immune competence by administration of GH and thyroxine.[7]

Conversely, the athymic "nude" mouse has been immunologically characterized as being T-cell deficient, thus lacking the ability to produce antibody to T-cell dependent antigens or respond to alloantigenic challenge. In addition to these immunologic abnormalities, these mice also have depressed levels of thyroxine and gonadotropins.[8] Attempts to restore endocrine function by transplanting thymus tissue, successfully returns estrogen concentrations to normal levels as well as restoring T-cell function. Hormone supplements with luteinizing hormone (LH), TSH, and adrenocorticotropic hormone (ACTH) will restore endocrine function. Collectively, these data indicate that a functional thymus must be present for the normal development of the hypothalamic-pituitary axis, and support the hypothesis that CNS and immune system interactions begin during development.

# III. CLINICAL CORRELATES OF CNS-IMMUNE INTERACTIONS

Individuals with a variety of psychiatric and neurological disorders also have abnormal immune responsiveness. Such diseases lend support to the theories that the CNS functions to modulate immune responsiveness. Patients with cerebral tumors display cutaneous anergy,[9] depressed antibody production,[10] impaired lymphocyte reactivity,[9] and decreased production of interleukin-2.[11] Similarly, patients with Huntingtons disease[12] and Guamanian Parkinson's-dementia complex[13] also have impaired *in vitro* correlates of cell-mediated and humoral immunity and diminished delayed-typed hypersensitivity. Impaired mitogen-induced lymphocyte proliferation and diminished antibody production is also present in individuals with bereavement-associated depression,[14] psychosis,[15] and emotional stress.[16]

These observations in man can also be extended to animal models in which stress induced by avoidance learning, overcrowding, and other stressful stimuli are found to have diminished delayed-type hypersensitivity,[17] prolonged survival of skin allografts,[18,19] and impaired lymphocyte reactivity to mitogens.[16,20]

# IV. EFFECTS OF ELECTROLYTIC LESIONS ON IMMUNITY

Such observations as presented in the two preceding sections thus help form the hypothesis that the CNS influences the immunologic status of an individual, and that stressful stimuli impair immune responsiveness. However, these studies do not directly address two questions. Can the CNS alter immune responsiveness? And if so, what areas of the brain are responsible for this modulation?

Initial studies to explore immunomodulatory centers in the CNS, used methods in which specific nuclei were electrolytically lesioned, and the effects of these lesions on a variety of both cell-mediated and humoral immune functions were examined. Preliminary studies in this area began by electrolytically lesioning a variety of areas in the brain, such as the hypothalamus, thalamus, reticular formation, superior colliculus, caudate nucleus, and several other sites both within and outside the limbic system.

Experiments by Isakovic and Jankovic examined the effects of these lesions on lymphoid tissue architecture,[21] Arthus reactivity,[22] antibody production,[22] and delayed-type hypersensitivity[22] in rats. These studies showed that lesions of the hypothalamus and reticular formation decreased the number of cells in the spleen and caused marked changes the architecture of the thymus. Such thymic changes included an overall loss of cells and a loss of the cortico-medullary junction.

Hypothalamic lesions also induced a marked decrease in lymphocyte functions. Humoral antibody production to bovine serum albumin was reduced, as well as the intensity of the Arthus and delayed-type hypersensitivity reactions. In support of these studies, Korneva and Khai also reported a decrease in antibody production in rabbits with hypothalamic lesions compared to controls.[23] Tyrey and Nalbandov extended this hypothesis by showing that hypophysectomy or adrenalectomy abrogated these effects.[24] Thus, a neural-endocrine-adrenal-immune pathway was hypothesized in which hypothalamic lesions induced changes in the pituitary that caused the adrenal glands to release corticosteroids. This output of steroids was proposed to be responsible for the observed decrease in antibody responsiveness.

In a series of papers by Schiavi and Stein, these investigations were continued by examining the effects of anterior, medial, or posterior hypothalamic lesions on histamine toxicity and hypersensitivity reactions in the guinea pig.[25] These studies showed that electrolytic lesions in the anterior hypothalamus (AHT) decreased the toxic effects of intravenously administered histamine. Lesions in other areas of the hypothalamus, however, did not have the same protective effects. These studies were extended by examining the protective effects of AHT lesions on the induction of lethal hypersensitivity. Such lesions protect animals from lethal anaphylaxis induced by the antigen picryl-chloride,[26] a phenomenon which was attributed to low levels of picryl antibody.

However, in a subsequent report, the same authors repeated this study using ovalbumin as the antigen.[27] Again, AHT lesions protected the mice from lethal anaphylaxis; however, the antibody titer to ovalbumin did not differ between the AHT- and sham-lesioned groups. The observed discrepancy in antibody titer may be due to differences in the two sensitizing antigens.

Studies to examine the effects of AHT lesions on *in vitro* correlates of cell-mediated immunity were performed by Keller and co-workers.[28] In these experiments, guinea pigs received AHT lesions, and suspensions of whole blood or purified peripheral blood lymphocytes (PBL) were cultured with the T-cell mitogen phytohemagglutinin (PHA). The results demonstrated that PHA-induced lymphocyte proliferation was inhibited in animals with AHT lesions. However, PHA responsiveness of purified PBLs were not affected by electrolytic lesions of the AHT. These authors hypothesize that humoral factors which are released following AHT lesions inhibit lymphocyte blastogenesis. Purification and washing of these cells removes this inhibitory factor and restores normal levels of cell activation.

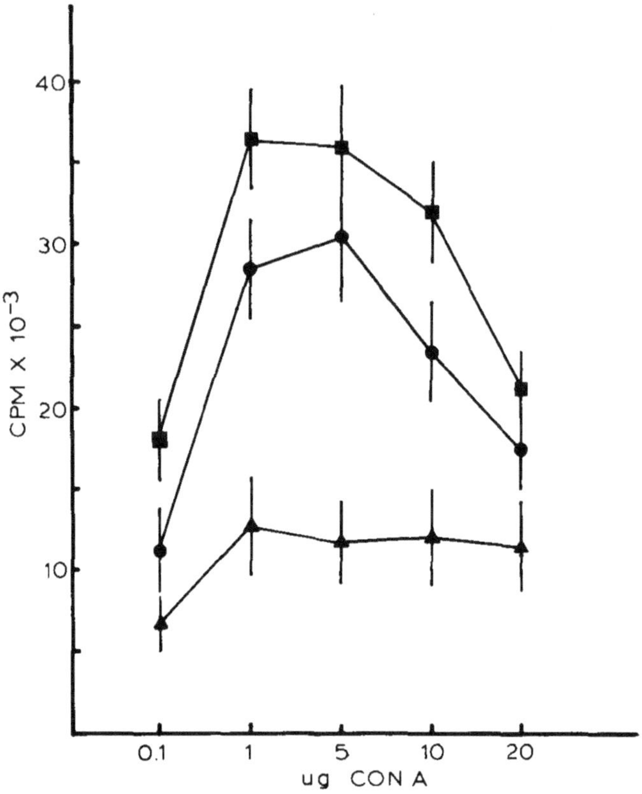


FIGURE 1. Effect of brain lesions on the blastogenic response of splenocytes to Con A. Results represent the mean ± S.E.M. of 7 to 14 separate experiments. Response of spleen cells from AHT-lesioned animals was *significantly different* ($p$ <0.05) from the response of normal animals to 1, 5, and 10 μg Con A. Values obtained with control-lesion animals were not significantly different from normal animals. ▲——▲, AHT-lesioned; ■——■ control-lesioned; ●——●, normal. (Reprinted from Cross, R. J., Markesbery, W. J., Brooks, W. H., and Roszman, T. L., Hypothalamic-immune interactions. I. The acute effect of anterior hypothalamic lesions on the immune response, *Brain Res.*, 196, 79, 1980. With permission.)

These studies reinforce the notion that humoral factors eminating from the CNS serve immunomodulators.

Studies examining the effects of CNS lesions on immune functions were begun in our laboratory during this same time. Thus, studies were initiated in rats to examine the effects of electrolytic lesions in the AHT on lymphoid tissue cell number and mitogen-induced lymphocyte proliferation.[29] As previously reported by Isokovic and Jankovic,[21,22] we also observed a marked involution of the thymus and a decrease in the number of cells in the spleen. Lymphocyte activation by the T-cell mitogen concanavalin A (Con A) was also impaired in rats with AHT lesions compared to animals with control lesions in the frontal cortex. These data (Figure 1) showed that AHT lesions impair the Con A-induced lymphocyte proliferation of spleen cells throughout a range of Con A concentrations. These studies also showed that AHT lesions do not alter the percentage of T-cells or B-cells in the spleen. The effects of AHT lesions are extremely acute and short-lived in that mitogen activation returns to normal levels 14 d after AHT lesion placement. The short-lived nature of the effect is in agreement with studies on neural plasticity which show that the CNS will recover from

**TABLE 1**
**The Effect of Glass Bead Adherent Cells from Normal and AHT-Lesioned Animals on the Response of Their Respective Nonadherent Cells to Con A (5 μg)**

| Cell source | Normal c.p.m. | Normal I (%)[a] | AHT-lesioned c.p.m. | AHT-lesioned I (%) |
|---|---|---|---|---|
| $1 \times 10^6$ Nonadherent | 28,569[b] | | 22,668 | |
| $1.5 \times 10^6$ Nonadherent | 25,799 | | 23,521 | |
| $1 \times 10^6$ Nonadherent + $1 \times 10^5$ adherent | 28,922 | 0 | 14,028 | 38 |
| $1 \times 10^6$ Nonadherent + $2 \times 10^5$ | 24,367 | 15 | 9,436 | 58 |
| $1 \times 10^6$ Nonadherent + $5 \times 10^5$ adherent | 15,002 | 47 | 4,644 | 80 |

[a] Percentage inhibition (I) = 100 × [(Cultures, c.p.m., with non-adherent + adherent cells)/Nonadherent cells, c.p.m.)] − 100.
[b] S.E.M. did not deviate 10%.

lesions as a result of dendritic sprouting within 10 to 14 days.[30] Measurement of serum corticosterone (CS) levels in rats with AHT or sham lesions indicate that these lesions markedly elevate CS levels. This response, however, was at its maximum 12 h after lesioning, and CS had returned to normal levels by 24 h. Significantly, both the AHT- and sham-lesioned animals displayed similar elevated levels of serum CS, indicating that the resulting impairment of mitogen-induced lymphocyte proliferation after the placement of AHT lesions does not result from increased levels of serum CS.

Subsequent studies examined the immunological mechanism(s) responsible for impairing the mitogen response of rats with AHT lesions.[31] Spleen cells from AHT-lesioned or normal rats were fractionated on glass bead columns prior to culture with Con A to remove macrophages. These data show that the passage of spleen cells through these columns removes the suppressive activity seen in unfractionated spleen cells from AHT-lesioned animals. It is possible that a glass bead adherent cell may be inducing a immunosuppressive state as the result of AHT lesion placement. Data presented in Table 1 shows that when the glass bead adherent cells from AHT-lesioned animals were removed from the glass bead columns with 0.02% ethylene diamine tetra acetate (EDTA) in phosphate-buffered saline and added to glass bead nonadherent cells in increasing concentrations, the resulting Con A response was suppressed in a dose-dependent manner. Adherent cells from normal rats were much less inhibitory. Further studies using antibody and complement treatment and irradiation indicate that the glass-adherent suppressor cells induced by lesions of the AHT are macrophages. Quantitation of splenic macrophages from AHT-lesioned and control animals shows that AHT lesions do not induce a quantitative change in the number of macrophages, but induce a qualitative change in suppressor macrophage activity. Such studies are in agreement with other studies of immunoregulation in the rat which have suggested the existence of macrophage-like suppressor cells.[32]

The experiments reviewed above demonstrate that the AHT is a primary immunoregulatory site within the CNS. Other investigators have lesioned other areas of the CNS with little indication that these areas of the brain can modulate immune function. However, in none of these studies were lesions placed in the limbic system or in areas which might be connected to the hypothalamus. We therefore investigated other areas of the hypothalamus and limbic system as possible CNS immunoregulatory centers.[33] Bilateral electrolytic lesions were placed in the ventro-medial hypothalamus (VMH), fornix (FX), mamillary bodies (MB), hippocampus (HC), and amygdaloid complex (AM), as well as the AHT with lesions in the frontal cortex serving as controls. Data presented in Table 2 shows that lesions in other areas of the CNS do have the ability to modulate spleen cell and thymocyte mitogen-

**TABLE 2**
**Effect of CNS Lesions on Con A Reactivity of Spleen Cells and Thymocytes[a]**

| | Days after lesioning | | | | | |
|---|---|---|---|---|---|---|
| | 4 | | 7 | | 14 | |
| Lesion | Spleen cells | Thymocytes | Spleen cells | Thymocytes | Spleen cells | Thymocytes |
| Normal | 64,670 ± 4,034 | 65,046 ± 1,977 | 64,670 ± 4,034 | 65,046 ± 1,977 | 64,670 ± 4,034 | 65,046 ± 1,977 |
| FC | 61,823 ± 2,166 | 59,071 ± 2,441 | 73,931 ± 5,138 | 67,927 ± 1,458 | 67,093 ± 4,764 | 79,901 ± 14,366 |
| AHT | 24,104 ± 5,614 ($p < 0.002$) | 62,612 ± 2,118 | 70,677 ± 6,401 | 66,732 ± 5,550 | 69,708 ± 9,198 | 72,313 ± 6,270 |
| VMH | 61,870 ± 2,807 | 63,315 ± 2,220 | 81,396 ± 19,464 | 97,466 ± 4,879 ($p < 0.01$) | 65,321 ± 4,352 | 88,711 ± 6,873 |
| FX | 67,349 ± 4,325 | 61,684 ± 2,470 | 74,539 ± 18,867 | 64,078 ± 9,601 | 58,053 ± 8,322 | 85,237 ± 6,230 |
| MB | 92,861 ± 8,904 ($p < 0.002$) | 58,005 ± 3,223 | 79,359 ± 18,927 | 69,371 ± 2,191 | 68,291 ± 9,545 | 77,008 ± 2,528 |
| HC | 100,110 ± 5,272 ($p < 0.001$) | 90,683 ± 4,210 ($p < 0.001$) | 77,629 ± 15,903 | 103,691 ± 12,953 ($p < 0.05$) | 64,849 ± 4,118 | 83,364 ± 7,864 |
| AM | 100,134 ± 6,276 ($p < 0.001$) | 76,412 ± 4,241 ($p < 0.01$) | 78,615 ± 3,623 | 61,653 ± 6,368 | 77,770 ± 4,054 | 72,706 ± 2,925 |

[a] Results are expressed as counts per minute × $10^6$ cells and represent the mean ± SEM of 4 to 12 separate experiments. Significance was determined by Student *t* test comparing values obtained from lesioned animals to those from control animals (normal and FC-lesioned animals did not differ significantly at any of the times).

induced lymphoid cell proliferation. As shown previously, lesions in the AHT inhibited spleen cell mitogen reactivity, with no effect on thymocyte reactivity. However, lesions in the MB, HC, and AM enchance lymphocyte responsiveness to Con A. In addition, thymocyte proliferation was also enchanced in rats with HC or AM lesions. Enhancement was short-lived as had been reported above, following lesion placement in the AHT. These studies show that electrolytic lesions in the CNS not only impair mitogen-induced lymphocyte proliferation, but can also enhance splenocyte and thymocyte responsiveness. In additional studies, we have shown that when lesions are placed in the HC or AM there is less suppressor activity than exhibited by normal splenic macrophages. These experiments therefore provide direct evidence that the CNS can modulate immune function in both an inhibitory and a facilitatory manner and that there is not one single immunoregulatory center in the brain.

The studies outlined above have focused on the immunologic mechanisms responsible for immune suppression or enhancement following the placement of lesions in the CNS. As yet, little effort had been placed on examining the pathway through which neural-immune interactions are manifest. Evidence supporting direct innervation of lymphoid tissue has been generated from a number of studies. Thus, it is possible that the immunomodulatory effects of CNS lesions are mediated via direct nerve pathways.[34,35] Conversely, other reports indicate that pituitary hormones can modulate immune function. Some hormones are known to be immunosuppressive (ACTH,[36] corticosteroids[37]), while others (insulin,[38] growth hormone,[39] prolactin[40]) are capable of enhancing immune responsiveness. Thus, it is important to determine the pathway through which destructive lesions in the CNS modulate immune responsiveness. We had previously shown that increases in serum CS levels appeared not to be a possible mechanism, but the role of other pituitary hormones as immunomodulators in this model need investigation.

To determine if pituitary hormones play a major role in mediating the effects of CNS lesions on immune responsiveness, hypophysectomized (Hx) rats were lesioned in either the AHT, HC, or AM.[41] In intact animals, these lesions would have either an inhibitory or facilitatory effect on Con A-induced spleen cell proliferation. However, when these lesions were placed in Hx rats, neither the suppressive nor enhancing effects were seen. These results suggest that mediators from the pituitary gland are responsible for the immune effects seen following CNS lesion placement.

Studies by Dann et al. have found that lesions in the tuberal region of the hypothalamus, which may correlate with our lesions placed in the MB, enhance skin graft rejection.[42] Intact rats with these lesions display a decrease in the mean survival time of allogeneic skin grafts. These data also show that rats which are hypophysectomized, followed 1 d later by lesioning of the tuberal hypothalamus, continue to reject skin grafts at a more rapid rate than controls. Thus, lesions in the posterior hypothalamus enhance cell-mediated immunity in both the rejection of skin grafts and in *in vitro* mitogen-induced proliferation. The discrepancy between the role of the pituitary in these systems may be due to the time in which the Hx was performed in relation to the time of lesion placement. Rats used in our study were Hx 2 weeks prior to lesioning, whereas those of Dann et al. were Hx 1 d prior.

## V. CNS MODULATION OF NATURAL KILLER CELL FUNCTION

Natural killer (NK) cells have been shown to be important effector cells in an immunosurveillance system against neoplastic cells,[43] virally transformed cells,[44] and solid tumors.[45] Likewise, evidence has shown that stress increases the incidence of neoplasia (reviewed in this monograph and Riley et al.[46]) Thus, several groups have provided evidence to show that lesions placed in the hypothalamus of mice or rats impairs NK function and promotes tumor growth. In a study by Forni et al., it was shown that mice which are lesioned in the medial basal hypothalamus have decreased numbers of large granular lymphocytes, the population of lymphocytes which possesses NK activity, and impaired NK activity against

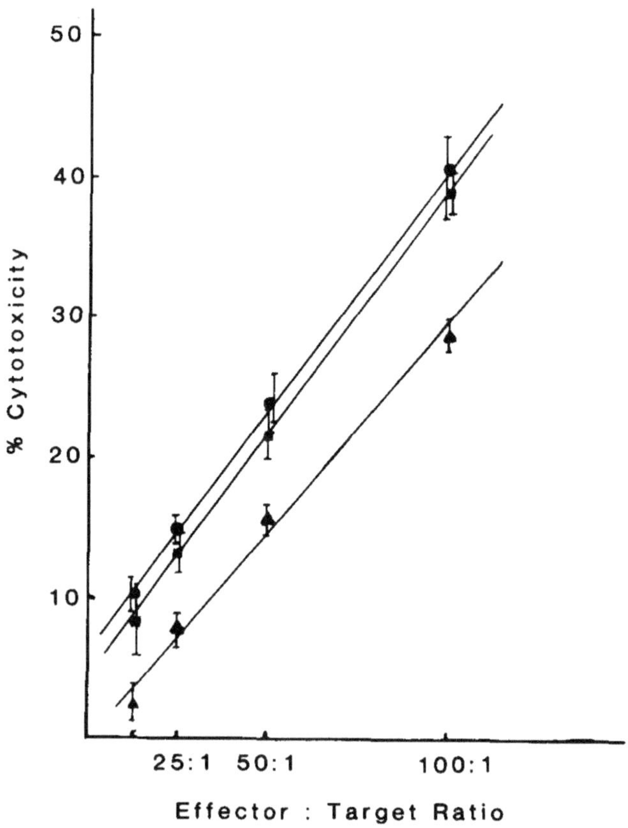


FIGURE 2. Splenic NK activity in AHT (▲), FC (■), and normal (●) animals at increasing effector: target cell ratios 4 d after lesion placement, NK activity in AHT-lesioned animals is significantly different from FC-lesioned ($p < 0.01$) and normal animals ($p < 0.001$). NK activity in FC-lesioned animals did not differ from normals. (Reprinted from Cross, R. J., Markesbery, W. R., Brooks, W. H., and Roszman, T. L., Hypothalamic-immune interactions: neuromodulation of natural killer activity by lesioning of the anterior hypothalamus, *Immunology*, 51, 399, 1984. With permission.)

YAC-1 lymphoma cell targets.[47] NK activity could be restored by treating lesioned mice with poly I:C, an inducer of interferon. These data implicate that AHT lesions induce a defect in NK cell maturation. The authors went on to show that hypothalamic lesions did not affect mixed lymphocyte responsiveness or impair the development of cytotoxic T lymphocytes (CTL). Thus, it appears that the CNS modulates the activity of endogenous cytotoxic NK cells, but does not affect the cytotoxic activity of CTL which must be induced by antigen. In a subsequent paper, these authors[48] showed that natural cytotoxic activity, (as described by Stutman[49]) was enhanced following the placement lesions in the anterior, medial, or posterior hypothalamus.

Studies in our laboratory have shown that electrolytic lesions in the AHT of the rat impaired NK activity against a virally transformed fibroblast cell line (F4) (Figure 2).[6] As shown by Forni et al. in their previous paper,[47] we also showed that impaired NK function was not due to the activation of suppressor cell mechanisms. Removal of macrophage suppressor cells by treatment with carbonyl iron did not abrogate impairment of NK function induced by AHT lesions (Table 3). Further studies also showed that an intact pituitary gland was necessary to mediate the neural effects of AHT lesions which result in impaired NK

**TABLE 3**
**NK Activity of Nonfractioned and Glass Bead Column Fractioned Spleen Cells from Normal FC- and AHT-Lesioned Animals**

| | | Lytic units/$10^6$ cells ± SEM | | |
|---|---|---|---|---|
| **Groups** | **Unfractionated** | **GBNA** | **GBA** | **Carbonyl iron-treated GBA** |
| AHT lesions (n = 16) | 1.46 ± 0.06 ($p$ <0.01) + | 1.59 ± 0.16 ($p$ >0.05) | 1.42 ± 0.19 ($p$ <0.05) | 1.45 ± 0.32 ($p$ <0.05) |
| FC lesions (n = 18) | 2.62 ± 0.62 | 1.61 ± 0.11 | 2.25 ± 0.25 | 2.75 ± 0.24 |
| Normal (n = 18) | 2.50 ± 0.30 | 1.71 ± 0.15 | 3.42 ± 0.67 | 2.86 ± 0.33 |

*Note:* Spleen cells obtained from animals 4 d after AHT or FC lesion placement. Significance was determined by Student *t* test comparing values obtained from lesioned animals to those from control animals. Normal and FC-lesioned animals did not differ significantly (n = number of animals/group).

activity. This is particularly important in light of the report by Saxena et al., in which NK activity in Hx mice was restored by GH treatment,[50] again implicating pituitary derived factors as immunomodulators in this system.

## VI. EFFECTS OF CNS NEUROTRANSMITTER DEPLETION ON IMMUNE FUNCTION

Over the last several years, our laboratory has developed a new model system to examine the role of CNS neurotransmitters as modulators immune function. Initial studies examined the effects of depleting CNS catecholamines by injection of 6-hydroxydopamine, into the cisterna magna, or by ablating CNS serotonergic neurons with 5,7 dihydroxytryptamine. These studies showed that depletion of CNS catecholamines markedly impaired both the primary humoral IgM and IgG humoral antibody response to sheep red blood cells (SRBC), whereas depletion of CNS serotonin had no effect.[51]

Further, 6-OHDA impaired the magnitude of the antibody response without altering the kinetics of the response (Figure 3).[52] To determine if 6-OHDA treatment was affecting the T-cell or B-cell component of the response, mice were treated with 6-OHDA, then 2 d later, immunized with either trinitrophenyl-keyhole limpet hemocyanin (TNP-KLH), a T-cell dependent antigen, or trinitrophenyl-lipopolysaccharide (TNP-LPS), a T-cell independent antigen. Data from these experiments showed that CNS catecholamine depletion impairs the TNP-KLH response, but has no effect on the TNP-LPS response, thus indicating that 6-OHDA treatment does not affect B-cells directly, but may impair the response by influencing the T-cell component of the response.[60] Immunofluorescence analysis also showed that 6-OHDA treatment does not alter the percentage of T-cells or B-cells in the spleen.

Experiments were designed to determine if 6-OHDA treatment impairs immune responsiveness by inhibiting the afferent or efferent phases of the response. Hitherto, mice received an injection of 6-OHDA at least 48 h prior to antigen, hereby affecting the afferent phase of the immune response. Data presented in Table 4 shows that when 6-OHDA is injected 2 h prior to antigen or 24 to 48 h after antigen (which would affect the efferent phase of the response), CNS catecholamine depletion has no effect on antibody responsiveness. Thus, the depletion of CNS catecholamines by 6-OHDA impairs immune responsiveness by inhibiting the afferent, early phase of the immune response. Further experiments also indicate that 6-OHDA treatment has no effect on the secondary humoral immune

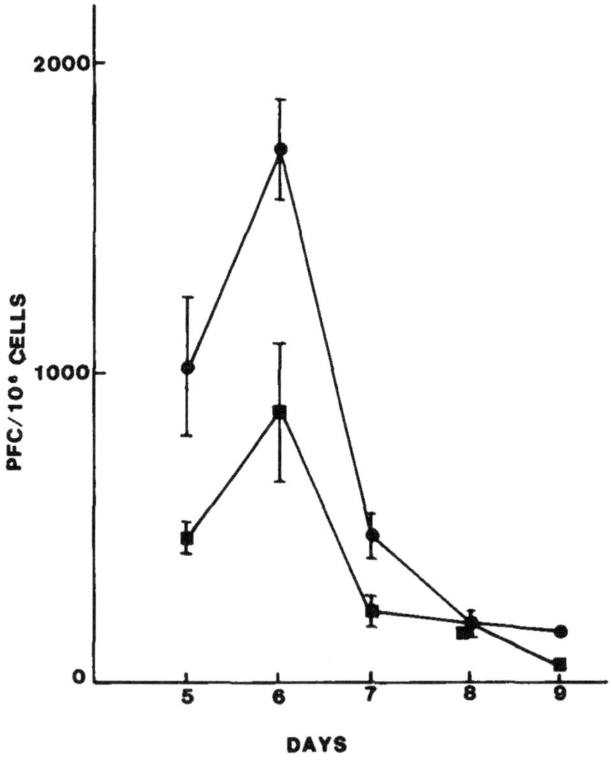


FIGURE 3. The effect of 6-OHDA on the kinetics of the IgG PFC response. Mice were injected with 6-OHDA (■) or saline (●). Two days later, the mice were immunized with SRBC and the IgG PFC response determined. The response of G-OHDA-injected mice is significantly depressed on days 5, 6, and 7 ($p < 0.05$). Each point represents the mean of five animals. (Reprinted from Cross, R. J., Jackson, J. C., Brooks, W. H., Sparks, D. L., Markesbery, W. R., and Roszman, T. L., Neuroimmunomodulation: impairment of humoral immune responsiveness by 6-hydroxydopamine treatment, *Immunology*, 57, 145, 1986. With permission.)

## TABLE 4
## Effect on the Primary SRBC Antibody Response of Treating Animals with 6-OHDA Before or After Immunization

| | **Treatment of animals** | |
|---|---|---|
| **6-OHDA injected before or after immunization**[a] | **Saline ($PFC/10^6$ cells)** | **6-OHDA ($PFC/10^6$ cells)** |
| 2 h before | 1170 ± 119 | 1480 ± 260 (NS) |
| 25 h after | 913 ± 133 | 918 ± 117 (NS) |
| 48 h after | 672 ± 61 | 926 ± 61 (NS) |

[a] Animals inoculated into the cisterna magna with 100 μg of 6-OHDA and immunized with $5 \times 10^8$ SRBC. There were five animals in each treatment. Number of indirect (IgG) PFC determined 6 d after immunization. Results represent the mean ± S.E.M. NS, not significant.

TABLE 5
Characterization of 6-OHDA-Induced Splenic Suppressor Cells

| Treatment of spleen cells transferred[a] | Antibody response[b] (DPFC/$10^6$ nucleated cells) 6-OHDA injected | Saline-injected | % Inhibition[c] |
|---|---|---|---|
| Whole spleen | 438 ± 43 ($p$ <0.01) | 647 ± 14 | 34 |
| anti-Thy 1.2 + C | 361 ± 58 (N.S.) | 405 ± 90 | 11 |
| G-10 column purified spleen | 325 ± 43 ($p$ <001) | 494 ± 30 | 34 |
| Nylon wool column purified spleen | 268 ± 32 ($p$ <0.01) | 563 ± 71 | 53 |

[a] Mice were administered either 6-OHDA or saline, then immunized 48 h later with SRBC. Two days following immunization, 5 × $10^6$ whole or treated spleen cells were transferred with SRBC (i.v.) into syngeneic recipients.
[b] Direct (IgM) PFC/$10^6$ nucleated cells were assayed on day 5 following cell transfer. The results represent the mean ± S.E.M. of values obtained from six mice/group.
[c] $$\% \text{ Inhibition} = 100 - \frac{\text{DPFC 6-OHDA}}{\text{DPFC Saline}} \times 100.$$

response, but does impair the development of immunological memory. These studies, in conjunction with those outlined above, indicate that catecholamine depletion impairs the function of virgin immunocompetent cells, but appears to have little affect on antigen activated or memory cells.

## VII. INDUCTION OF SUPPRESSOR CELLS BY 6-OHDA TREATMENT

To examine the immunological mechanisms underlying the suppressive effects of CNS catecholamine depletion, experiments were designed to determine if suppressor activity could be transferred from 6-OHDA injected mice to normal naive recipients.[53] As outlined in Table 5, whole or fractionated spleen cells from 6-OHDA-treated, antigen-primed mice were transferred to syngeneic recipients with additional antigen. These studies showed that suppressor cell activity induced following 6-OHDA treatment can be transferred to other mice. The suppressor cell was characterized as a T-cell based on its sensitivity to anti-T-cell antibody and complement treatment and the ability to enrich for the activity by passing the cells through a nylon wool column. Further studies also showed that the suppressor cells were not induced in 6-OHDA treated animals unless they were immunized. Thus, transfer of T-cells from nonantigen primed 6-OHDA treated mice does not inhibit the response of the recipient animals (Table 6).

## VIII. THE IMPORTANCE OF PITUITARY HORMONES IN MEDIATING THE SUPPRESSIVE EFFECTS OF 6-OHDA TREATMENT

Previously, we showed that hypophysectomized rats did not display either suppressed or enhanced immune function following the placement of lesions in the AHT or HC, re-

**TABLE 6**
**The Effect of Immunization on the Suppressor Activity of Splenic T-Cells from Mice Administered 6-OHDA or Saline in the Cisterna Magna**

| | Source of splenic T-cells | |
|---|---|---|
| **Treatment** | **Nonimmunized animals**[a] | **Immunized animals**[b] |
| | **DPFC/10⁶ nucleated cells** | |
| 6-OHDA injected | 616[c] ± 97 | 268 ± 32 |
| Saline injected | 641 ± 154 | 563 ± 71 ($p$ <0.01) |

[a] Mice were treated with 6-OHDA or saline, and 2 days later splenic T-cells were recovered from nylon wool columns and $5 \times 10^6$ T-cells and $1 \times 10^8$ SRBC were transferred to syngeneic recipients. DPFC were determined 5 d layer.

[b] Mice were treated with 6-OHDA or saline, and 2 days later the animals were immunized with SRBC. Two days after immunization splenic T-cells were recovered from nylon wool columns and $5 \times 10^6$ T-cells and $1 \times 10^8$ SRBC were transferred to recipients.

[c] Direct (IgM) plaque forming cells (PFC)/$10^6$ spleen cells ± S.E.M. Five mice were employed in each treatment group.

spectively.[33] Other studies have also shown that injection of 6-OHDA into the CNS markedly alters the output of pituitary hormones, particularly GH,[54] prolactin, and LH.[55] Likewise, suppressor cell activity induced by 6-OHDA treatment in mice can only be transferred to recipients when the suppressor cells are antigen induced in the environment which results from CNS catecholamine depletion. Therefore, it was of interest to determine if pituitary hormones mediate the suppressive effect 6-OHDA treatments in mice. To address this question, mice were hypophysectomized, and 2 weeks later, received 6-OHDA or saline into the cisterna magna. All animals were immunized with SRBC 2 d following 6-OHDA treatment. These studies show that an intact pituitary gland appears essential to the activation of suppressor mechanisms by 6-OHDA. In addition, we have recently shown that spleen cells from 6-OHDA-treated mice cultured *in vitro* with mitogens or alloantigens in a hormone-rich, serum-supplemented culture media respond normally.[60] Therefore, it appears that CNS 6-OHDA treatment alters the output of pituitary hormones which can then impinge on the immune response and result in suppression of the immune response.

Extrapolating from the data presented above, two possibilities for CNS modulation of immune function exist. It is possible that the CNS presets the gain or the magnitude of the immune response, such as in the situations of stress or emotional disorders, or neural pathology. Likewise, it is also possible, as shown by Besedovsky and co-workers,[56] that the CNS reacts to factors produced by lymphocytes during the immune response. In either case, changes in CNS catecholamines can alter the output of pituitary hormones.[54,55] These hormones may then interact with specific hormone receptors[57-59] on the lymphocyte membrane and induce a number of second messenger systems, resulting in either up or down regulation of lymphocyte function. Thus the immune system is no longer a self-regulating autonomous entity, but one that is continually sending signals to and responding to input from the CNS.

## ACKNOWLEDGMENTS

Supported by U.S. Public Health Service Grants NS17423 and NS22512. The author also wishes to thank Judy Chism for help in preparing the manuscript.

## REFERENCES


1. **Asherson, G. L., Colizzi, V., and Zembala, M.,** An overview of T-suppressor cell circuits, *Annu. Rev. Immunol.*, 4, 37, 1986.
2. **Hoffman, G. W.,** A theory of regulation and self-nonself discrimination, *Eur. J. Immunol.*, 5, 638, 1975.
3. **Jerne, N. K.,** Towards a network theory of the immune system, *Ann. Immunol. (Inst. Pasteur)*, 125C, 373, 1974.
4. **Benacerraf, B. and McDevitt, H. O.,** Histocompatibility-linked immune response genes, *Science*, 175, 273, 1973.
5. **McDevitt, H. O.,** Regulation of the immune response by the histocompatibility system, *N. Engl. J. Med.*, 303, 1514, 1980.
6. **Cross, R. J., Markesbery, W. R., Brooks, W. H., and Roszman, T. L.,** Hypothalamic-immune interactions: neuromodulation of natural killer activity by lesioning of the anterior hypothalamus, *Immunology*, 51, 399, 1984.
7. **Fabris, N., Pierpaoli, W., and Sorkin, E.,** Hormones and the immunological capacity. IV. Restorative effects of developmental hormones or of lymphocytes on the immunodeficiency syndrome of the dwarf mouse, *Clin. Exp. Immunol.*, 9, 227, 1971.
8. **Pierpaoli, W. and Besedovsky, H. O.,** Role of the thymus in programming neuroendocrine function, *Clin. Exp. Immunol.*, 20, 323, 1975.
9. **Brooks, W. H., Netsky, M. G., Normansell, D. E., and Horowitz, D. A.,** Depressed cell-mediated immunity in patients with primary intracranial tumors, *J. Exp. Med.*, 136, 1631, 1972.
10. **Mahaley, M. S., Brooks, W. H., Roszman, T. L., Bigner, D. D., Dudka, L., and Richardson, S.,** Immunobiology of primary intracranial tumors. I. Studies of the cellular and humoral general immune competence of brain-tumor patients, *J. Neurosurg.*, 46, 467, 1977.
11. **Elliott, L., Brooks, W., and Roszman, T.,** Role of interleukin 2 (IL-2) and IL-2 receptor expression in the proliferative defect observed in mitogen-stimulated lymphocytes from patients with gliomas, *J. Natl. Cancer Inst.*, 78, 919, 1987.
12. **Morrell, R. M.,** Antibody-dependent cytotoxicity in Huntington's disease, *Adv. Neurol.*, 23, 443, 1979.
13. **Hoffman, P. M., Robbins, P. S., Nolte, M. T., Gibbs, C. J., and Gajdusek, D. C.,** Cellular immunity in Guamanians with amyotrophic lateral sclerosis Parkinson dementia, *N. Engl. J. Med.*, 299, 680, 1978.
14. **Bartrop, R. W., Lazarus, E., and Luckhurst, L. G.,** Depressed lymphocyte function after bereavement, *Lancet*, 1, 834, 1977.
15. **Kovaleva, E. S., Bonartsev, P. D., and Prilipko, L. L.,** Lymphocyte reaction in schizophrenic patients to the phytomitogens, concanavalin A and phytohemogglutinin, *Bull. Exp. Biol. Med.*, 84, 1136, 1977.
16. **Solomon, G. F.,** Emotions, stress, the central nervous system and immunity, *Ann. N.Y. Acad. Sci.*, 164, 335, 1969.
17. **Rassmussen, A. F., Spencer, E. S., and Marsh, J. T.,** Decreased susceptibility of mice to passive anaphylaxis following avoidance learning stress, *Proc. Soc. Exp. Biol. Med.*, 100, 878, 1959.
18. **Solomon, G. F.,** Stress and antibody response in rats, *Int. Arch. Allergy Appl. Immunol.*, 36, 97, 1969.
19. **Wistar, R. and Hildemann, W. H.,** The effect of stress on skin transplantation immunity in mice, *Science*, 131, 159, 1960.
20. **Marsh, J. T. and Rassmussen, A. F.,** Response of adrenals, thymus, spleen, and leukocytes to shuttle box and confinement stress, *Proc. Soc. Exp. Biol. Med.*, 104, 108, 1960.
21. **Isakovic, K. and Jankovic, B. D.,** Neuroendocrine correlates of immune response. II. Changes in the lymphatic organs of brain lesioned rats, *Int. Arch. Allergy Appl. Immunol.*, 45, 373, 1973.
22. **Jankovic, B. D. and Isakovic, K.,** Neuroendocrine correlates of immune response. I. Effects of brain lesions on antibody production, Arthus reactivity, and delayed-hypersensitivity in the rat, *Int. Arch. Allergy Appl. Immunol.*, 45, 360, 1973.
23. **Korneva, E. A. and Khai, L. M.,** Effect of destruction of hypothalamic areas on immunogenesis, *Fed. Proc.*, 23, T88, 1964.
24. **Tyrey, L. and Nalbandov, A. V.,** Influence of anterior hypothalamic lesions on circulating antibody titers in the rat, *Am. J. Physiol.*, 222, 179, 1972.
25. **Schiavi, R. C., Adams, J., and Stein, M.,** Effect of hypothalamic lesions on histamine toxicity in the guinea pig, *Am. J. Physiol.*, 211, 1269, 1966.

26. **Macris, N. T., Schiavi, R. C., Camerino, M. S., and Stein, M.,** Effect of hypothalamic lesions on immune processes in the guinea pig, *Am. J. Phsyiol.*, 219, 1205, 1970.
27. **Schiavi, R. C., Macris, N. T., and Camerino, M. S., and Stein, M.,** Effect of hypothalamic lesions on immediate hypersensitivity, *Am. J. Physiol.*, 228, 596, 1975.
28. **Keller, S. E., Stein, M., Camerino, M. S., Schleifer, S. J., and Sherman, J.,** Suppression of lymphocyte stimulation by anterior hypothalamic lesions in the guinea pig, *Cell. Immunol.*, 52, 334, 1980.
29. **Cross, R. J., Markesbery, W. R., Brooks, W. H., and Roszman, T. L.,** Hypothalamic-immune interactions. I. The acute effect of anterior hypatholamic lesions on the immune response, *Brain Res.*, 196, 79, 1980.
30. **Goldowitz, D., Scheff, S., and Cotman, C.,** The specificity of reactive synaptogenesis: comparative study in the rat hippocompal formation, *Brain Res.*, 170, 427, 1979.
31. **Roszman, T. L., Cross, R. J., Brooks, W. H., and Markesbery, W. R.,** Hypothalamic-immune interactions. II. The effect of hypothalamic lesions on the ability of adherent spleen cells to limit lymphocyte blastogenesis, *Immunology*, 45, 737, 1982.
32. **Bash, J. A. and Waksman, B. H.,** The suppressive effect of immunization on the proliferative responses of rat T cells *in vitro*, *J. Immunol.*, 114, 782, 1975.
33. **Brooks, W. H., Cross, R. J., Roszman, T. L., and Markesbery, W. R.,** Neuroimmunomodulation: neural anatomical basis for impairment and facilitation, *Ann. Neurol.*, 12, 56, 1981.
34. **Williams, J. M., Peterson, R. G., Shea, P. A., Schmedtje, J. F., and Bauer, D. C.,** Sympathetic innervation of murine thymus and spleen: evidence for a functional link between the nervous and immune systems, *Brain Res. Bull.*, 6, 83, 1981.
35. **Felton, D. L., Ackerman, K. D., Wiegand, S. J., and Felton, S. Y.,** Noradrenergic sympathetic innervation of the spleen. I. Nerve fibers associate with lymphocytes and macrophages in specific compartments of the splenic white pulp, *J. Neurosci. Res.*, 18, 28, 1987.
36. **Johnson, H. M., Smith, E. M., Torres, B. A., and Blalock, J. E.,** Regulation of the *in vitro* antibody response by neuroendocrine hormones, *Proc. Natl. Acad. Sci. U.S.A.*, 79, 4171, 1982.
37. **Vischer, T. L.,** Effect of hydrocortisone on the reactivity of thymus and spleen cells to *in vitro* stimulation, *Immunology*, 23, 777, 1972.
38. **Snow, E. C., Feldbush, T. L., and Oaks, J. A.,** The role of insulin in the response of murine T lymphocytes to mitogenic stimulation *in vitro*, *J. Immunol.*, 124, 739, 1980.
39. **Pandian, M. R. and Talwar, G. P.,** Effect of growth hormone on the metabolism of thymus and on the immune response against SRBC, *J. Exp. Med.*, 134, 1095, 1971.
40. **Berczi, I., Nagy, E., Kovacs, K., and Horvath, E.,** Regulation of humoral immunity in rats by pituitary hormones, *Acta Endocrinol.*, 98, 506, 1981.
41. **Cross, R. J., Brooks, W. H., Roszman, T. L., and Markesbery, W. R.,** Hypothalamic-immune interactions. Effect of hypophysectomy on neuroimmunomodulation, *J. Neurol. Sci.*, 53, 557, 1982.
42. **Dann, J. A., Wachtel, S. S., and Rubin, A. L.,** Possible involvement of the central nervous system in graft rejection, *Transplantation*, 27, 223, 1979.
43. **Heberman, R. B. and Ortaldo, J. R.,** Natural killer cells: their role in defense against disease, *Science*, 214, 24, 1981.
44. **Williams, R. M., Leifer, J., and Moore, M. J.,** Hybrid effect in natural cell-mediated cytotoxicity of SV-40 transformed fibroblasts by rat spleen cells, *Transplantation*, 23, 283, 1977.
45. **Brooks, C. G., Flanner, G. R., Webb, P. J., and Baldwin, R. W.,** Quantitative studies of natural cytotoxicity to solid tumors in rats. The nature of the killer cell depends on the type of assay, *Immunology*, 41, 673, 1980.
46. **Riley, V., Fitzmaurice, M. A., and Spackman, D. H.,** Psychoneuroimmunologic factors in neoplasia: studies in animals, in *Psychoneuroimmunology*, Ader, R., Ed., Academic Press, New York, 1981, 31.
47. **Forni, G., Bindoni, M., Santoni, A., Belluardo, N., Marchese, A. E., and Giovarelli, M.,** Radio frequency destruction of the tuberoinfundibular region of hypothalamus permanently abrogates NK cell activity in mice, *Nature (London)*, 306, 181, 1983.
48. **Belluardo, N., Mudo, G., Cella, S., Santoni, A., Forni, G., and Bindoni, M.,** Hypothalamic control of certain aspects of natural immunity in the mouse, *Immunology*, 62, 321, 1987.
49. **Bykowsky, M. J. and Stutman, O.,** The cell responsible for murine natural cytotoxicity (NC) activity: a multi-lineage system, *J. Immunol.*, 173, 1120, 1986.
50. **Saxena, Q. B., Saxena, R. K., and Adler, W. H.,** Regulation of natural killer activity *in vivo*. *III*. Effect of hypophysectomy and growth hormone treatment on the natural killer activity of the mouse spleen cell population, *Int. Arch. Allergy Appl. Immunol.*, 67, 169, 1982.
51. **Jackson, J. C., Cross, R. J., Walker, R. F., Markesbery, W. R., Brooks, W. H., and Roszman, T. L.,** Influence of serotonin on the immune response, *Immunology*, 54, 505, 1935.
52. **Cross, R. J., Jackson, J. C., Brooks, W. H., Sparks, D. L., Markesbery, W. R., and Roszman, T. L.,** Neuroimmunomodulation: impairment of humoral immune responsiveness by 6-hydroxydopamine treatment, *Immunology*, 57, 145, 1986.

53. **Cross, J. R., Brooks, W. H., and Roszman, T. L.**, 'Modulation of T-suppressor cell activity by central nervous system catecholomine depletion, *J. Neurosci. Res.*, 18, 75, 1987.
54. **Willoughby, J. O. and Day, T. A.**, Central catecholamine depletion: effects on physiological growth hormone and prolactin secretion, *Neuroendocrinology*, 32, 65, 1981.
55. **Fenske, M. and Wuttke, W.**, Effects of intraventricular 6-hydroxydopamine injections on serum prolactin and LH levels: absence of stress-induced pituitary prolactin release, *Brain Res.*, 104, 63, 1976.
56. **Besedovsky, H. O., Del Rey, A. E., and Sorkin, E.**, What do the immune system and brain know about each other?, *Immunol. Today*, 4, 342, 1984.
57. **Russell, D. H., Kibler, R., Matrision, L., Larson, D. F., Paulos, B. F., and Magun, B. E.**, Prolactin receptors on human T and B lymphocytes: antagonism of prolactin binding by cyclosporine, *J. Immunol.*, 134, 3027, 1985.
58. **Arrenbrecht, S.**, Specific binding of growth hormone to thymocytes, *Nature (London)*, 252, 255, 1974.
59. **Harrison, L. C., Flier, J., and Itin, A.**, Radioimmunoassay of the insulin receptor: a new probe of receptor structure and function, *Science*, 203, 544, 1979.
60. **Cross, R. J. and Roszman, T. L.**, Central catecholamine depletion impairs in vivo immunity but not in vitro lymphocyte activation, *J. Neuroimmunol.*, 19, 33, 1988.

Chapter 10

# IMMUNOREGULATION BY LEUKOTRIENES AND PLATELET-ACTIVATING FACTOR

**Marek Rola-Pleszczynski**

# I. INTRODUCTION

In most instances when the host is asked to mount a specific immune response, the initial event or the subsequent response, or both, are associated with nonspecific inflammation. In recent years, with the identification of numerous molecular and cellular components of the inflammatory response, investigators have initiated studies of potential interactions between the latter and the protagonists of more specific immunological reactions. In particular, soluble mediators of inflammation, produced by phagocytes, endothelial cells, or nerves have been studied in regard to their possible modulation of lymphocyte and monocyte-macrophage functions. In the present chapter, two lipid mediators, leukotriene (LT) $B_4$ and platelet-activating factor (PAF), which are very potent proinflammatory molecules, will be discussed as potentially powerful immunoregulatory agents.

# II. LEUKOTRIENE $B_4$

Leukotriene $B_4$ ($LTB_4$) is derived from the oxidative metabolism of arachidonic acid.[1-3] Following membrane perturbations and activation of phospholipase $A_2$, arachidonic acid is released from membrane phospholipids. It may undergo oxidative changes either through cyclooxygenase which results in the formation of the various prostaglandins, or through one of several lipoxygenases. Leukotrienes are derived from the action of 5-lipoxygenase, with initial formation of 5-hydroperoxyeicosatetraenoic (5-HPETE) acid, followed by $LTA_4$. $LTB_4$, which is an enzymatic, hydrolytic product of $LTA_4$ differs from the other leukotrienes $LTC_4$, $LTD_4$, and $LTE_4$ in that it has no peptidic component. It is a more stable molecule than the peptido-leukotrienes, but can be rapidly degraded through omegaoxidation by polymorphonuclear leukocytes (PMN) to the relatively inactive metabolites, 20-OH-$LTB_4$ and 20-COOH-$LTB_4$.

While sharing some of the myotropic properties of the other leukotrienes, $LTB_4$ has been found to exert very strong leukocytotropic activities. It is one of the most powerful chemokinetic[4] and chemotactic[5] agents, and it can induce cation fluxes,[6] hexose transport,[7] neutrophil aggregation[4] and degranulation,[8] and their enhanced binding to endothelial cells.[9] It can also synergize with prostaglandins $E_1$ and $E_2$ in causing macromolecule leakage in the skin through increased vascular permeability.[10]

$LTB_4$ is rapidly synthesized by phagocytic cells, principally neutrophilic $PMN_3$ and alveolar macrophages,[11] upon challenge with a variety of stimuli such as microbial pathogens, toxins, particulate material, or ionophores. Its synthesis can also be induced by platelet-derived 12-HPETE[12] or PAF-acether.[13,14] Effective cell separation techniques suggest that $LTB_4$ is not a product of lymphocytes,[15,16] in spite of several earlier reports to the contrary.[17,18] It has been detected in significant concentrations in inflammatory synovial exudates,[19] psoriatic skin lesions,[20] and inflammatory bowel disease,[21] and is thought to be involved in the adult respiratory distress syndrome.[22]

Globally, $LTB_4$ thus appears to be a potent proinflammatory agent. Several years ago, we and others became interested in its potential immunoregulatory functions (reviewed in Reference 23). It became progressively evident that $LTB_4$ could effect many regulatory functions through interactions with a variety of lymphocyte and other cell subpopulations, as presented in the next sections.

## A. EFFECTS ON LYMPHOCYTES

When $LTB_4$ was added to human peripheral blood mononuclear leukocyte (PBML) cultures stimulated with the mitogenic lectins concanavalin A (Con A) or phytohemagglutinin (PHA), a modest inhibition of the proliferative response could be observed.[24] Similar findings were also reported for $LTD_4$ and $LTE_4$,[25] and for 15-HETE[26] in mice. These initial observations suggested the possibility that suppressor cell activation could be involved in these responses. When suppressor cells are generated *in vitro,* their suppressive activity can be subsequently tested by adding them to fresh PBML cultures, in the absence of the inducing agent, and measuring the proliferative response of the latter to mitogens or antigens. Using such a coculture system, normal human PBML were preincubated for 24 h with graded concentrations of $LTB_4$. A significant suppressor cell activity was generated at concentrations of $LTB_4$ ranging from $10^{-14}$ to $10^{-10}$ *M*.[24] $LTD_4$ induced no significant suppressor cell activity. The magnitude of the suppressor cell activity generated by $LTB_4$ was comparable to that seen with Con A or histamine, but the concentrations at which it occurred were orders of magnitude lower than those needed with the latter. Although a 24-h preincubation with $LTB_4$ was optimal, as little as 3 h were sufficient for generating suppressor activity.[24]

$LTB_4$-induced suppressor cells were shown to be T lymphocytes which did not require proliferation to exert their activity, as mitomycin C treatment of preincubated cells before addition to fresh PBML cultures did not affect the suppressor function.[24] In contrast, irradiation with 2000 rad abolished all suppressor activity. This also suggested that carry over of $LTB_4$ on the surface of preincubated lymphocytes was not responsible for the subsequent suppression. Similar findings were later reported by Payan and Goetzl,[27] and Atluru and Goodwin,[28] in regard to lymphokine production and regulation of immunoglobulin synthesis, respectively.

More specifically, when T-cells were fractionated into $CD4^+$ and $CD8^+$ T-cells using panning, complement-mediated lysis, or fluorescence-activated cell sorting techniques, $LTB_4$ could induce the appearance of suppressor cells in either subset, but the effective suppressor cells finally bore the $CD8^+$ phenotype.[26,29] Proliferation of $CD4^+$ cells was inhibited by $LTB_4$, while the proliferation of $CD8^+$ cells was enhanced.[29,30] Further analysis of the cellular requirements for $LTB_4$-induced suppressor cell activity revealed that, while only T-cells needed to interact with $LTB_4$, the expression of their suppressor activity required the presence of monocytes in the responding cell population.[24,29,31] Only purified $CD8^+$ T-cells could bypass this requirement.[29] For preincubated $CD4^+$ T-cells or unfractionated T-cells, little suppression was seen in the absence of monocytes, while a helper effect was evident in the presence of monocytes plus the cyclooxygenase inhibitor indomethacin.[29] We postulated that this phenomenon could be explained by a dual stimulation of monocytes to produce both PGs and interleukin-1 (IL-1) by $LTB_4$ preincubated T-cells. We recently demonstrated that $LTB_4$-pulsed T-cells released a soluble factor that stimulated monocytes to produce $PGE_2$ and/or IL-1 and that most of the activity of such T-cell supernatants was attributable to interferon-gamma (IFN-$\gamma$).[32]

Indeed, when lymphokine production was studied following interaction of $LTB_4$ with T-cell subsets, it was found that production of both IL-2 and IFN-$\gamma$ by $CD4^+$ T-cells was enhanced by nanomolar concentrations of $LTB_4$, while their production by $CD8^+$ cells was impaired.[33,34] This was observed with human cells with a variety of stimuli including lectins, phorbol ester plus ionomycin, or poly I:C. In the mouse, $LTB_4$ (as well as other leukotrienes and arachidonic acid itself) could replace helper cells in IFN-$\gamma$ production.[35]

In addition to interacting with T-cells, $LTB_4$ was shown to modulate the activity of natural killer (NK) and natural cytotoxic (NC) lymphocytes. NK-cell activity was measured using the NK-sensitive K562 erythroleukemia cell line as target while NC activity was measured using target cells (MA-160) infected with herpes simplex virus type 1 during a 4-h coculture. At concentrations of $10^{-12}$ to $10^{-8}$ *M*, $LTB_4$ and, to a lesser extent, $LTA_4$ markedly augmented NK and NC activities.[36,37] The lipoxygenase inhibitor nordihydroguaiaretic acid (NDGA) inhibited NC activity and this inhibition could be effectively reversed by $LTB_4$, indicating that exogenous lipoxygenase products could compensate for endogenous lipoxygenase inhibition.[38] $LTB_4$ acted on effector NK or NC lymphocytes rather than on their target cells, as demonstrated by preincubation experiments.[39] $LTB_4$ was also shown to increase both the binding of effector lymphocytes to their target cells and the rate of target cell killing. These effects were rapid, within 30 min for binding and 2 to 4 h for lysis of target cells. No IL-2 or IFN-γ was detectable in these cultures.[39]

### B. EFFECTS ON MONOCYTES

Many of the effects of $LTB_4$ on lymphocyte functions, described above, could be amplified or modulated by interactions of this mediator with accessory cells, such as monocytes. Since these phagocytic cells can produce $LTB_4$ themselves, it can also play a role of feedback regulator or internal messenger.

Indirect evidence that lipoxygenase metabolites of arachidonic acid would possibly be involved in IL-1 production was provided by use of metabolic inhibitors of lipoxygenation.[40] Subsequently, we provided more direct evidence that $LTB_4$ and, to a lesser extent, $LTD_4$ enhanced IL-1 production by lipopolysaccharide (LPS)-stimulated human monocytes two- to threefold.[41] Similar effects were seen when monocytes were stimulated with adherence, silica, zymosan, or muramyl dipeptide. Optimal enhancement of IL-1 production occurred when LPS and $LTB_4$ were present together at the initiation of the 24-h monocyte culture, but significant enhancement was also observed when monocytes were either preincubated with $LTB_4$ before addition of LPS or cultured with LPS alone for 3 h before addition of $LTB_4$.[42]

An analogous effect was observed when the production of tumor necrosis factor (TNF-α) by human monocytes was assessed in the presence of graded concentrations of $LTB_4$.[42] Significant enhancement of TNF production was evident at $LTB_4$ concentrations of $10^{-12}$ to $10^{-8}$ *M*. TNF production was markedly impaired by culture of monocytes in the presence of the lipoxygenase inhibitors NDGA or AA-861 (Takeda, Japan), and exogenous $LTB_4$ was capable of reversing the inhibition.[43]

The observations may be quite relevant to the concomitant findings that $LTB_4$ significantly enhanced monocyte-mediated tumor cell killing in an 18-h cytotoxicity assay.[43] This enhanced cytotoxic activity was associated with augmented monocyte-to-tumor target cell binding and with subsequent higher rate of lysis in a single cell cytotoxicity assay. Although monocytes can kill tumor cells through several mechanisms in addition to TNF, $LTB_4$ was also found to augment the production of $H_2O_2$ by monocytes, while the production of neutral proteases was unaffected.[43]

As can be seen, an increasing body of evidence suggests that $LTB_4$ may be an important modulator of many immune cell functions, either as an internal messenger within some cells or as an external signal, linking inflammatory cells to lymphocytes.

## III. PLATELET-ACTIVATING FACTOR

A factor released by stimulated leukocytes which could activate platelets was described in the late 1960s and early 1970s and termed platelet-activating factor (PAF).[44-46] Subsequent structural elucidation demonstrated that PAF is 1-alkyl-2(R)-acetyl-glycero-3-phosphorylcholine,[47,48] and it is now interchangeably referred to as PAF-acether or AGEPC. PAF-acether is a potent mediator of anaphylaxis and hypersensibility and may be involved in shock, graft rejection, postischemic disorders, ovoimplantation, and certain neurological perturbations (reviewed in Reference 49). Because both PAF-acether and leukotrienes can derive from the same membrane phospholipid, alkyl-arachidonyl-GPC, they may be produced simultaneously or in sequence, and some of the effects of PAF appear to be mediated by subsequent LT production.

Cellular sources of PAF are numerous and include various leukocyte populations, endothelial cells, and probably other cell types.[49] Resting lymphocytes are unable to produce PAF, possibly due to a lack of acetyl transferase.[50] However, several T- and B-cell lines[51] and large granular lymphocytes[52] can be activated to produce PAF under certain circumstances. Many of the cells or tissues which produce PAF are themselves targets of PAF-induced bioactions, and the recent discovery and development of many PAF receptor antagonists has opened the way for more detailed research in this field.

### A. EFFECTS ON LYMPHOCYTES

When PAF was added to human PBML stimulated with PHA or Con A, a concentration-dependent inhibition of lymphocyte proliferation occurred.[53] The $IC_{50}$ was between $10^{-12}$ and $10^{-10}$ *M* and PAF needed to be present during the first 3 h of the 72-h culture for suppression to be observed. Lyso-PAF or (S)-PAF were inactive, while a similar effect was seen with the nonhydrolyzable PAF analog, ethoxy-PAF. PAF-induced inhibition of PBML proliferation was prevented by the use of the PAF receptor antagonist BN52021. Suppression was also reversed by indomethacin, suggesting that PGs were involved as second messengers in the PAF-induced inhibition of lymphocyte proliferation. IL-2 production by mitogen-stimulated human PBML was also inhibited by PAF and ethoxy-PAF and this effect was similarly blocked by BN52021. Recently, Dulioust et al.[54] reported that PAF suppressed $CD4^+$ T-cell proliferation, but without effect on IL-2 production: the effect occurred at high concentrations of PAF and was seen when PAF was added late during culture. In contrast, Behrens and Goodwin[55] reported augmented $CD4^+$ T-cell proliferation in reponse to PAF, which could be blocked by BN52021.

Because inhibition of proliferation and IL-2 production could be due to the generation or activation of suppressor cells, we investigated the effects of PAF in a coculture system. PBML preincubated with PAF acquired the ability to suppress the proliferation of fresh autologous lymphocytes stimulated with mitogens.[56] Hence, PAF could activate suppressor cells which subsequently inhibited lymphocyte function in the absence of any further contact with PAF. This induction of suppressor cell activity was accompanied by an increase in $CD8^+$ T-cell numbers and a slight decrease in $CD4^+$ T-cell numbers.[57] $CD2^+$ (sheep erythrocyte receptor-positive) T-cells were also increased in numbers, but the most striking increase was seen in T-cells bearing DR antigens. No effect was seen on numbers of IL-2 receptor (Tac)-positive T-cells. In contrast, Vivier et al. have found discording results regarding T-cells phenotypes following incubation with PAF.[58]

When specific PBML subsets were isolated and purified *before* preincubation with PAF, a remarkable spectrum of activities emerged.[57] PAF-preincubated monocytes exerted suppression via an indomethacin-sensitive mechanism, while PAF-preincubated $CD8^+$ T-cells exerted suppression via an indomethacin-resistant mechanism, and at 10- to 100-fold lower concentrations of PAF. In contrast, PAF-preincubated $CD4^+$ T-cells markedly enhanced lymphocyte proliferation. In a different system, PAF and two nonhydrolyzable PAF analogs increased the proliferation of IL-2-stimulated human lymphoblasts, while some PAF receptor antagonists (RO-193704, CV-3988 and L-652,731, but not BN52021 or WEB-2086) and the PAF synthesis inhibitor L-648,611 inhibited such proliferation.[59] Corroboration of some of these *in vitro* observations was provided by *in vivo* infusion of PAF into rats via implanted miniosmotic pumps. After 7 d of infusion, rat splenocytes showed enchanced IL-2 production in response to Con A, *in vitro*.[60,61]

Lymphocyte-mediated cytotoxicity was also shown to be modulated by PAF.[62,63] NI-cell activity was markedly augmented by low ($10^{-13}$ to $10^{-10}$ *M*) concentrations of PAF while higher concentrations ($>10^{-8}$ *M*) were inhibitory. The inhibition was blocked by BN52021, but this antagonist (as well as WEB-2086) did not prevent the augmented NK activity at low concentrations of PAF. Although impossible to verify at the present time, these findings could be explained by the presence of two PAF receptors on NK cells: one with a lower affinity, which would be linked with inhibition of cell function and could be blocked by BN-52021, and another with high affinity, linked with augmentation of cell function, and not affected by BN52021. Preliminary data concerning allo-specific cytotoxic T-cells (CTL) also suggest PAF may have inhibitory effects, since BN52021 enhances the generation of CTL when included in the mixed lymphocyte culture.[64] Conversely, BN52021 increases cardiac allograft survival in rats, acting synergistically with azathioprine and cyclosporine A.[65] The mechanism(s) and cells involved in these phenomena in relation to PAF are essentially unknown at this time.

### B. EFFECTS ON MONOCYTES-MACROPHAGES

Since accessory cells are essential to many T-cell functions, it was interesting to study the effect of PAF on monocyte-macrophage cytokine production. IL-1 production by rat splenic monocytes was modulated in a biphasic manner, with low concentrations ($10^{-12}$ to $10^{-10}$ *M*) of PAF causing a significant enhancement, with higher concentrations ($10^{-8}$ to $10^{-6}$ *M*) being inhibitory.[66] Human monocytes also produced augmented quantities of IL-1 when incubated with PAF and stimulated with LPS or MDP;[70] other investigators have also found enhanced IL-1 production with PAF.[67] Under baseline conditions, (R) PAF and the PAF analogs, PR 1501 and PR 1502, but not (S)PAF, have been reported to induce IL-1 production,[68] although to a lesser extent than LPS. Continuous low-rate administration of PAF *in vivo,* in rats, via osmotic mini-pumps also revealed a biphasic effect on subsequent *ex vivo* IL-1 production by rat splenic monocytes.[61] Cells from rats which received 28 μg of PAF during 7 d showed impaired IL-1 (both released and cell associated) production in response to LPS stimulation. In contrast, cells from rats having received 1 to 9 μg of PAF showed augmented IL-1 production. These effects were abolished in animals having received a concomitant treatment with BN52021, given orally.

Macrophages and natural cytotoxic cells mediate their cytotoxic activities, at least in part, through synthesis and release of TNF-α. When preincubated or coincubated with PAF, LPS-treated human monocytes and rat alveolar macrophages produced significantly higher quantities of TNF.[69-71]

## IV. CONCLUSIONS

$LTB_4$ and PAF are known protagonists of numerous inflammatory processes. As reviewed in this chapter, they can also mediate varied and powerful effects on many cells involved in immunological reactions. Many ongoing studies will probably establish even more clearly the interactions between inflammation and immune responses, thus opening new venues for research and eventual therapeutic intervention.

## ACKNOWLEDGMENTS

Studies by the author presented in this chapter were supported by grants from the National Cancer Institute and the Medical Research Council of Canada. The author is a Research Scholar of the Fonds de la Recherche en Santé du Québec.

## REFERENCES

1. **Borgeat, P. and Samuelsson, B.,** Metabolism of arachidonic acid in polymorphonuclear leukocytes: unstable intermediate in formation of dihydro acids, *Proc. Natl. Acad. Sci. U.S.A.*, 76, 3213, 1979.
2. **Sirois, P. and Borgeat, P.,** Leukotrienes: a new approach to the biochemistry of hypersensitivity, *Surv. Immunol. Res.*, 1, 279, 1982.
3. **Borgeat, P. and Samuelsson, B.,** Transformation of arachidonic acid by rabbit polymorphonuclear leukocytes: formation of a novel dihydro-eicosa-tetraenoic acid, *J. Biol. Chem.*, 254, 2643, 1979.

4. **Ford-Hutchinson, A. W., Bray, M. A., Doig, M. V., Shipley, M. E., and Smith, M. J. H.,** Leukotriene B, a potent chemokinetic and aggregating substance released from polymorphonuclear leucocytes, *Nature (London),* 286, 264, 1980.
5. **Palmblad, J., Malmsten, C. L., Uden, A. M., Radmar, K. O., Engstedt, L., and Samuelsson, B.,** Leukotriene $B_4$ is a potent stereospecific stimulator of neutrophil chemotaxis and adherence, *Blood,* 58, 658, 1981.
6. **Molski, T. F. P., Naccache, P. H., Borgeat, P., and Sha'afi, R. I.,** Similarities in the mechanisms by which formylmethionyl-leucyl-phenulalamine, arachidonic acid and leukotriene $B_4$ increase calcium and sodium influxes in rabbit neutrophils, *Biochem. Biophys. Res. Commun.,* 103, 227, 1981.
7. **Bass, D. A., Thomas, M. J., Goetzl, E. J., DeChatelet, E. R., and McCall, C. E.,** Lipoxygenase-derived products of arachidonic acid mediate stimulation of hexose uptake in human polymorphonuclear leukocytes, *Biochem. Biophys. Res. Commun.,* 100, 1, 1981.
8. **Showell, H. J., Naccache, P. H., Borgeat, P., Picard, S., Valerand, P., Becker, E. L., and Sha'afi, R. I.,** Characterization of the secretory activity of $LTB_4$ toward rabbit neutrophils, *J. Immunol.,* 128, 811, 1982.
9. **Bray, M. A., Ford-Hutchinson, A. W., and Smith, M. J. H.,** Leukotriene $B_4$: an inflammatory mediator *in vivo, Prostaglandins,* 22, 213, 1981.
10. **Bray, M. A., Cunningham, F. M., Ford-Hutchinson, A. W., and Smith, M. J. H.,** Leukotriene $B_4$: a mediator of vascular permeability, *Br. J. Pharmacol.,* 72, 483, 1981.
11. **Fels, A. O., Pawlowski, N. A., Cramer, E. B., King, T. K., Cohen, A. Z., and Scott, W. A.,** Human alveolar macrophages produce leukotriene $B_4$, *Proc. Natl. Acad. Sci. U.S.A.,* 79, 7866, 1982.
12. **Maclouf, J., Fruteau de Laclos, B., and Borgeat, P.,** Stimulation of leukotriene biosynthesis in human blood leukocytes by platelet-derived 12-hydroperoxy-icosatetraenoic acid, *Proc. Natl. Acad. Sci. U.S.A.,* 79, 6042, 1982.
13. **Chilton, F. H., O'Flaherty, J. T., Walsh, C. E., Thomas, M. J., Wykle, R. L., DeChatelet, L. R., and Waite, B. M.,** Stimulation of the lipoxygenase pathway in polymorphonuclear leukocytes by 1-*O*-alkyl-2-*O*-acetyl-*SN*-glycero-3-phosphocholine, *J. Biol. Chem.,* 257, 5402, 1982.
14. **Lin, A. H., Morton, D. R., and Gorman, R. R.,** Acetyl glyceryl ether phosphorylcholine stimulates leukotriene $B_4$ synthesis in human polymorphonuclear leukocytes, *J. Clin. Invest.,* 70, 1058, 1982.
15. **Goldyne, M. E., Burrish, G. F., Poubelle, P., and Borgeat, P.,** Arachidonic acid metabolism among human mononuclear leukocytes. Lipoxygenase related pathways, *J. Biol. Chem.,* 259, 8815, 1984.
16. **Poubelle, P., Borgeat, P., and Rola-Pleszczynski, M.,** Assessment of leukotriene $B_4$ synthesis in human lymphocytes using high performance liquid chromatography and radioimmunoassay methods, *J. Immunol.,* 139, 1273, 1987.
17. **Goetzl, E. J.,** Selective feed-back inhibition of the 5-lipoxygenation of arachidonic acid in human T-lymphocytes, *Biochem. Biophys. Res. Commun.,* 1901, 344, 1981.
18. **Atluru, D., Lianos, E. A., and Goodwin, J. S.,** Arachidonic acid inhibits 5-lipoxygenase in human T cells, *Biochem. Biophys. Res. Commun.,* 135, 670, 1986.
19. **Davidson, E. M., Rae, S. A., and Smith, M. J. H.,** Leukotriene $B_4$ in synovial fluid, *J. Pharm. Pharmacol.,* 34, 410, 1982.
20. **Brain, S. D., Camp, R. D. R., Dowd, P. M., Black, A. K., Woolard, P. M., Mallet, A. I., and Greaves, M. W.,** Psoriasis and leukotriene $B_4$, *Lancet,* 2, 762, 1982.
21. **Sharon, P. and Stenson, W. F.,** Production of leukotrienes by colonic mucosa from patients with inflammatory bowel disease, *Gastroenterology,* 84, 1306, 1983.
22. **Goetzl, E. J., Payan, D. G., and Goldman, D. W.,** Immunopathogenetic roles of leukotrienes in human diseases, *J. Clin. Immunol.,* 4, 79, 1984.
23. **Rola-Pleszczynski, M.,** Immunoregulation by leukotrienes and other lipoxygenase metabolites, *Immunol. Today,* 6, 302, 1985.
24. **Rola-Pleszczynski, M., Borgeat, P., and Sirois, P.,** Leukotriene $B_4$ induces human suppressor lymphocytes, *Biochem. Biophys. Res. Commun.,* 108, 1531, 1982.
25. **Webb, D. R., Nowowiejski, I., Healy, C., and Rogers, T. J.,** Immunosuppressive properties of leukotrienes $D_4$ and $E_4$ in vitro, *Biochem. Biophys. Res. Commun.,* 108, 1531, 1982.
26. **Bailey, J. M., Bryant, R. W., Low, C. E., Pupilla, M. B., and Vanderhoeck, J. Y.,** Regulation of T-lymphocyte mitogenesis by the leucocyte product, 15-HETE, *Cell Immunol.,* 67, 112, 1982.
27. **Payan, D. G. and Goetzl, E. J.,** Specific suppression of human lymphocyte function by leukotriene $B_4$, *J. Immunol.,* 131, 551, 1983.
28. **Atluru, D. and Goodwin, J. S.,** Control of polyclonal immunoglobulin production from human lymphocytes by leukotrienes: leukotriene $B_4$ induces an $OKT8^+$, radiosensitive suppressor cell from resting human $OKT8^-$ T cells, *J. Clin. Invest.,* 74, 1444, 1984.
29. **Rola-Pleszczynski, M.,** Differential effects of leukotriene $B_4$ on $T4^+$ and $T8^+$ lymphocyte phenotype and immunoregulatory functions, *J. Immunol.,* 135, 1357, 1985.
30. **Gualde, N., Atluru, D., and Goodwin, J. S.,** Effect of lipoxygenase metabolites of arachidonic acid on proliferation of human T cells and T cell subsets, *J. Immunol.,* 134, 1125, 1985.

31. **Rola-Pleszczynski, M. and Sirois, P.**, Leukotrienes as immunoregulators, in *Leukotrienes and Other Lipoxygenase Products*, Piper, P. J., Ed., John Wiley & Sons, London, 1983, 234.
32. **Rola-Pleszczynski, M., Bouvrette, L., Gingras, D., and Girard, M.**, Identification of interferon-γ as the lymphokine which mediates leukotriene $B_4$-induced immunoregulation, *J. Immunol.*, 139, 513, 1987.
33. **Rola-Pleszczynski, M., Chavaillaz, P.-A., and Lemaire, I.**, Stimulation of interleukin-2 and interferon gamma production by leukotriene $B_4$ in human lymphocyte cultures, *Prostagl. Leukotr. Med.*, 23, 207, 1986.
34. **Rola-Pleszczynski, M., Gagnon, L., and Chavaillaz, P.-A.**, Immune regulation by leukotriene $B_4$. The biology of the leukotrienes, *Ann. N.Y. Acad. Sci. U.S.A.*, 524, 218, 1988.
35. **Johnson, H. M. and Torres, B. A.**, Leukotrienes, positive signals for regulation of interferon-γ production, *J. Immunol.*, 132, 413, 1984.
36. **Rola-Pleszczynski, M., Gagnon, L., and Sirois, P.**, Leukotriene $B_4$ augments human natural cytotoxic cell activity, *Biochem. Biophys. Res. Commun.*, 113, 531, 1983.
37. **Rola-Pleszczynski, M., Gagnon, L., Rudzinska, P., Borgeat, P., and Sirois, P.**, Human natural cytotoxic cell activity: enhancement by leukotrienes (LT) $A_4$, $B_4$, and $D_4$ but not by stereoisomers of $LTB_4$ or HETEs, *Prostagl. Leukotr. Med.*, 13, 113, 1984.
38. **Rola-Pleszczynski, M., Gagnon, L., and Sirois, P.**, Natural cytotoxic cell activity enhanced by leukotriene $B_4$: modulation by cyclooxygenase and lipoxygenase inhibitors, in *Icosanoids and Cancer*, Thaler-Dao, H., Ed., Raven Press, New York, 1984, 235.
39. **Gagnon, L., Girard, M., Sullivan, A. K., and Rola-Pleszczynski,M.**, Augmentation of human natural cytotoxic cell activity by leukotriene $B_4$ mediated by enhanced effector-target cell binding and increased lytic efficiency, *Cell. Immunol.*, 104, 243, 1987.
40. **Dinarello, C. A., Bishai, I., Rosenwasser, L. J., and Coceani, F.**, The influence of lipoxygenase inhibitors on the *in vitro* production of human leukocytic pyrogen and lymphocyte activating factor (interleukin-1), *Int. J. Immunopharmacol.*, 1, 43, 1984.
41. **Rola-Pleszczynski, M. and Lemaire, I.**, Leukotrienes augment interleukin-1 production by human monocytes, *J. Immunol.*, 135, 3958, 1985.
42. **Gagnon, L., Fillion, L., and Rola-Pleszczynski, M.**, Leukotriene $B_4$ augments tumor necrosis factor production by human monocytes, *Fed. Proc.*, 46, 6253, 1987.
43. **Gagnon, L., Fillion, L., Dubois, C., and Rola-Pleszczynski, M.**, Leukotrienes and macrophage activation: augmented cytotoxic activity and enhanced interleukin-1, tumor necrosis factor and hydrogen peroxide production, *Agents Actions*, 26, 141, 1989.
44. **Barbaro, J. F. and Zvailer, N. J.**, Antigen-induced histamine release from platelets of rabbits producing homologous PCA antibody, *Proc. Soc. Exp. Biol. Med.*, 122, 1245, 1966.
45. **Siraganian, R. P. and Bowman, B. J.**, Stimulation of human neutrophil degranulation with 1-*O*-octadecyl-2-*O*-acetyl-*sn*-glyceryl-3-phosphorylcholine: modulation by inhibitors of arachidonic acid metabolism, *Biochem. Biophys. Res. Commun.*, 104, 1495, 1982.
46. **Benveniste, J., Le Couedic, J. P., Polonsky, J., and Tencé, M.**, Structural analysis of purified platelet activating factor by lipases, *Nature (London)*, 269, 170, 1977.
47. **Benveniste, J., Tencé, M., Varenne, P., Bidault, J., Boullet, C., and Polonsky, J.**, Semi-synthesis and proposed structure of platelet-activating factor (PAF): PAF-acether and alkyl ether analog of lysophosphatidylcholine, *C.R. Acad. Sci.*, 289D, 1037, 1979.
48. **Demopoulos, C. A., Pinckard, R. N., Hanahan, D. J.**, Platelet-activating factor. Evidence for 1-*O*-alkyl-2-acetyl-*sn*-glyceryl-3-phosphorylcholine as the active component (a new class of lipid chemical mediators), *J. Biol. Chem.*, 254, 9355, 1979.
49. **Braquet, P., Touqui, L., Shen, T. Y., and Vargaftig, B. B.**, Perspectives in platelet activating factor research, *Pharmacol. Rev.*, 39, 97, 1987.
50. **Jouvin-Marche, E., Ninio, E., Beauvain, G., Tencé, M., Niaudet, P., and Benveniste, J.**, Biosynthesis of PAF-acether (platelet-activating factor). VII. Precursors of PAF-acether and acetyl-transferase activity in human leukocytes, *J. Immunol.*, 133, 892, 1984.
51. **Bussolino, F., Foa, R., Malavasi, F., Ferrando, M. L., and Camussi, G.**, Release of platelet-activating factor (PAF)-like material from human lymphoid cell lines, *Exp. Hematol.*, 12, 688, 1984.
52. **Malavasi, F., Tetta, C., Funaro, A., Bellone, G., Ferrero, E., and Caligaris-Cappio, F.**, $F_c$ receptor triggering induces expression of surface activation antigens and release of platelet-activating factor in large granular lymphocytes, *Proc. Natl. Acad. Sci. U.S.A.*, 83, 2443, 1986.
53. **Rola-Pleszczynski, M., Pignol, B., Pouliot, C., and Braquet, P.**, Inhibition of human lymphocyte proliferation and interleukin-2 production by platelet-activating factor (PAF-acether): reversal by a specific antagonist: BN52021, *Biochem. Biophys. Res. Commun.*, 142, 754, 1987.
54. **Dulioust, A., Vivier, E., Salem, P., Deryckz, S., Benveniste, J., and Thomas, Y.**, Inhibition of human $T4^+$ cell proliferation by PAF-acether (platelet-activating factor), *Fed. Proc.*, 46, 784, 1987.
55. **Behrens, T. and Goodwin, J. S.**, Effect of PAF on B and T cell proliferation, *Prostaglandins*, 34, 154, 1987.

56. **Rola-Pleszczynski, M. and Turcotte, S.,** Activation of suppressor lymphocytes and monocytes by PAF-acether, *Prostaglandins,* 34, 148, 1987.
57. **Rola-Pleszczynski, M., Pouliot, C., Turcotte, S., Pignol, B., Braquet, P., and Bouvrette, L.,** Platelet activating factor induces human suppressor cell activity with a concomitant increase in $CD8^+$ and decrease in $CD4^+$ T cell numbers, *J. Immunol.,* 140, 3547, 1988.
58. **Vivier, E., Salem, P., Dulioust, A., Metezeau, P., Benveniste, J., and Thomas, Y.,** PAF-acether (platelet-activating factor) alters the expression of T cell surface antigens, *Fed. Proc.,* 46, 1219, 1987.
59. **Ward, S. G., Lewis, G. P., and Westwick, J.,** Differential effect of platelet activating factor receptor antagonists on human T lymphocyte proliferation, *Prostaglandins,* 34, 149, 1987.
60. **Pignol, B., Hénane, S., Sorlin, B., Mencia-Huerta, J. M., and Braquet, P.,** Effect of long-term treatment with platelet-activating factor (PAF-acether) on interleukin-1 (IL-1) and interleukin-2 (IL-2) production in the rat, *Prostaglandins,* 34, 150, 1987.
61. **Pignol, B., Hénane, S., Sorlin, B., Rola-Pleszczynski, M., Mencia-Huerta, J. M., and Braquet, P.,** Effect of long-term treatment with platelet-activating factor (PAF) on interleukin-1 (IL-1) and interleukin-2 (IL-2) production by rat spleen cells, in *New Trends in Lipid Mediators Research,* Vol. 1, Braquet, P., Ed., S. Karger, Basel, 1988, 38.
62. **Rola-Pleszczynski, M., Pouliot, C., Pignol, B., and Braquet, P.,** PAF-acether augments human natural killer cell activity, *Prostaglandins,* 34, 156, 1987.
63. **Rola-Pleszczynski, M., Turcotte, S., Gagnon, L., Pignol, B., Braquet, P., Bolduc, D., and Bouvrette, L.,** Enhancement of natural killer (NK) cell activity by platelet-activating factor, in *New Trends in Lipid Mediators Research,* Vol. 1, Braquet, P., Ed., S. Karger, Basel, 1988, 89.
64. **Gebhardt, B. M., Braquet, P., Bazan, H. E. P., and Bazan, N. G.,** The effect of platelet-activating factor on cellular immune reactions *in vitro, Prostaglandins,* 34, 153, 1987.
65. **Foegh, M. and Ramwell, P.,** PAF and transplant immunology, *Prostaglandins,* 34, 186, 1987.
66. **Pignol, B., Hénane, S., Mencia-Huerta, J. M., Rola-Pleszczynski, M., and Braquet, P.,** Effect of platelet-activating factor (PAF-acether) and its specific receptor antagonist, BN 52021, on interleukin-1 (IL-1) release and synthesis by rat spleen adherent monocytes, *Prostaglandins,* 33, 931, 1987.
67. **Salem, P., Dulioust, A., Deryckz, S., Vivier, E., Benveniste, J., and Thomas, Y.,** PAF-acether (Platelet-activating factor) increases interleukin-1 (IL-1) secretion by human monocytes, *Fed. Proc.,* 46, 922, 1987.
68. **Ward, S. G., Lewis, G. P., and Westwick, J.,** Further characterisation of platelet-activating factor-induced IL 1 production by human adherent monocytes, *Prostaglandins,* 34, 151, 1987.
69. **Rola-Pleszczynski, M., Bossé, J., Bissonnette, E., and Dubois, C.,** Augmentation of tumor necrosis factor production by monocytes and alveolar macrophages in response to PAF-acether, *Prostaglandins,* 35, 802, 1988.
70. **Rola-Pleszczynski, M., Gingras, D., and Poubelle, P.,** PAF-acether enhances the production of tumor necrosis factor and interleukin 1 by human monocytes, *NER Allergy Proc.,* 9, 263, 1988.
71. **Dubois, C., Bissonnette, E., and Rola-Pleszczynski, M.,** Platelet-activating factor stimulates tumor necrosis factor production by alveolar macrophages: prevention by PAF receptor antagonists and lipoxygenase inhibitors, *J. Immunol.,* 143, in press, 1989.

Chapter 11

# IMMUNOREGULATORY ACTIVITY OF ENDOGENOUS OPIOIDS

**Deborah V. Harbour and Eric M. Smith**

## TABLE OF CONTENTS

# I. INTRODUCTION

There appear to be many ways that the immune and nervous systems can communicate. In general, it seems that this may be through the action of neurotransmitters in innervated lymphoid organs or by soluble mediators such as neuroendocrine hormones. Many of the chapters in this volume consider these separately. There is one class of factors, the endogenous opiates, that links both categories with the immune system. They are found in neurons and glandular tissues. Furthermore, endogenous opiates appear to be synthesized by the immune system as well. This chapter will review the studies suggesting that endogenous opiates can modulate immune functions, not only as regulation by the nervous/neuroendocrine systems, but as autoregulatory factors derived from lymphoid cells themselves.

# II. ENDOGENOUS OPIATES

Endorphins were discovered in the 1970s and along with enkephalins have been referred to as endogenous opioids.[1-3] They are a family of related opioid peptides that were originally isolated from brain tissue and found to have potent analgesic and euphoric activities. The endorphins and enkephalins are synthesized from different genes as large inactive polypeptide precursors termed proopiomelanocortin, proenkephalin, and prodynorphin.[3-5] These large parent molecules are further processed through specific proteolytic cleavages into smaller peptides. The opioid activity is present in the N-terminal amino acids which are common to all the neuroactive enkephalins and endorphins. Proopiomelanocortin (POMC), a 31,000-Da polypeptide encodes adrenocorticotropin (ACTH) from its central portion and beta ($\beta$) lipotropin ($\beta$-LPH) in the amino terminus of POMC (see Figure 1). The $\beta$-LPH is a 91 amino acid inactive precursor and contains the sequences for $\beta$-endorphin ($\beta$-END) ($\beta$-LPH 61-91); gamma ($\gamma$)-END ($\beta$-LPH 61-77); delta ($\delta$)-END ($\beta$-LPH 61-87); and alpha ($\alpha$)-END ($\beta$-LPH 61-76). The processing of POMC occurs differently depending on the tissue location seemingly as a result of tissue specific proteases.[6,7] For example, in the anterior pituitary, the major opioid product is $\beta$-LPH which as an opioid is biologically inactive. In the mouse intermediate lobe, this product is rapidly cleaved into $\beta$-endorphin which contains the biologically active Tyr-Gly-Gly-Phe at its N-terminal sequence. In addition to differential cleavages there is also differential acetylation of the $\beta$-endorphin in different tissues (for review see Reference 6). The acetylated $\beta$-endorphin in the intermediate lobe is 1/1000, as active an opiate as the unacetylated form. POMC synthesis has been reported in other tissues besides pituitary and brain. These include placenta, ovarian granulosa and testicular cells, gastrointestinal tract, adrenal glands, mouse spleen and human peripheral blood lymphocytes, and lung cells.[6] Processing, when it occurs, in extrapituitary tissues generally appears to resemble the intermediate lobe where the major opiate product is $\beta$-endorphin.

Pro-enkephalin A is processed to Met-enkephalin (Tyr-Gly-Gly-Phe-Met) and longer Met-enkephalin (ENK) species, as well as Leu-ENK (Tyr-Gly-Gly-Phe-Leu) as the major products.[3-5] Besides the pituitary and the brain, another major source of ENKs is the adrenal medulla (but generally in a different distribution than END) and lymphocytes.[8,9] As with POMC, there is differential processing of the proenkephalin precursor and its products dpending on the tissue of location.[7]

Prodynorphin or pro-enkephalin B is the most recently discovered member of the opioid family and contains the sequence of Leu-ENK, dynorphins A and B, and $\alpha$-neo-END.[3-5] The peptides are stored and coreleased with vasopressin in the posterior pituitary and in gonadotrophs of the anterior lobe. Various cleavage products are found in brain and adrenal medulla, again implicating tissue specific proteases in the ultimate bioactive end products.[7]

The differential processing provides a family of peptides capable of bringing about diverse biological effects, not all of which are opioid in nature. The opioid peptides appear

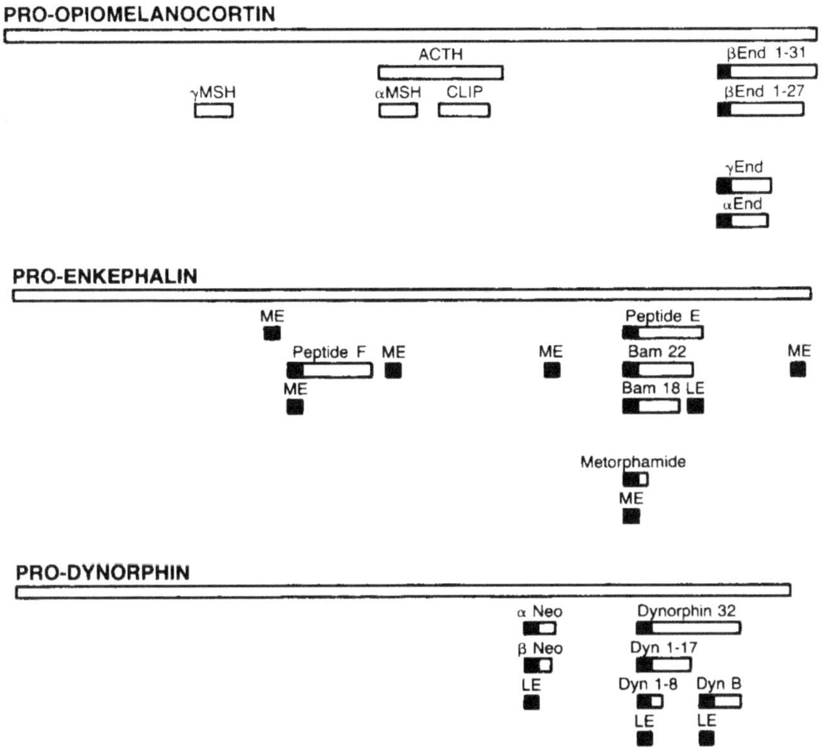


FIGURE 1. The opioid family of prohormones: proopiomelanocortin, proenkephalin, and prodynorphin. The opioid active $NH_2$ terminal Tyr-Gly-Gly-Phe is in solid regions. MSH = melanocyte stimulating hormone; ACTH = adrenocorticotropic hormone; END = endorphin; ME = methionine enkephalin; Leu = leucine enkephalin; Dyn = dynorphin; Neo = neoendorphin. Based on information from Reference 5.

to have various modes of action on the target tissue including paracrine, endocrine, neurotransmitter, and neuroendocrine pathways.[6,10] On a molecular basis this may be brought about by interactions with a single specific receptor from a heterogeneous family of opiate receptors present on various target cells including those of the immune system.[11] This will be discussed in detail in a later section. The opioid family of peptides function directly and indirectly in a regulatory capacity in peripheral and central homeostatic mechanisms.[4,6,8,10] This is particularly true with regards to the physiologic response to stress in which circulating levels of endorphins and enkephalins are both elevated.[12] Their pharmacologic role in analgesia is well documented, particularly with the potent opioid β-END which appears to mediate the majority of its effects through the mu (μ-morphine) and epsilon (ε) opiate receptors.[6,10,13] The other opioid peptides are less active. In addition, the μ-receptor has also been shown to mediate euphoria, respiratory depression, changes in body temperature, and catalepsy.[10,12] The delta (δ) receptor mediates similar actions and Leu-ENK appears to bind preferentially here.[10,11] Dynorphin on the other hand is the proposed endogenous ligand for the kappa (κ) opiate receptor.[10,13] These preferences with regard to binding are not absolute, as the endogenous opioids appear to bind to all the opiate receptors, but with varying affinity.

The END and ENK have been shown to facilitate memory by aquisition of avoidance behavior and maintenance of that learned behavior, plus they have neuroleptic or excitory behavioral effects.[6,14,15] Increased endogenous opioid activity has been implicated in the pathophysiology of schizophrenia.[6] In addition to regulatory homeostatic pathways, pain modulation, and behavioral effects, there is a well-documented endogenous opioid component that mediates in part the pathophysiologic response to shock, in particular endotoxic shock.[10,12] Thus it appears that the endogenous opioid families have a multitude of effects, both beneficial and detrimental, mediated through a heterogeneous family of opiate receptors. Therefore, in light of these multitude of locations and activities it is not surprising that the opioid peptides have in recent years been shown to act on the immune system as well (Table 1).

## III. ACTIONS OF ENDOGENOUS OPIATES IN THE IMMUNE SYSTEM

### A. HUMORAL IMMUNE FUNCTION

Some of the earliest studies on the effects of various opioid peptides on immune responses were reported by Johnson et al. in which they measured the *in vitro* plaque forming response (PFC) to T-dependent and -independent antigens.[16] The results showed α-END ($5 \times 10^{-7}$ *M*) was a potent suppresser of the PFC response (≥80%), and the inhibition was reversed by naloxone and $2 \times 10^{-6}$ *M* β-END implicating an opiate receptor mediated event. Met- and Leu-ENK were less active (60% at $2 \times 10^{-6}$ to $2 \times 10^{-7}$ *M*, respectively) in the PFC suppression and β- and γ-END were only weakly suppressive. Heijnen et al.[17] confirmed the immunosuppressive effects of α-END on the PFC response at concentrations of $5 \times 10^{-6}$ to $5 \times 10^{-8}$ *M*. β-END was minimally active only at $5 \times 10^{-6}$ *M*.[17] These authors propose the α-END inhibition of the PFC response is mediated through two pathways. First, α-END acts on T-cells by inhibiting the secretion of T helper factor. Second, it acts on B-cells (plasmacytes) by inhibiting antibody secretion. In that same study, Heijnen et al. showed enhancement of PFC with low concentrations of β-END. Thus there was a bimodal response, the end result being dose dependent.[17] Other investigators have shown that high concentrations ($10^{-4}$ to $10^{-6}$ *M*) of morphine, β-END, and Met-ENK modulate IgE-mediated release from mast cells through opioid receptors.[18] Although there are no *in vivo* studies of opiates on antibody production, studies of drug addicts suggest that antibody production is affected, either directly or through action on T lymphocytes.[19,20]

### B. CELL-MEDIATED IMMUNE FUNCTION

#### 1. Lymphocyte Mitogenesis

The role of opioid peptides in modulation of mitogen proliferative responses by lymphocytes[21-24] is a complex and controversial topic. Some investigators found β-END enhanced Con A- and PHA-induced proliferation,[22] while others showed $10^{-7}$ *M* β-END inhibited lymphocyte PHA proliferation *in vitro*.[23] In elucidating the mechanism of suppression of T-cell blastogenic response to mitogens, McCain et al.[25] have recently postulated that β-END ($10^{-7}$ *M*) functions similarly to Con A through activation of T-suppressor cells. They further showed that Con A and β-END acted synergistically in inducing the T-suppressor populations in human peripheral blood cells. Additionally, these authors showed that the carboxy terminal portion of β-END, glycyl-L-glutamine, inhibits Con A induction of suppressor T-cells.[25] We will return to this observation in the following sections. α-END appears to have no effect on T-cell mitogenesis.[26] Met- and Leu-ENK have been reported to enhance the PHA proliferation response of human peripheral blood lymphocytes,[26] but Leu-ENK inhibited an optimal dose PHA response.[27] Paradoxically, morphine and naloxone decrease T-cell proliferation to PHA at very high doses ($10^{-3}$ and $10^{-4}$ *M*, respectively) and both increase the response at lower doses ($10^{-6}$ to $10^{-14}$ *M* and $10^{-7}$ to $10^{-15}$ *M*, respectively).[28] The diverse *in vitro* responses were paralleled by diverse *in vivo* responses. Berkenbosch et al. have shown that the responses to the mitogens Con A (T-cell mitogen) and LPS (B-cell mitogen) are different with different forms of stress.[29] Using different stress paradigms they showed that stress leading to a conflict situation resulted in decreases in

## TABLE 1
## Opioid Peptide Modulation of Immune Function

| Immune function | Opioid peptide | Effect | Ref. |
|---|---|---|---|
| Humoral | | | |
| 1. PFC (ab response) | α-END (5 × $10^{-7}$ *M*) | − − − | 16, 17 |
| | Met-ENK (2 × $10^{-6}$ *M*) | − − | 16 |
| | Leu-ENK (2 × $10^{-7}$ *M*) | − − | 16 |
| | β-END (5 × $10^{-6}$ *M*) | −[w] | 16 |
| | β-END (5 × $10^{-8}$ *M*) | + | 17 |
| 2. IgE-mediated serotonin release | β—END | + | 18 |
| | Met-ENK | + | 18 |
| Cellular | | | |
| 1. Mitogenesis | β-END ($10^{-7}$ *M*) | − − | 22, 23 |
| | β-END ($10^{-7}$ *M*) | + | 24 |
| | Glycyl-L-Glutamine | NE | 25 |
| | α-END | NE | 24 |
| | Met-, Leu-ENK | + | 26 |
| | Leu-ENK | − | 27 |
| | $10^{-3}$ Morphine, $10^{-4}$ Naloxone | − − | 28 |
| | $10^{-6}$ to $10^{-14}$ Morphine | + | 28 |
| | $10^{-7}$ to $10^{-15}$ Naloxone | + | 28 |
| 2. Receptor modulation E-rosette | Morphine ($10^{-9}$ *M*) | − | 30 |
| | Met-ENK | − | 30 |
| | Leu-, Met-ENK | + | 31, 32 |
| IL-2 | β-END | NE | 33 |
| | Met-ENK | + | 34 |
| 3. Lymphokine modulation | | | |
| IFN (*in vivo*) | Morphine | − | 35 |
| IFN (*in vitro*) | Morphine | − | 36 |
| IFN (*in vitro*) | β-END | + | 37 |
| IFN (*in vitro*) | β-END | + | 38 |
| IFN (*in vitro*) | Met—ENK | + | 38 |
| LCF | β-END | − | 39 |
| LCF | Met—ENK | − | 39 |
| LCF | α-END | NE | 39 |
| 4. Cellular effector function | | | |
| NK *in vivo* | Inescapable shock | − − | 42—45 |
| NK *in vitro* | Inescapable shock | − − | 42—45 |
| NK *in vitro* | Morphine | NE | 45, 46 |
| NK *in vitro* | Levorphanol, detrophan | − − | 45—47 |
| NK *in vitro* | β-END, γ-END | + + | 24, 48—53 |
| NK *in vitro* | Met-END | + + | 24, 48—537 |
| NK *in vitro* | α-END | NE | 24, 48—537 |
| CTL | β-END | + | 51 |
| 5. Macrophage, granulocyte | | | |
| Chemotaxis *in vitro* | β-END ($10^{-8}$ to $10^{-12}$ M) | + + | 53, 54 |
| | Met-ENK ($10^{-8}$ to $10^{-12}$ M) | + + | |
| Chemotaxis *in vivo* | β-END | + | 55 |
| Phagocytosis | Met-ENK (low dose) | + | 56, 57 |
| ADCC | Met-ENK (low dose) | + | 56, 57 |
| Phagocytosis | Met-ENK (high dose) | − | 56, 57 |
| ADCC | Met-ENK (high dose) | − | 56, 57 |
| Peritoneal macrophage killing | dynorphin | + | 54, 58 |

*Note:* Effects are as follows: ([w]) — weak (−) — slightly inhibited (− −) — moderately inhibited (− − −) — strongly inhibited (+) — slightly enhanced (+ +) — moderately enhanced (+ + +) — strongly enhanced both T- and B-cell compartments, whereas a rapid adverse reaction may result in rapid increase in immune reactivity, particularly the T-cell compartment. This type of experimentation points to the fact that the nature and duration of the stress may result in the release of different neurochemicals (ENK, END, catecholamines, ACTH, etc.) which account for the diverse *in vivo* effects.

### 2. T-Cell Surface Receptor Modulation

In addition to the modulation of effector cell growth in response to mitogens, there is also some evidence that opiates may play a role in the regulation and modulation of E-rosette receptors on cells. Donahoe and colleagues have done elegant experiments showing down regulation of the E-rosette receptor through capping and endocytosis in response to opioid ligand binding (for review see Reference 30). This effect with morphine ($10^{-9}$ *M*) and Met-ENK are both reversible by naloxone and may be mediated by the effect of opioid peptides on E-rosette receptor recycling mechanisms. They point out that response to stress, in which the endogenous opioids may be elevated, may alter receptor expression and lead to homeostatic changes through a change in T-cell effector function. They propose this could happen since the E-rosette receptor is functionally linked to expression of the T-cell antigen receptor, Ia antigens, Fc receptors, IL-2 receptors, and γ-interferon receptors.[30] Thus one can readily see how opioid mediated modulation of the E-rosette receptor could have significant effects on immune function. Other investigators have reported that Leu- and Met-ENK enhance T-rosettes[31,32] and possibly IL-2 and Ia receptor expression during lymphocyte activation. Again the effects are reversible with naloxone and probably occur through opiate receptors.[34]

### 3. Modulation of Lymphokines

The lymphokine studied most extensively to date has been interferon. Hung et al. originally showed that the narcotics morphine, dihydromorphine, and methadone effected a decrease in circulating interferon (IFN) levels.[35] Further, they showed that polyinosine polycytotidylic acid (poly I:C)-induced IFN induction was suppressed by morphine, and this suppression was reversed by naloxone.[36] As with other immune parameters modulated by opioid peptides there are contradictory results with α-IFN modulation in that another opioid, β-END, will enhance PHA or poly I:C-induction of IFN.[37] Consistent with these reports, Brown and Van Epps were able to show modulation of *in vitro* γ-IFN production by Met-ENK and β-END.[38] Additionally, the authors showed Met-ENK and β-END, but not α-END suppressed the production of T lymphocyte chemotactic factor (LCF) produced by Con A-stimulated peripheral blood mononuclear cells. They postulated prostaglandins are the effector molecules in this pathway.[39] It is interesting that IFN has been shown to have opiate-like activities[40] and so it may not be just fortuitous that opiates can modulate IFN induction.

### 4. T-Cell and NK Cell Effector Mechanisms

The fact that the perception of stress is important with regards to immune response is evidenced by the "escapable vs. inescapable" foot shock animal model. Inescapable shock activates the endogenous opioid system and has a high incidence of tumor development associated with it, whereas the escapable foot shock does not cause activation of the opioid cascade and has no increased tumor development.[41-44] In addition to tumor development, rats given inescapable shock had a significantly more suppressed response to the mitogens phytohemagglutinin (PHA) and concanavalin A (Con A) than those rats unstressed or stressed with escapable shock.[21]

The role of endogenous opiates in these described processes are varied and at times seem contradictory. Shavit and co-workers have proposed that inhibition of natural killer

(NK) cell cytotoxicity is one mechanism whereby stress involving endogenous opioid elevations cause an enhancement of tumor development.[45] They tested this hypothesis by exposing rats to intermittent footshock (opioid releasing) and continuous footshock (non-opioid-induced analgesia). After the stress paradigm, the investigators found that *in vitro* spleen NK function from the opioid-releasing stressed rats was significantly inhibited. In addition, they found a consistent suppression of NK function in rats treated with morphine in doses high enough to enhance mammary tumor growth. They showed opioid specificity by reversing the NK suppressive effects of both the footshock and the morphine by prior *in vivo* injection of naltrexone, an opiate antagonist.[46] So these authors concluded that enhanced tumor growth may be mediated through opioid suppression of NK function which may be dependent on dose and route of entry of the morphine. This *in vivo* immunosuppressive effect is not necessarily verified by *in vitro* studies. In the author's own hands, *in vitro* exposure of NK cells to high dose morphine had no effect on NK function. However, levorphanol and dextrorphan, opiate agonists and antagonists, respectively, both suppressed *in vitro* NK function.[47] Other investigators have shown enhancement of *in vitro* NK function with β-endorphin, Met-ENK, and γ-endorphin, but not α-END,[48-50] and these effects were reversible with naloxone. In addition, recent evidence shows enhancement of cytotoxic T-cell (CTL) activity by β-END.[51] These authors postulated that mechanisms for the enhancement include increased recycling of effector cells and increased generation of NK and/or CTL effectors. It seems clear that the effects of opioids on the natural killing effector function of cellular immunity and *in vivo* tumor growth are very complex and are probably at least dependent on the nature of the stress, the type and location of the tumor, and the concentration and type of endogenous opiate involved (i.e., β endorphin vs. α-endorphin vs. Met-ENK, etc.). This last point is important since the opiates obviously mediate at least some of their effects on immune cells through specific opiate receptors on these cells. Recent experimentation using fragments of des-tyr endorphin (Tyrosine at position 1 removed) have mapped the site of maximum NK activity to amino acid fragment 6 to 9 the site of α-helix formation.[52] Of course different opiates may bind to different receptors with various affinities thereby resulting in diverse effects. There is some evidence that not all effects of opioid peptides on immune function are mediated through opiate receptors, but may function at different receptors. As recently reviewed by Solomon et al.,[24] the fact that Met-ENK (1 to 5) and γ-endorphin (1 to 17) enhance NK function, but α-endorphin (1 to 16) does not, plus the fact that not all effects are reversible with naloxone, provides evidence for a nonopiate receptor mechanism, possibly mediated through the carboxy terminal end of the species.

### 5. Macrophage and Granulocyte Cell Function

Several investigations have shown that endogenous opioid peptides stimulate a number of phagocytic functions. Van Epps and Saland reported that β-END and Met-ENK stimulated in doses of $10^{-12}$ and $10^{-8}$ *M*, human peripheral blood mononuclear chemotaxis to 80% of that observed with formyl-methionyl-leucyl-phenylalanine (FMLP), a classical chemoattractant.[53] Higher doses and lower doses of the immunomodulators had different results, suggesting a multiple receptor effect. The effects were reversible with naloxone and could be mimicked *in vivo* by injections of β-END into rat cerebral ventricles, resulting in migration of macrophages.[55] The neutrophil chemotaxis induced by Met-ENK and β-END was only 30% of FMLP's induction. However, β-END enhanced neutrophil migration to FMLP and opioid peptides mediate (through opiate receptor mechanisms) morphologic changes in these cells.[54] This suggests that FMLP and opiates modulate chemotaxis through separate receptors. Met-ENK has also been shown to enhance phagocytosis, IgG2a-mediated antibody dependent cellular cytotoxicity (ADCC) (for review see Reference 56), and chemiluminescence.[57] At higher concentrations Met-ENK inhibits phagocytic and ADCC functions. This is probably due to effects on different second messengers.[56] Recently, Foster and Moore[58] showed that

**TABLE 2**
**Receptors for Opioid Peptides on Cells of the Immune System**

| Cell | Radioligand | Kd or competor | Ref. |
|---|---|---|---|
| | | **Classical opiate receptors** | |
| Mouse spleen cells | Met-ENK | $5.9 \times 10^{-10}$ *M* | 59 |
| Human PBL and platelets | Naloxone | Morphine | 60 |
| Human granulocytes | Dihydromorphine | $1 \times 10^{-8}$ *M* | 61 |
| Human monocytes | Dihydromorphine | $8 \times 10^{-8}$ *M* | 61 |
| T-cells | β-END | Naloxone | 33 |
| Human PMN | β-END | β-END | 62 |
| Mouse spleen cells | β-END | Naloxone | 49, 63 |
| Mouse spleen cells | α-END | Naloxone or modification of amino terminus | 19<br>19 |
| | | **Nonopiate receptors** | |
| Cultured lymphocytes (B-cell) | β-END | $3 \times 10^{-9}$ *M* | 64 |
| Thymoma cells | β-END | β-END only | 65, 66 |

dynorphin and related peptides enhanced cytolytic function in activated peritoneal macrophages. This was reversible by naloxone implicating a κ, μ or δ opiate receptor on the effector cell.

These data taken together verify that the endogenous opioid peptides have definite and specific effects on various compartments of the immune response. The opiates may be immunoenhancing, immunosuppressive, or have no effect. The results at times appear contradictory but a closer examination of the experimental parameters such as effector concentration, *in vivo* vs. *in vitro* experiments, and source of cells may elucidate the specific immunoregulatory pathways. Further, the exact mechanisms whereby such strikingly similar peptides produce diverse immunomodulatory activities may be related to the type of receptor on the target cell (opiate vs. nonopiate), the cell type of the target (T, B, macrophage, or neutrophil) and the second messenger system activated on the target (calcium, cyclic nucleotide, phosphotidyl inositol, etc.), or the presence of other neurotransmitters in the microenvironment.

## C. MOLECULAR MECHANISMS OF OPIOID PEPTIDE CONTROL

### 1. Opiate Receptors on Cells of the Immune System

As stated previously, the immunomodulation effected by neuropeptides appears to be mediated in part through specific receptors on leukocytes (see Table 2). Wybran et al.[59] and Hazum et al.,[64] both in 1979, were the first to report opioid receptors on lymphocytes. Wybran and others related that Met- and Leu-ENK enchanced active T-cell rosettes with cells from normal and lymphoma patients.[31,32] The proposal was made that the ENK receptor was very close to or similar to the E-rosette receptor on human T lymphocytes which allowed for steric mechanisms to bring about the E-rosette effect.[59] It has subsequently been shown by Johnson et al. that Met-ENK binds to mouse spleen cells at a specific opiate receptor (Kd $5.9 \times 10^{-10}$ *M*) which might be a δ-receptor (since Met-ENK binds with highest affinity to δ-receptors).[16] Numerous other investigators have reported modulation of NK and T-cell function by the enkephalins, however, no radioligand binding studies were performed to verify specific binding of the enkephalins.[26,32,67] Mehrishi and Mills[60] showed specific opioid receptor binding on human peripheral blood lymphocytes and platelets with $^3$H-naloxone, displaceable 43 to 57% by morphine. The authors concluded that μ-opiate receptors are present on at least a subpopulation of leukocytes. Other investigators have demonstrated saturable binding of $^3$H-dihydromorphine to human granulocytes (Kd 10 n$M$, 1000 sites per cell) and monocytes (Kd 8 n$M$, 3000 sites per cell).[61] These findings are further supported by the leukocyte migration studies of Van Epps et al.[68] in which they showed that naloxone inhibited chemotaxis to β-END and Met-ENK. Falke and Fisher have recently studied the molecular mechanisms of the early events of $^{125}$I-β-END binding to polymorphonuclear leukocytes (PMN).[62] Using ultrastructural autoradiography the investigators were able to show that β-END specifically bound to the membrane was internalized within 2 min at 37°C which paralleled the rapid change in morphology in these cells after treatment with β-END. Farrar has reviewed the results of binding studies on T-, B-, and NK cell lines using ligands specific for different opiate receptors.[33] The results show that T- and NK cells showed similar patterns of binding since β-endorphin was able to compete with $^3$H-etorphin for sites on the cells of perhaps ε, μ, or κ types. The B-cell line on the other hand demonstrated a different pattern of binding since broad spectrum diprenorphin competed effectively for binding to these cells and β-END did not. The conclusion was the B-cells demonstrated a classical opiate receptor that was quite different than that on T-cells. The results also showed a parallel increase in IL-2 receptors with the opiate-like receptors suggesting concomitant immunoregulatory activities of opioid peptides with IL-2. Therefore, it appears that leukocytes possess classical opiate receptors, μ and δ, with possibly ε and κ. These receptors may be differentially expressed on cells from different compartments of the immune system. Carr et al. has recently purified an opiate receptor present on murine splenocytes.[63] *In situ* cross-linking allowed purification by electrophoresis and overlaying with $^{125}$I-β-END allowed the authors to identify a major binding chain of 46 kDa and a minor binding chain of 31 kDa. They further showed complete loss of binding if the labeling was performed in the presence of excess naloxone which verified the specificity of opiate receptors on immune cells resembles that of neural cells.

Cells of the immune system have also been shown to possess nontypical opiate receptors. This was suggested in the previous section by several studies showing that effects of several opioid peptides were not reversed by naloxone or morphine. Hazum and co-workers were the first to show specific binding of radiolabeled β-END (Kd $3 \times 10^{-9}$ $M$) to a B lymphocyte cell line.[64] This binding could not be inhibited by naloxone, morphine, or α-endorphin, and only weakly by Met- and Leu-ENK. Complete inhibition in dose dependent fashion was only by β-END and a β-END derivative. The suggestion was that the carboxyterminal end of the β-END molecule was important for binding, thus implicating a non opioid receptor for β-END on cells of the immune system. Thus, it appears the immune system interacts with opioid peptides in the classical opioid method as do nerve cells, but it also has a unique method for interaction (at least for β-END) that nerve cells apparently do not use. Further evidence for this nonopioid receptor came from Schweigerer and co-workers.[65] These authors demonstrated β-END binding to a 72,000 MW protein on the surface of EL4 thymoma cells. The specificity of binding appeared to reside in the carboxyterminal portion of β-END. The β-END was subsequently internalized and the 72K extracellular protein was down regulated suggesting that the β-END may regulate immune function at an internal site. Previously, these authors had shown specific β-END binding to complement complexes.[68] Again this binding was dependent on the carboxyterminal end of the β-END since it was not blocked by naloxone. Taken together this evidence points to the fact in some cases β-END may preferentially bind to specific receptors on some immune cells, unique from the brain opioid receptors, μ and δ.

α-END apparently binds to a classical opiate receptor, probably μ. Indirect evidence for this postulate was provided by Johnson et al., when they showed that α-END's suppression of ab response was reversed by naloxone.[16] Heijnen and colleagues further strengthened this postulate when they showed that modification of the amino terminal end of α-END abrogated its suppressive function in the PFC assay.[17] A rationale for these diverse effects

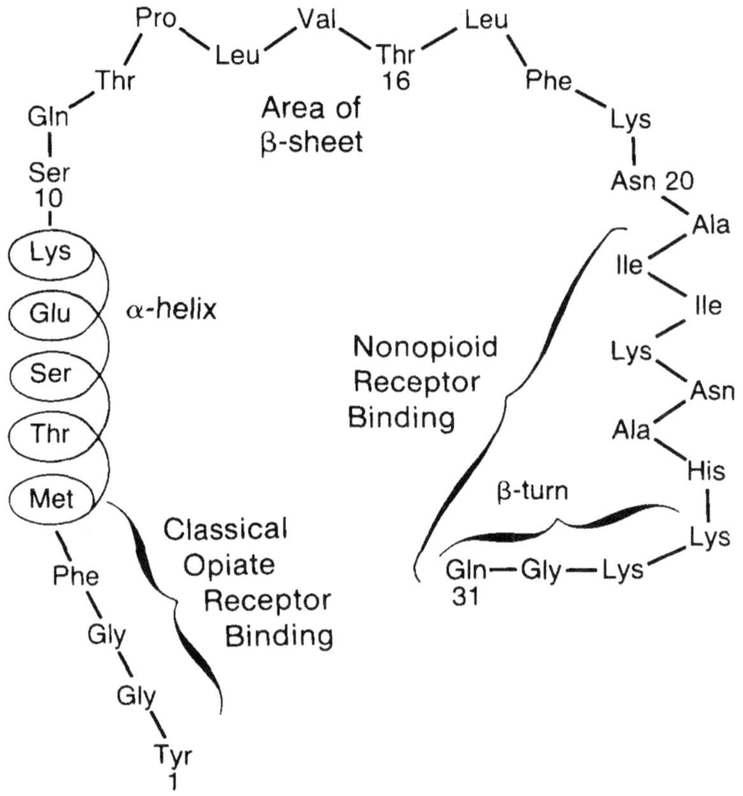


FIGURE 2. Structure of the β-END molecule with the two amino putative binding regions. The terminal region shows classical opiate binding region. The carboxyl terminal region shows the non opiate binding region. Based on information in Reference 24.

may be postulated in that β-END preferentially binds to a unique nonopioid receptor and enhances antibody response, whereas α-END, and under certain conditions β-END, bind to a classical opioid receptor suppressing the antibody response.

All of the above studies have led investigators to propose several theories of opioid peptide-receptor immunomodulation. Wybran and co-workers in 1980 published the two hit opioid peptide lymphocyte receptor hypothesis. They suggested that an opioid peptide, i.e., β-End, recognized T- and B-cells by two different determinants of itself — one opioid and one nonopioid[69] (see Figure 2). This mechanism, they postulated, may provide a type of chemical bridge between T and B lymphocytes, thereby providing for cooperation between the two cell types necessary in order for an adequate immune response. This hypothesis was followed by the Ying-Yang hypothesis by Plotnikoff in which he proposed that there are two different opioid receptors on T lymphocytes.[70] This hypothesis provided a mechanism whereby different forms of opioid peptides could cause diverse effects on immune cells. For example, Met- and Leu-ENK and β-END when bound to δ- and ε-receptors were mediators of upregulation of T-cells. Contrary to this, the larger opioid fragments and morphine (and probably α-END) bind to μ- and κ-receptors to function as down regulators of T-cell function. Plotnikoff further proposed that receptor cooperativity may occur to bring about heterogeneous responses as a result of different types of fragments being released in response to stress. In view of the experimental data, one can readily see that both postulates are certainly applicable and we can expand them now to include a direct opioid peptide action on B-cells through a specific nonopioid receptor.[64]

### 2. Opioid Activation of Subcellular Mechanisms in Cells of the Immune System

Met-enkephalin has been shown to modulate its effect on neural cells through specific δ-receptors negatively coupled to adenylate cyclase. The overall result is an elevation of cGMP.[71] This effect appears to be active over the range of concentrations $10^{-7}$ to $10^{-9}$ *M*. However, at very high concentrations of $10^{-6}$ *M*, the result is reversed and intracellular cAMP levels increase with clustering of the δ-receptor.[71] These same results, as well as $Ca^{2+}$ uptake (only at high concentrations $10^{-6}$ *M*) have been shown *in vitro* for peritoneal macrophages.[72] These diverse intracellular mechanisms occurred with different effector functions, providing a molecular mechanism for concentration specific effects of opioid peptides on immune response. Furthering this concept, at lower Met-ENK concentrations ADCC activity was enhanced through Fc receptors, whereas phagocytosis activity decreased.[56] Conversely, at higher concentrations ($>10^{-6}$ *M* Met-ENK) both ADCC and phagocytic intracellular killing were diminished. The inhibitory effects were reversed by calmodulin, suggesting that there were several intracellular pathways involved in the immunosuppressive actions of Met-ENK.

β-endorphin has also been shown to inhibit adenylate cyclase elevations of cAMP in T-cells lines. Presumably, this was through opiate receptors since the effect was reversed with naloxone.[33] Endogenous prostaglandins have been implicated as intracellular mediators of the extracellular binding of β-endorphin to an opiate receptor on T-cells. One report provides evidence of β-endorphin abrogation of $PGE_2$, resulting in enhanced IL-2 dependent proliferation.[33] Another report suggests β-END enhances the $PGE_2$ suppressive effects of T-cells by inhibiting release of lymphocyte chemotactic factor (LCF).[39]

Opioid peptide-mediated intracellular signaling system studied in NK cells have shown calmodulin to be an important regulator of the β-END enhancement of NK activity since the calmodulin inhibitor naphthelene sulfonamide W13 reversed the β-END effect.[73] Others have shown dynorphin mediated enhancement of tumoricidal activity of peritoneal macrophages may occur as a result of modulation of calcium fluxes.[58]

Lastly, investigators have examined oxygen metabolism and superoxide ($O_2^-$) production and release in effector cells. One report[74] studied the effect of naloxone on neutrophil $O_2^-$ release. They measured $^3$H-naloxone binding and release of $O_2^-$ finding that naloxone at $10^{-6}$ *M* inhibited $O_2^-$ release through ligand-specific opiate receptor binding. In another study, β-end was found to stimulate $O_2^-$ release at physiologic concentrations ($10^{-14}$ to $10^{-8}$ *M*).[75] This effect was blocked by a (−)-naloxone stereoisomer. *N*-acetyl β-end was ineffective in inducing $O_2^-$ formation, again suggesting an opiate receptor mediated effector mechanism. Recent investigations by Carr et al. studied the influence of $K^+$ ion channel effects mediated by β-end.[63] The investigators used the patch clamp technique and showed that β-end at $7 \times 10^{-7}$ *M* concentration resulted in reduction of $K^+$ conductance by decreasing the potassium channel openings in human T-cells. This effect was mediated through the opiate receptor since channel openings reappeared after treatment with naloxone.[63] Therefore, it seems clear that the immune and neuroendocrine systems utilize similar intracellular second messenger systems for the mediation of target cell effects. It also seems plausible that as data in this area accumulate, unique pathways will be discovered for cells of the immune system, possibly through direct regulation of effector genes.[65]

## IV. OPIATE PRODUCTION BY THE IMMUNE SYSTEM

The opioid peptides have heterogenous actions depending in part upon the target cell type as well as the concentration of the effector.[10] The endogenous opiates function as neurotransmitters, defined by their release from neurons and subsequent actions on neurons. They also function as classical hormones by being released from the adrenal medulla and pituitary endocrine glands into the circulation and acting on remote target tissues. The opioid peptides have also been shown to have paracrine and possibly autocrine actions. For example, the ENK and β-END are released from neural cells in the brain and bind at opiate receptors in adjacent cells in a paracrine fashion, probably to modulate pain pathways. Recent evidence shows that enkephalins released from the adrenal medulla feed back and inhibit release of new enkephalins, therefore providing a mechanism of autoregulation (for review see Reference 10). The action of endogenous opiates from the neuroendocrine system on lymphocytes seems to fit an endocrine type of regulation. Our laboratory and others have found that the immune system provides a source of endogenous opioids, enkephalins, and endorphins, and recent evidence has shown that these peptides are capable of exerting biologic effects probably through similar pathways as those described above. It is important to consider these potential paracrine and autocrine pathways during the following discussion of opiate production by lymphoid cells.

## A. HYPOTHESIS FOR LPS INDUCTION OF LEUKOCYTE-DERIVED ENDORPHINS

Endogenous opioids have been implicated as possible mediators of some of the pathophysiological changes induced during endotoxic shock and Gram-negative sepsis.[76,77] This implication was a result of the ability of the potent opiate antagonist, naloxone, to alleviate endotoxin-induced hypotension and changes in body temperature by apparently blocking the effector molecule.[78,79] Additionally, naloxone treatment improved survival rates,[80] and plasma endorphin levels have been shown to rise significantly in sheep injected with *Escherichia coli* endotoxin (lipopolysaccharide [LPS]).[81,82] Holaday and co-workers have shown that this endogenous opioid component specifically binds to δ-opiate receptors which then induced the pathophysiologic responses.[10,83] They have shown that these responses can be mimicked *in vivo* by treatment with DADLE, a δ-agonist.

Blalock and Smith reported the production of ACTH and END-like molecules from Newcastle disease virus infected and corticotropin releasing factor (CRF) treated leukocytes.[84,85] In addition, Lolait and co-workers reported a similar immunoreactive (ir)-END present in mouse spleen macrophages.[86,87] This data is important when one considers that depletion studies indicate that leukocytes may also mediate some of the aforementioned endotoxin effects.[88,89] Therefore, we postulated that leukocytes may serve as an extrapituitary source of END-like molecules that are produced in response to bacterial endotoxin.

## B. LPS-INDUCED MONONUCLEAR CELL SYNTHESIS OF ir-ENDORPHINS

To test this hypothesis, we initially determined whether human mononuclear cells or mouse spleen cells were able to synthesize ir-END *in vitro* in response to LPS.[90] Radiolabeled material from LPS-treated cultures was purified by affinity chromatography on an anti-γ-endorphin antibody affinity column. Subsequent sizing of the LPS-induced radiolabled ir-endorphin by gel filtration showed a peak of radioactivity which comigrated with α- and γ-endorphin at approximately 1800 Da. A similar radiolabeled peak was not observed from control cultures. Therefore, LPS seemed to induce human mononuclear and mouse spleen cell production of a molecule that is antigenically and structurally related to α- or γ-endorphin. The fact that the ir-endorphin was intrinsically radiolabeled shows that it was synthesized *de novo*. As a means to assess its potential biological activity, the ir-endorphin was tested for the ability to bind opiate receptors. In an *in vitro* opiate receptor binding assay,[91] LPS-induced ir-endorphin, but not control culture material specifically inhibited $^3$H-dihydromorphine binding to opiate receptors from mouse brain tissue.[90] Thus, in addition to a similar molecular weight, the LPS-induced ir-endorphin has a biological activity similar to that of α- or γ-endorphin.

In order to determine whether the leukocyte-derived endorphin was produced concomitantly with ACTH and therefore probably a cleavage product from pro-opiomelanocortin,

**TABLE 3**
**Binding of Ir-Endorphin to the δ-Opiate Receptor**

| Sample | | Specifically bound $^3$H-DADLE CPM ± SD | % Inhibition of specific $^3$H-DADLE binding |
|---|---|---|---|
| B | Mock | 1025 ± 68 | 18.9 |
| | LPS-treated | 782 ± 121 | 99.3 |
| T | Mock | 1067 ± 90 | 4.7 |
| | LPS-treated | 1054 ± 137 | 9.1 |
| β-Endorphin | $10^{-6}$ | 833 ± 64 | 83.5 |
| | $10^{-8}$ | 940 ± 135 | 52.5 |
| | $10^{-10}$ | 1035 ± 157 | 15.5 |
| DADLE | (δ-Agonist) | 920 ± 49 | 54.2 |
| DAGO | (μ-Agonist) | 1139 ± 73 | 0 |
| Media | | 995 ± 184 | 28.9 |

*Note:* Sample ligands are as follows: T or B lymphocyte culture supernates at a 1:2 dilution, culture media alone, 50 μg/ml DAGO (D-Ala$^2$, N-Me-Phe$^4$, Met (O)$^5$-OL)-Enkephalin, 50 μg/ml DADLE (D-Ala$^2$-D-Leu$^5$)-Enkephalin, $10^{-5}$ M Naloxone, β-endorphin. The labeled ligand was $^3$H-DADLE, (Amersham specific activity 76.8 mCi/ng). NG 108 cells were plated in 96 well polystyrene microtiter plates at a concentration of $1 \times 10^5$ cells per well. The sample ligands were added in 100 μl volumes in triplicate and incubated for 30 min at 4°C. The cells were washed with phosphate buffered saline supplemented with 1% fetal bovine serum followed by incubation with the labeled ligand for 15 min at 4°C. The cells were washed three times and each well was counted in liquid scintillation fluid. Table is reproduced with permission from Reference 92.

we also tested the culture supernatant fluid for the presence of ACTH. Our results strongly suggested concomitant *de novo* synthesis of a bioactive ACTH (1 to 22 to 1 to 26)-like molecule produced *in vitro* by LPS treated mononuclear cells.[90] Thus, mononuclear cell production of these pro-opiomelanocortin derived peptides in response to LPS seemed analagous to that observed in the pituitary gland.

## C. B LYMPHOCYTE PRODUCTION OF BIOACTIVE ir-ENDORPHIN

In order to determine the cell type(s) producing the LPS induced ir-endorphin, we fractionated the cells into macrophage depleted enriched populations of T- and B-cells prior to treatment with LPS. Immunofluorescent analysis of cultured cells demonstrated that significant staining with the monospecific antibody to γ-endorphin occurs only in LPS treated B lymphocyte, but not T-cell enriched populations.[92]

Since ligand binding to the δ-opiate receptor has been shown to elicit similar pathophysiologic responses, hypotension and hypothermia, as seen in endotoxic shock, the T- and B-cell supernatant culture fluids were tested for the ability to bind δ-opiate receptors.[10,93] Radioreceptor assays were performed with the agonist DADLE on the neuroblastoma cell line NG108 which carries the δ-opiate receptor. The results are presented in Table 3 showing a significant inhibition of the labeled δ-agonist with the supernate of the B-cell enriched population treated with LPS and no significant inhibition with the T-cell-derived material.[92] Thus, it appears that B-cells are the major cell type producing the LPS induced ir-endorphin and this can bind the δ-opiate receptor, providing a mechanism whereby B-cell derived endorphins could effect pathophysiologic responses of endotoxic shock.

### D. *IN VIVO* PRODUCTION OF ir-ENDORPHIN PRECEDES PATHOPHYSIOLOGIC RESPONSES

We chose to further study the production of ir-endorphin by using C3H/HeJ, LPS resistant and C3HeB/FeJ, LPS sensitive imbred mice. It has been shown that C3H/HeJ mice do not experience the pathophysiologic responses seen in endotoxic shock.[94] Such resistance increases their survival time when exposed to endotoxin as compared to the C3HeB/FeJ mice. We postulated that lack of production of leukocyte endorphin may contribute to the lack of response and increased survival of the LPS resistant mice. To test this, we injected groups of LPS-sensitive and -resistant mice with LPS and monitored pathophysiologic changes (i.e., body temperature).[95] The *in vivo* response of the two strains of inbred mice became significantly different after 1 h. The LPS injected, but not media injected, C3HeB/FeJ mice presented with marked hypothermia and respiratory alterations. These pathophysiologic responses were preceded by positive immunofluorescence of the spleen cells stained with anti-γ-END from the C3HeB/FeJ mice (peak immunofluorescence was 30 min) suggesting that the *in vivo* production of ir-END may in part mediate some of LPS-induced pathophysiologic responses. In contrast, the LPS resistant C3H/HeJ mice did not manifest significant pathophysiologic changes nor did their splenocytes fluoresce when stained with antisera to γ-END.

### E. B LYMPHOCYTE-DERIVED ir-ENDORPHIN ELICITS ENDOTOXIC SHOCK-LIKE RESPONSES

It was necessary to demonstrate that the *in vivo* pathophysiologic effects observed by the administration of LPS were mediated by the released leukocyte-derived ir-endorphin. In order to determine this, affinity purified ir-endorphin produced by the *in vitro* stimulation of C3HeB/FeJ B cells with LPS was injected intraperitoneally into C3H/HeJ and C3HeB/FeJ mice.[95] As a control for *de novo* synthesis of the B-cell derived ir-endorphin, the mice were also injected with affinity purified material from the supernatant fluid of mock treated C3HeB/FeJ B-cells. The results are depicted in Figure 3 showing that the injected LPS-treated B-cell derived END, but not the mock elicits a similar hypothermic response pattern as seen when mice are injected with LPS. This hyperthermia was coupled with lethargy, an observed reduction in respiration rate, catatonia, and discharge from the eyes. The fact that the LPS resistant mice are capable of presenting with similar pathophysiologic responses as those observed in the LPS sensitive mice indicates that these mice are capable of responding to the endorphins, but they probably do not synthesize the eliciting opioid components when injected with the LPS. This evidence strongly suggests that the lymphocyte-derived ir-endorphin mediates some of the pathophysiologic responses seen in endotoxic shock.

### F. ALTERNATIVE PROCESSING OF POMC IN THE IMMUNE SYSTEM

An interesting finding from our results of purification of the LPS induced ir-endorphin was the possibility of alternative processing of the POMC-derived peptides as shown in Table 4. NDV or CRF elicit the production of mRNA, and full-length ACTH 1 to 39 and β-endorphin from cultured leukocytes.[84,85,96] LPS elicits the production of ACTH and endorphins which correspond to the molecular weight of ACTH 1 to 22 to 26 and α- or γ-endorphin from B lymphocytes.[90] We have hypothesized that the truncated ACTH and endorphins represent novel proteolytic cleavage products from ACTH 1 to 39 and β-endorphin.[90]

This alternative processing was confirmed in the LPS mouse model in which we showed that CRF induced the LPS-sensitive mouse spleen cell production of β-END, whereas LPS induced a smaller species, α- or γ-End size and ACTH (1 to 22 to 29) as the major species with ACTH 1 to 39 as the minor species (Reference 97 and 108). Taken together, these data suggest the existence of stimulus dependent processing of POMC by splenocytes. This

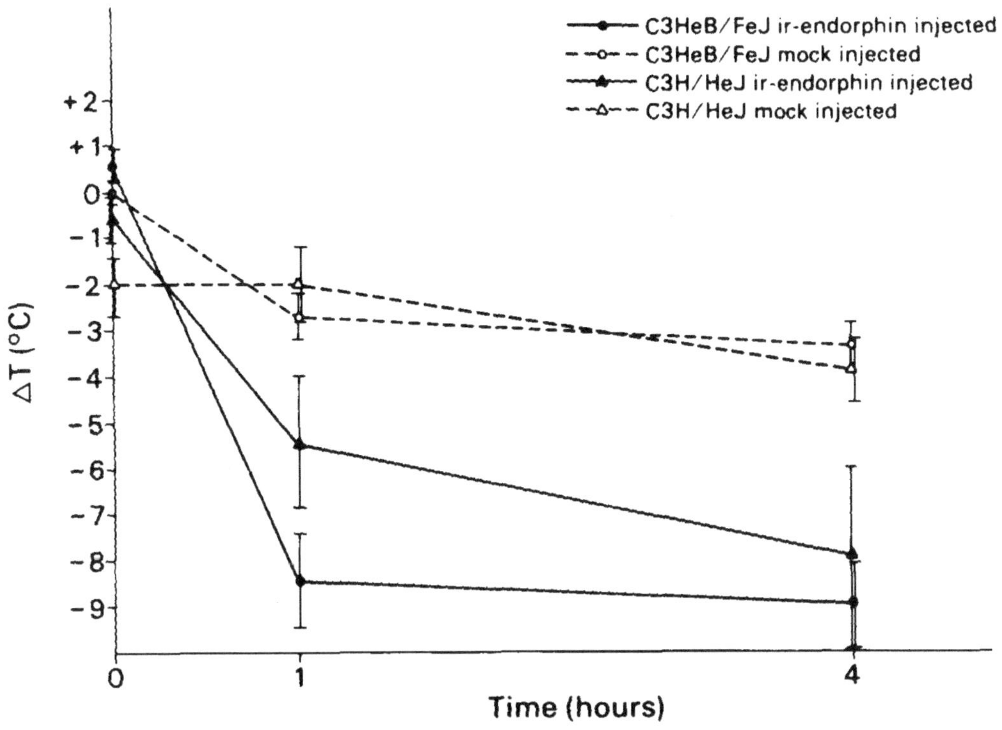


FIGURE 3. B-cell-derived ir-endorphin elicits body temperature changes in LPS-sensitive and -resistant mice. C3H/HeJ or C3HeB/FeJ mice were each injected with 0.2 ml of concentrated fractions bound to the anti-γ-endorphin affinity column from either C3HeB/FeJ B-cells stimulated with LPS- or mock-induced cultures. Rectal temperatures in each group of mice were recorded in the same order/time point. Total number of mice in each group was 4. Data points represent average temperatures ± standard deviation.[93] Reproduced with permission from Reference 95.

**TABLE 4**
**Stimulus-Dependent Production and Processing of Leukocyte-Derived Endorphins and Enkephalins**

| Cell source | Stimulus | Major product | Reference |
|---|---|---|---|
| Unfractionated cells | NDV, CRF | β-END | 84, 85, 96 |
| Macrophages | Constitutive | β-END | 86, 87 |
| Mouse B-cells | LPS | α- or γ-END | 90, 92, 97 |
| Mouse B-cells | CRF | β-END | 97 |
| T-cells | Constitutive | Pro-ENK | 98 |

statement can be expanded to include a stimulus-dependent production of opioid peptides in the immune system since Con A-stimulated mouse spleen cells produced ENK mRNA (Table 4). In contrast, LPS treatment of the LPS resistant, C3H/HeJ mouse spleen cells induced the production of a large molecular weight, approximately a 31,000-Da molecule that was biologically inactive in opiate receptor binding.[97] It is tempting to speculate that the larger molecular weight material produced by the LPS-treated C3H/HeJ leukocytes is POMC, the 31,000 Da molecular weight precursor to ACTH and endorphins. Thus, it appears that while LPS may induce POMC synthesis, the C3H/HeJ splenocytes lack the ability to process the precursor into an ir-endorphin of the molecular weight of α- or γ-endorphin.

## G. LPS INDUCES A NOVEL PROTEASE ACTIVITY

Since induction of the POMC peptides appeared to be cell and stimulus dependent, we

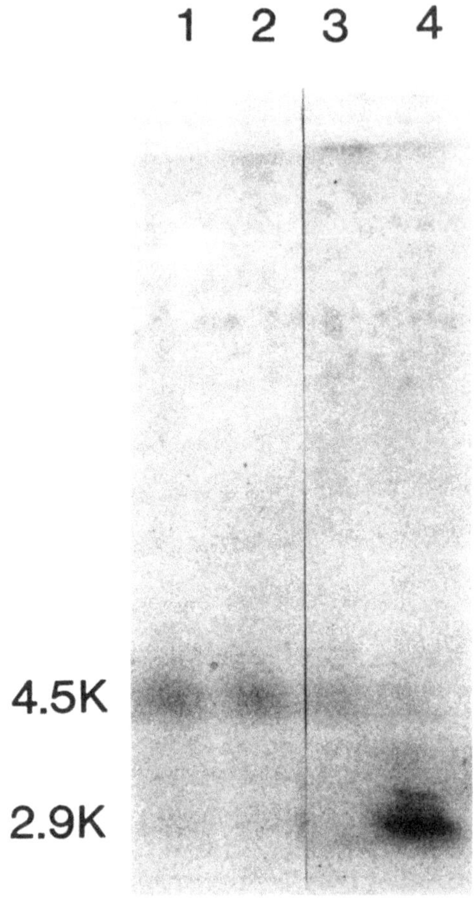


FIGURE 4. Comparison of LPS-induced proteolytic activity of B lymphocytes from C3H/HeJ and C3HeB/FeJ mice. Procedures are described in detail in Reference 98. Briefly, the cells were sonicated, and incubated with $I^{125}$ ACTH, sonicates were clarified by centrifugation and applied to 15% polyacrylamide gels. Molecular weight markers are ACTH (1 to 39) 4.5K and ACTH (1 to 24) 2.9K. Lane 1 is mock-treated C3H/HeJ B-cell sonicate incubated with $I^{125}$ ACTH at pH 5.0. Lane 2 is LPS-treated C3H/HeJ B-cell sonicate incubated with $I^{125}$ ACTH at pH 5.0. Lane 3 is mock-treated C3HeB/FeJ B-cell sonicate incubated with $I^{125}$ ACTH at pH 5.0 Lane 4 is LPS-treated B-cell sonicate incubated with $I^{125}$ ACTH at pH 5.0. Reproduced with permission from Reference 97.

determined whether the LPS would induce or activate an enzymatic activity that would process the POMC derived peptide ACTH (1 to 39). We incubated $^{125}I$ ACTH (1 to 39) with the cell sonicates from LPS or mock treated B lymphocytes of C3HeB/FeJ and C3H/HeJ mice.[97] Figure 4 lane 4 shows that LPS-treated C3HeB/FeJ B lymphocyte sonicates possess an enzymatic activity at pH 5 (lane 4) which cleaves ACTH 1 to 39 into a 2900 Da species. This same activity was not seen at pH 7 nor did the mock treated sensitive B-cells possess significant proteolytic activity (lane 3). In contrast, neither mock nor LPS-treated B lymphocytes from C3H/HeJ (LPS-insensitive) mice express proteolytic activity (lanes 1 and 2) at pH 5. Thus, this novel proteolytic enzyme does not appear to be induced or activated in the LPS-resistant C3H/HeJ B-cells, and its absence may well be the reason for the lack of precursor processing.

These data provide evidence that LPS treated B lymphocytes from C3HeB/FeJ (LPS-sensitive), but not C3H/HeJ (LPS-resistant) mice synthesize and process ACTH and β-endorphin-like molecules into smaller peptides. Whereas, CRF elicits the production and processing of ACTH 1 to 39 (4500 Da) and β-endorphin (3500 Da) from POMC, LPS causes the activation of a novel proteolytic activity which further cleaves these molecules into ACTH 1 to 22 to 26 (approximately 2900 Da) and α- or γ-endorphin (approximately 1800 Da), respectively. The processing of the precursor POMC and its products appears to be a major pathway for production of different families of bioactive hormones and has led researchers to postulate that different enzymes may be responsible for such differential processing.[7-10] This type of regulation, at the level of posttranslational cleavage, is seen with many hormones and particularly in several families of endogenous opioids including enkephalins and endorphins.[99] Numerous membrane bound enzymes have been reported to be responsible for many of these activities; however, these enzymes often function at neutral pH[100] and some investigators believe prohormone cleaving enzymes must be active at the pH of secretory vesicles which is around 5 to 6.[101] Interestingly, the enzyme activity we have demonstrated appears to fulfill such a criteria since it has a pH optimum of approximately 5.

Thus, as with the neuroendocrine system, there appears to be processing of POMC, but unique to the immune system, the data suggest a possible stimulus specific B lymphocyte processing mechanism for the POMC as outlined in Table 4. This alternative processing results in the production of various compounds with different immunoregulatory activities. As has been reviewed in this chapter (see Table 1), β-END is an enhancer of *in vitro* antibody production, CTL, NK function, and mitogenesis, while α-END is a potent suppressor of *in vitro* antibody response. ACTH (1 to 39) suppresses antibody response and macrophage function, but ACTH (1 to 24) has no effect. Therefore, one can readily see that the production of different neurohormone peptides by various stimuli or with different kinetics may result in not only a variety of immune responses, but also in the overall physiologic response (e.g., hypothermia vs. hyperthermia, etc.).

## V. IMPLICATIONS OF A TOTALLY INTEGRATED IMMUNE-NEUROENDOCRINE SYSTEM

This chapter has emphasized that there are numerous similarities between the endogenous opioid systems of the immune and neuroendocrine systems. These include production of similar opioid peptides (endorphins and enkephalins),[101-104] presence of classical opiate receptors coupled to similar intracellular signaling pathways, and cell-dependent mechanisms of processing large prohormones thereby yielding a variety of peptides with diverse bioactivities. There are some properties of the opioid components of the immune system that are unique from the neuroendocrine system including: stimulus dependent processing of prohormones, production of unique ACTH (1 to 22 to 1 to 26), presence of a nonclassical opiate receptor.

Therefore, based on these data, it seems important to alter our view of the immune and the neuroendocrine systems as separate entities to that of a completely integrated neuroimmunoendocrine system.[102-105] This knowledge of the totally integrated circuit has several implications. First, is the understanding that can be gained of the natural role of endogenous opiates and of the regulatory interactions that control the neuroendocrine and immune systems. The possibilities are great as evidenced by the variety of laboratories and interests reviewed above.

These data further substantiate the hypothesis that the immune system and the neuroendocrine system are linked via common peptide hormones and receptors.[102-105] A possible function suggested for this is that the immune system serves a sensory function. It perceives stimuli which are not directly recognized by the CNS, such as viruses and bacteria.[101-104] Subsequent to this recognition, the immune system signals the neuroendocrine system through chemical mediators such as hormone-like peptides, that are recognizable by the neuroendocrine system. An example of this is the data reviewed here. In this system, cells of the immune system recognize a bacterial stimulus (LPS) and respond by synthesizing and secreting an endorphin molecule. The endorphin may in turn influence the hypothalamus and vascular system to alter body temperature or blood pressure or both, thereby resulting in some of the detrimental effects seen in endotoxic shock.[95,105] This hypothesis is strengthened by two points. First, the mice resistant to the *in vivo* effects of LPS do not appear to express the enzymes necessary to process the POMC molecule in the lymphocyte system. This suggests that the lymphocyte's enzymatic cleavage of the opioid component from POMC is an important mediator in the pathophysiology of LPS-induced shock. Second, a possible pathophysiological role for their *in vivo* production of this leukocyte-derived peptide seems a distinct possibility when one considers the levels of ir-endorphin that are produced and released by LPS-treated spleen cells. Analysis by an enzyme-linked immunosorbent assay with monospecific antiserum against $\gamma$-endorphin shows extracellular levels of approximately 60 ng/$10^8$ spleen cells of ir-endorphin.[90] When one considers that there are approximately $1 \times 10^8$ to $2 \times 10^8$ mononuclear cells per mouse spleen, then the total capacity of this organ would be 60 to 120 ng. Since the blood volume of a mouse is only about 3 ml, only 10 to 20% of the total would need to be released to achieve the level of 4 ng of $\beta$-endorphin per milliliter that is observed in rats that are stressed by electric shock.[105] Thus, this one lymphoid organ apparently has the capacity to produce enough endorphin to have physiological or pathophysiological consequences.

A second implication is a practical, clinical one. This is probably best illustrated by several problems in which both opioids and the immune system have been implicated. One is the observation that the immune response of heroin addicts is typically suppressed.[20,106] The other is the controversial findings that in stressed (or bereaved) individuals there is an increase in circulating endogenous opiates and also an inhibited immune response with possibly a greater potential for death from cancer or infectious disease.[24] Could manipulation of the opiate system help these problems? Or could blockage of lymphocyte opiate receptors with a novel ligand prevent this immunosuppression? Or, in the case of acquired immunodeficiency syndrome, can opiates be used to enhance the failing immune response? In addition to these questions, do bioactive opioid peptides, i.e., endorphins (putatively behaviorally active neurotransmitters) from stimulated cells of the immune systems provide a mechanism whereby a psychological response is associated with infectious disease? Can we use this knowledge to further our understanding and approach to holistic medicine? Fortunately, the prospects for answering these questions appear excellent.

## ACKNOWLEDGMENTS

The authors thank Dr. J. E. Blalock for his central and essential role in those studies of ours that were discussed by this review. The authors are supported in part by the James W. McLaughlin Fellowship Fund (D.V.H.) and the Office of Naval Research (N00014-84-K-0486).

# REFERENCES


1. **Li, C. H. and Chung, D.,** Isolation and structure of an untriakantapeptide with opiate activity from camel pituitary glands, *Proc. Natl. Acad. Sci. U.S.A.*, 73, 1145, 1976.
2. **Guillemin, R.,** Beta-lipotropin and endorphins: implications of current knowledge, *Hosp. Pract.*, 13, 53, 1978.
3. **Marx, J. L.,** Synthesizing the opioid peptides, *Science*, 220, 395, 1983.
4. **Morley, J. S.,** Chemistry of opioid peptides, *Br. Med. Bull.*, 39, 5, 1983.
5. **Evans, C. J., Erdelyi, E., and Barchas, J. S.,** Candidate opioid peptides for interaction with the immune system, in *Enkephalins and Endorphins: Stress and the Immune System*, Plotnikoff, N. P., Faith, R. E., Murgo, A. J., and Good, R. A., Eds., Plenum Press, New York, 1986.
6. **Krieger, D. T.,** Brain peptides: what, where, and why?, *Science*, 222, 975, 1983.
7. **Douglass, J. O., Civelli, O., Binberg, N., Comb, M., Uhler, M., Lissitzky, J.-C., and Herbert, E.,** Regulation of expression of opioid peptide genes, *Ann. Neurol.*, Suppl. 16, S22, 1984.
8. **Rossier, J. and Bloom, F.,** Central neuropharmacology of endorphins, in *Neurochemical Mechanisms of Opiates and Endorphins*, Advances in Biochemistry and Psychopharmacology, Vol. 20, Loh, H. H. and Ross, D. H., Eds., Raven Press, New York, 1979, 165.
9. **Zurawski, G., Benadik, M., Kamb, B. J., Abrams, J. S., Zurawski, S. M., and Lee, F. D.,** Activation of mouse T-helper cells induces abundant preproenkephalin mRNA synthesis, *Science*, 232, 772, 1986.
10. **Holaday, J. W.,** Neuropeptides in shock and traumatic injury: sites and mechanisms of action, in *Neuroendocrine Perspective*, Vol. 3, Muller, E. E. and MacLeod, R. M., Eds., Elsevier, Amsterdam, 1984, 161.
11. **Chang, K.-J. and Cuatrecacas, P.,** Multiple opiate receptors, enkephalins and morphine bind to receptors of different specificity, *J. Biol. Chem.*, 254, 2610, 1979.
12. **Bernton, E. W., Long, J. B., and Holaday, J. W.,** Opioids and neuropeptides: mechanisms in circulatory shock, *Fed. Proc.*, 44, 290, 1985.
13. **Lee, N. M.,** Is dynorphin a K ligand?, *Proc. West. Pharmacol. Soc.*, 26, 309, 1983.
14. **Barchas, J. D. and Sullivan, S.,** Opioid peptides as neuroregulators: potential areas for the study of genetic-behavioral mechanisms, *Behav. Genet.*, 12, 69, 1982.
15. **Kiraly, I. and Van Ree, J. M.,** Non-opiate β-endorphin fragments and depamine. VI. Behavioral analysis of the interaction between gamma-type endorphins and dopaminergic systems in the nucleus accumbens of rats, *Neuropharmacology*, 23, 511, 1984.
16. **Johnson, H. M., Smith, E. M., Torres, B. A., and Blalock, J. E.,** Neuroendocrine hormone regulation of *in vitro* antibody production, *Proc. Nat. Acad. Sci. U.S.A.*, 79, 4171, 1982.
17. **Heijnen, C. J., Bevers, C., Kaveliars, A., and Ballieux, R. E.,** Effect of alpha-endorphin on the antigen induced primary antibody response of human blood B cells *in vitro*, *J. Immunol.*, 136, 213, 1986.
18. **Yamasaki, Y., Shimamura, O., Kizu, A., Nakagawa, M., and Ijichi, H.,** IgE mediated $^{14}$C-serotonin release from rat mast cells modulated by morphine and endorphins, *Life Sci.*, 31, 471, 1982.
19. **Brown, S. M., Stimmel, B., Taub, R. N., Kochura, S., and Rosenfeld, R. E.,** Immunologic dysfunction in heroin addicts, *Arch. Intern. Med.*, 134, 1001, 1974.
20. **McDonough, R. J., Madden, J. J., Falek, A., Falek, A., Shafer, D. A., Pline, M., Gordon, D., Bokos, P., Kuehnle, J. C., and Mendelson, J.,** Alteration of T and null lymphocyte frequencies in the peripheral blood of human opiate addicts: *in vivo* evidence for opiate receptor sites on T lymphocytes, *J. Immunol.*, 125, 2539, 1980.
21. **Laudenslager, M. L., Ryan, S. M., Drugan, R. C., Hyson, B. L., and Maier, S. F.,** Coping and immunosuppression: inescapable but not escapable shock suppresses lymphocyte proliferation, *Science*, 221, 568, 1983.
22. **Gilman, S. C., Schwartz, J. M., Milner, R. J., Bloom, F. E., and Feldman, J. D.,** β-Endorphin enhances lymphocyte proliferative responses, *Proc. Nat. Acad. Sci. U.S.A.*, 79, 4226, 1982.
23. **McCain, H. W., Lamster, I. B., Bozzone, J. M., and Gribic, J. T.,** β-Endorphin modulates human immune activity via non-opiate receptor mechanisms, *Life Sci.*, 31, 1619, 1982.
24. **Solomon, G. S., Kay, N., and Morley, J. E.,** Endorphins: a link between personality, stress emotions, immunity and disease?, in *Enkephalins and Endorphins: Stress and the Immune System*, Plotnikoff, N. P., Faith, R. E., Murgo, A. J., and Good, R. A., Eds., Plenum Press, New York, 1986, 129.
25. **McCain, H. W., Lamoter, I. B., and Bilotta, J.,** Modulation of human T cell suppressor activity by beta endorphin and glycyl-L-glutamine, *Int. J. Immunopharmacol.*, 8, 443, 1986.
26. **Plotnikoff, N. and Miller, G. E.,** Enkephalins as immunodilators, *Int. J. Immunopharmacol.*, 5, 437, 1983.
27. **Wybran, J.,** Enkephalins and endorphins as modifiers of the immune system: present and future, *Fed. Proc.*, 44, 92, 1985.
28. **Bocchini, G., Bonanno, G., and Canevari, A.,** Influence of morphine and naloxone on human peripheral blood T-lymphocytes, *Drug Alcohol Depend.*, 11, 233, 1983.

29. **Berkenbosch, F., Heijnen, C. J., Croiset, G., Revers, C., Ballieux, R. E., Binnekado, R., and Tilders, F. J. H.,** Endocrine and immunological responses to acute stress, in *Enkephalins and Endorphins: Stress and the Immune System,* Plotnikoff, N. P., Faith, F. E., Murgo, A. J., and Good, R. A., Eds., Plenum Press, New York, 1986, 109.
30. **Donahoe, R., Madden, J., Hollingsworth, F., Shafer, D., and Falek, A.,** T cell E-receptor modulation by opiates opioids and other behaviorally active substances, in *Enkephalins and Endorphins: Stress and the Immune System,* Plotnikoff, N. P., Faith, R. E., Murgo, A. J., and Good, R. A., Eds., Plenum Press, New York, 1986, 289.
31. **Wybran, J., Appelbloom, T., Famaey, J. P., and Govaerts, A.,** Suggestive evidence for morphine and met-enkephalin receptors on normal blood T lymphocytes, *J. Immunol.,* 123, 1070, 1979.
32. **Miller, G. C., Murgo, A. J., and Plotnikoff, N. P.,** Enkephalins: enhancement of active T cell rosettes from normal volunteers, *Clin. Immunol. Immunopathol.,* 31, 132, 1984.
33. **Farrar, W. L.,** Relationship between lymphokine and opiatergic modulation of lymphocyte proliferation, in *Enkephalins and Endorphins: Stress and the Immune System,* Plotnikoff, N. P., Faith, R. E., Murgo, A. J., and Good, R. A., Eds., Plenum Press, New York, 1986, 241.
34. **Plotnikoff, N. P., Wybran, J., Nimeh, N. F., and Muller, G. C.,** Methionine enkephalin: enhancement of T cells in patients with Kaposis' sarcoma, AIDS and lung cancer, in *Enkephalins and Endorphins: Stress and the Immune System,* Plotnikoff, N. P., Faith, R. E., Murgo, A. J., and Good, R. A., Eds., Plenum Press, New York, 1986, 425.
35. **Hung, C. Y., Lefkowitz, S. S., and Geber, W. F.,** Interferon inhibition by narcotic analgesics, *Proc. Soc. Exp. Biol. Med.,* 142, 106, 1973.
36. **Geber, W. F., Lefkowitz, S. S., and Hung, C. Y.,** Action of naloxone on the interferon lowering activity of morphine in the mouse, *Pharmacology,* 14, 322, 1976.
37. **Mandler, R. N., Biddison, W. E., Mandler, R., and Serrate, S. A.,** β-Endorphin augments the cytolytic activity and interferon production of natural killer cells, *J. Immunol.,* 135, 934, 1986.
38. **Brown, S. L. and Van Epps, D. E.,** Opioid peptides modulate production of interferon gamma by human mononuclear cells, *Cell. Immunol.,* 103, 19, 1986.
39. **Brown, S. L. and Van Epps, D. E.,** Suppression of T lymphocyte chemotactic factor production by the opioid peptides β-endorphin and met-enkephalin, *J. Immunol.,* 134, 3384, 1985.
40. **Blalock, J. E. and Smith, E. M.,** Human leukocyte interferon (HuIFN-α): potent endorphin-like opioid activity, *Biochem. Biophys. Res. Commun.,* 101, 472, 1981.
41. **Maier, S. F., Drugan, R. C., and Grace, J. W.,** Controllability, coping behavior and stress induced analgesia in the rat, *Pain,* 12, 47, 1982.
42. **Shavit, Y., Ryan, S. M., Lewis, J. W., Laudenslager, M. L., Terman, G. W., Maier, S. F., Gale, R. P., and Liebeskind, J. C.,** Inescapable but not escapable stress alters immune function, *Physiologist,* 26, A-64, 1983.
43. **Shavit, Y., Terman, G. W., Martin, F. C., Lewis, J. W., Liebeskind, J. C., and Gale, R. P.,** Stress, opioid peptides, the immune system and cancer, *J. Immunol.,* Suppl. 135, 834S, 1985.
44. **Visintainer, M. A., Volpicelli, J. R., and Seligman, M. E. P.,** Tumor rejection in rats after inescapable or escapable shock, *Science,* 216, 437, 1982.
45. **Shavit, Y., Lewis, J. W., Terman, G. W., Gale, R. P., and Liebeskind, J. C.,** Opioid peptides mediate the suppressive effect of stress on natural killer cell cytotoxicity, *Science,* 223, 188, 1984.
46. **Shavit, Y., Terman, G. W., Martin, F. C., Gale, R. P., and Liebskind, J. C.,** Naltrexone sensitive suppression of the immune systems natural cells by morphine, *Soc. Neurosci. Abstr.,* 10, 726, 1984.
47. **Martin, F. C., Shavit, Y., Terman, G. W., Pechnick, R. N., Oh, C., and Liebeskind, J. C.,** Stereo-specificity of opiate immunosuppression in rats, *Soc. Neurosci. Abstr.,* 11, 524, 1985.
48. **Matthews, P. M., Froelich, C. J., Sibbitt, J. R., Sibbitt, W. L., and Bankhurst, A. D.,** Enhancement of natural cytotoxicity by β-endorphin, *J. Immunol.,* 130, 1658, 1983.
49. **Kay, N., Allen, J., and Morley, J. E.,** Endorphins stimulate normal human peripheral blood lymphocyte natural killer activity, *Life Sci.,* 35, 53, 1984.
50. **Faith, R. E., Liang, H. J., Murgo, A. J., and Plotnikoff, N. P.,** Neuroimmunomodulation with enkephalins: enhancement of natural killer (NK) cell activity in vitro, *Clin. Immunol. Immunopathol.,* 31, 412, 1984.
51. **Carr, D. J. and Klimpel, G. R.,** Enhancement of the generation of cytotoxic T cells by endogenous opiates, *J. Neuroimmunol.,* 12, 75, 1986.
52. **Li, C. H., Yamashiro, D., Tseng, L. F., Chang, W. C., and Ferraro, P.,** Beta endorphin omission analogs: dissociation of immunoreactivity from other biological activities, *Proc. Natl. Acad. Sci. U.S.A.,* 77, 3211, 1980.
53. **Van Epps, D. E. and Saland, L.,** β-Endorphin and met enkephalin stimulate human peripheral blood mononuclear cell chemotaxis, *J. Immunol.,* 132, 3046, 1984.
54. **Simpkins, C. O., Dickey, C. A., and Fink, M. P.,** Human neutrophil migration is enhanced by Beta-endorphin, *Life Sci.,* 34, 2251, 1984.

55. **Saland, L. C., Van Epps, D. E., Ortiz, E., and Samora, A.,** Acute injections of opiate peptides into the rat cerebral ventricle: a macrophage-like cellular response, *Brain Res. Bull.*, 10, 523, 1983.
56. **Foris, G., Medgyesi, E., Gyimesi, E., and Hauck, E.,** Met-enkephalin induced alterations of macrophage functions, *Mol. Immunol.*, 21, 747, 1984.
57. **Miller, G. C., Murgo, A. J., and Plotnikoff, N. P.,** The influence of leucine and methionine enkephalin on immune mechanisms, *Int. J. Immunopharmacol.*, 4, 367, 1982.
58. **Foster, J. S. and Moore, R. N.,** Dynorphin and related opioid peptides enhance tumoricidal activity mediated by murine peritoneal macrophages, *J. Leukocyte Biol.*, 42, 171, 1987.
59. **Wybran, J. and Dupont, E.,** The active T rosette: an early marker for T cell activation, *Ann. Immunol. (Paris)*, 133, 211, 1982.
60. **Mehrishi, J. N. and Mills, I. H.,** Opiate receptors on lymphocytes and platelets in man, *Clin. Immunol. Immunopathol.*, 27, 240, 1983.
61. **Lopker, A., Abood, L. G., Hoss, W., and Lionetti, F. J.,** Stereoselective muscarinic, acetylcholine and opiate receptors in human phagocytic leukocytes, *Biochem. Pharmacol.*, 29, 1361, 1980.
62. **Falke, N. E. and Fischer, E. G.,** Opiate receptor mediated internalization of $^{125}$I-β-End in human polymorphonuclear leukocytes, *Cell Biol. Int. Rep.*, 19, 429, 1986.
63. **Carr, D. J., Bubein, J. K., Woods, W. T., and Blalock, J. E.,** Opioid receptors on murine splenocytes: possible coupling to $K^+$ channels, *Ann. N.Y. Acad. Sci.*, in press 1987.
64. **Hazum, E., Chang, K.-J., and Cuatrecacas, P.,** Specific nonopiate receptors for beta endorphin, *Science*, 205, 1033, 1979.
65. **Schweigerer, L., Schmidt, W., Teschemacher, H., and Gramsch, C.,** β-Endorphin: surface binding and internalization in thymoma cells, *Proc. Natl. Acad. Sci. U.S.A.*, 82, 5751, 1985.
66. **Schweigerer, L., Bhakdi, S., and Teschemacher, H.,** Specific non-opiate binding sites for human β-END on the terminal complex of human complement, *Nature (London)*, 296, 5857, 1982.
67. **Faith, R. E., Liang, H. J., Murgo, A. J., and Plotnikoff, N. P.,** Neuroimmunomodulation with enkephalins: enhancement of human natural killer cell activity in vitro, *Clin. Immunol. Immunopathol.*, 31, 412, 1984.
68. **Van Epps, D. E., Saland, L., Taylor, C., and Williams, R. C., Jr.,** *In vitro* and *in vivo* effects of Beta endorphin and met enkephalin on leukocyte locomotion, *Prog. Brain Res.*, 59, 361, 1983.
69. **Wybran, J., Applebloom, T., Famaey, J. P., and Govaerts, A.,** Receptors for morphine and met-enkephalin on human T lymphocytes: the two hits opioid lymphocyte receptor hypothesis, in *New Trends in Human Immunology and Cancer Immunotherapy*, Serrou, B. and Rosenfield, C., Eds., Doin Publishers, Paris, 1980, 48.
70. **Plotnikoff, N. P.,** The Ying-Yang hypothesis of opioid peptide immunomodulation, *Psychopharmacol. Bull.*, 21, 489, 1985.
71. **Gilbert, J. A. and Richelson, E.,** Function of delta opioid receptors in cultured cells, *Mol. Cell. Biochem.*, 55, 83, 1983.
72. **Foris, G., Medgyesi, G. A., and Szekely, J. I.,** Significant role of receptor coupling in the neuropeptide-induced alterations of macrophage cytotoxicity, in *Enkephalins and Endorphins: Stress and the Immune System*, Plotnikoff, N. P., Faith, R. E., Murgo, A. J., and Good, R. A., Eds., Plenum Press, New York, 1986, 303.
73. **Moon, T. D., Morley, J. E., Vesella, R. L., and Lange, P. H.,** The role of calmodulin in human NK activity, *Scand. J. Immunol.*, 18, 255, 1983.
74. **Simpkins, C. O., Alailima, S. T., and Tate, E. A.,** Inhibition by naloxone of neutrophil superoxide release: a potentially useful anti-inflammatory effect, *Circ. Shock*, 20, 181, 1986.
75. **Sharp, B. M., Tsukayama, D. T., Gekker, G., Keane, W. F., and Peterson, P. K.,** β-Endorphin stimulates human polymorphonuclear leukocyte superoxide production via a stereoselective opiate receptor, *J. Pharmacol. Exp. Ther.*, 242, 579, 1987.
76. **Faden, A. I. and Holaday, J. W.,** Opiate antagonists: a role in the treatment of hypovolemic shock, *Science*, 205, 317, 1979.
77. **Holaday, J. W. and Faden, A. I.,** Naloxone reversal of endotoxin hypotension suggests role of endorphins in shock, *Nature (London)*, 275, 450, 1978.
78. **Faden, A. I. and Holaday, J. W.,** Naloxone treatment of endotoxin shock: stereospecificity of physiologic and pharmacologic effects in the rat, *J. Pharmacol. Exp. Ther.*, 212, 441, 1979.
79. **Mamazza, J., Hinckey, E. J., and Chu-Jeng, C. R.,** The pulmonary effects of opiate blockade in septic shock, *J. Surg. Res.*, 36, 625, 1984.
80. **Reynolds, D. G., Gwill, N. V., Vargish, T., Lechner, R. B., Faden, A. I., and Holaday, J. W.,** Blockade of opiate receptors with naloxone improves survival and cardiac performance in canine endotoxic shock, *Circ. Shock*, 7, 39, 1980.
81. **Carr, D. B., Bergland, R., Hamilton, A., Blume, H., Kasting, N., Arnold, M., Martin, J. B., and Rosenblatt, M.,** Endotoxin-stimulated opioid peptide secretion: two secretory pools and feedback control in vivo, *Science*, 217, 845, 1982.

82. **Traber, D. L., Thompson, P. D., Blalock, J. E., Smith, E. M., Adams, T., and Sziebert, L. A.,** Action of an opiate receptor blocker on ovine cardiopulmonary responses to endotoxin, *Am. J. Physiol.*, 245, H189, 1983.
83. **D'Amato, R. J. and Holaday, J. W.,** Multiple opiate receptors in endotoxic shock: evidence for delta involvement and mu-delta interactions in vivo, *Proc. Natl. Acad. Sci. U.S.A.*, 81, 2898, 1984.
84. **Smith, E. M., Meyer, W. J., III, and Blalock, J. E.,** Virus induced increases in corticosterone in hypophysectomized mice: a possible lymphoid-adrenal axis, *Science*, 218, 1311, 1982.
85. **Smith, E. M., Morrill, A. C., Meyer, W. J., III, and Blalock, J. E.,** Corticotropin releasing factor induction of leukocyte derived immunoreactive ACTH and endorphins, *Nature (London)*, 321, 881, 1986.
86. **Lolait, S. J., Lim, A. T. W., Toh, B. H., and Funde, J. W.,** Immunoreactive β-endorphin in a subpopulation of mouse spleen macrophages, *J. Clin. Invest.*, 73, 277, 1984.
87. **Lolait, S. J., Clements, J. A., Markwick, A. J., Cheng, C., McNally, M., Smith, A. I., and Funde, J. W.,** Pro-opiomelanocortin messenger ribonucleic acid and posttranslational processing of beta endorphin in spleen macrophages, *J. Clin. Invest.*, 77, 1776, 1986.
88. **Bohs, C. T., Fish, J. C., Miller, T. H., and Traber, D. L.,** Pulmonary vascular response to endotoxin in normal and lymphocyte depleted sheep, *Circ. Shock*, 6, 13, 1979.
89. **Traber, D. L., Jinkins, T., Rice, K., Eziebert, L., Adams, T., Jr., Henriksen, N., Traber, L. D., and Thompson, P. D.,** Role of lymphocytes in the ovine response to endotoxin, *Adv. Shock Res.*, 9, 257, 1983.
90. **Harbour-McMenamin, D., Smith, E. M., and Blalock, J. E.,** Endotoxin induction of leukocyte-derived proopiomelanocortin related peptides, *Infect. Immun.*, 48, 813, 1985.
91. **Simon, E. J., Hiller, J. M., and Edelman, I.,** Stereospecific binding of the potent narcotic analgesic ($^3$H)-etorphine to rat brain homogenate, *Proc. Natl. Acad. Sci. U.S.A.*, 70, 1947, 1973.
92. **Harbour-McMenamin, D., Smith, E. M., and Blalock, J. E.,** B lymphocyte enriched populations produce immunoreactive endorphin that binds to delta opiate receptors and may play a role in endotoxic shock, in *Advances in Gene Technology: Molecular Biology of the Endocrine System*, ICSU Short Reports, Vol. 4, 1986, 378.
93. **Holaday, J. W. and Totella, F. C.,** Multiple opioid receptors: possible physiological functions of μ and δ binding sites *in vivo*, in *Central and Peripheral Endorphins: Basic and Clinical Aspects*, Muller, E. E. and Genzani, A. R., Eds., Raven Press, New York, 1984, 237.
94. **Morrison C. and Rifan, J. L.,** Bacterial endotoxins and host immune responses, *Adv. Immunol.*, 28, 293, 1979.
95. **Harbour, D. V., Smith, E. M., and Blalock, J. E.,** Splenic lymphocyte production of an endorphin during endotoxic shock, *Brain Behav. Immun.*, 1, 123, 1987.
96. **Westly, H. J., Kleiss, A. J., Kelley, K. W., Wong, P. K. Y., and Yuen, P.,** Newcastle disease virus-infected splenocytes express the proopiomelanocortin gene, *J. Exp. Med.*, 163, 1589, 1986.
97. **Harbour, D. V., Smith, E. M., and Blalock, J. E.,** Novel processing pathway for pro-opiomelanocortin in lymphocytes: endotoxin induction of a new prohormone cleaving enzyme, *J. Neurosci. Res.*, 18, 95, 1987.
98. **Zurawski, G., Benedik, M., Kamb, B. J., Abrams, J. S., Zurawski, S. M., and Lee, F. D.,** Activation of mouse T-helper cells induces abundant preproenkephalin mRNA synthesis, *Science*, 232, 772, 1986.
99. **Herbert, E.,** Discovery of pro-opiomelanocortin — a cellular polyprotein, *Trends Biochem. Sci.*, 6, 184, 1981.
100. **Benuck, M., Berg, N. J., and Marks, N.,** Membrane-bound enzymes and their role in processing of the dynorphins and of the proenkephalin octapeptide met-enkephalin-Arg-Gly-Leu, *Neurochem. Res.*, 9, 733, 1987.
101. **Loh, Y. P., Parish, D. C., and Tuteja, R.,** Purification and characterization of a paired basic residue-specific pro-opiomelanocortin converting enzyme from bovine pituitary intermediate lobe secretory vesicles, *J. Biol. Chem.*, 260, 7194, 1985.
102. **Blalock, J. E. and Smith, E. M.,** A complete regulatory loop between the immune and neuroendocrine systems, *Fed. Proc.*, 44, 108, 1985.
103. **Blalock, J. E., Harbour-McMenamin, D. V., and Smith, E. M.,** Peptide hormones shared by the neuroendocrine and immunologic systems, *J. Immunol.*, Suppl. 135, 858s, 1985.
104. **Blalock, J. E.,** The immune system as a sensory organ, *J. Immunol.*, 132, 1067, 1984.
105. **Smith, E. M., Harbour, D. V., and Blalock, J. E.,** Leukocyte production of endorphins, *Ann. N.Y. Acad. Sci.*, 496, 192, 1987.
106. **Rossier, J., French, E. D., Rivier, C., Ling, N., Guillemin, R., and Bloom, F. E.,** Foot-shock induced stress increases β-endorphin levels in blood but not brain, *Nature (London)*, 270, 618, 1977.
107. **Cherubin, C. E. and Brown, J.,** Systemic infections in heroin addicts, Lancet, 1, 298, 1968.
108. **Smith, E. M., et al.,** manuscript in preparation, 1989.

Chapter 12

# STRESS NEUROPEPTIDES, IMMUNITY, AND NEOPLASIA

**Yehuda Shavit, Raz Yirmiya, and Benzion Beilin**

## TABLE OF CONTENTS

# I. STRESS, IMMUNITY, AND CANCER

## A. STRESS AND INFECTION

Behavioral medicine has long been interested in the interaction of emotions, stress, and susceptibility to infection. Tobach and Bloch were among the first to study the relationship between stress and susceptibility to acute tuberculosis.[1] Their data indicated that the stress of crowding significantly reduced survival time of mice challenged with acute tuberculosis infection.

Pursuing this line of research, numerous stressors (such as handling, crowding, isolation, separation from mother, loud noise, electric shock, forced swimming, and rotation) have been evaluated in behavioral epidemiological studies. No direct measures of immune function were taken in these early studies, but the effect of the stressors on the susceptibility to bacterial and viral infectious agents was studied across species ranging from fish to human beings. In general, stress was found to reduce the resistance of animals to infectious agents. However, several reports found that stress has a protective effect against infection. The nature and chronicity of the stress, the time at which the infectious agent is introduced relative to the stress, and the housing and social conditions of the animals appear to be important factors in determining the outcome of this type of study. Several reviews of this subject have recently appeared.[2-4]

## B. STRESS AND CANCER

Reviewing the literature on stress and tumors reveals a picture similar to that relating stress and infection. Stress can alter the incidence and development of experimental tumors in animals. In general, stress appears to enhance tumor induction and development, although stress-induced retardation of tumor growth has also been reported. These inconsistent findings may be due to differences in stressors, tumor lines, timing of the stress relative to tumor administration, and species utilized in the different studies. Nevertheless, several major factors fundamental in determining the effects of stress on tumorigenicity have been identified. These include: chronic *vs.* acute stress, where chronic stress reportedly causes retardation of tumor growth and acute stress tumor enhancement; coping ability of the animal, where animals given control over the stress are less affected by it; and housing conditions (for recent reviews, see References 5 and 6).

## C. STRESS AND IMMUNE FUNCTION

Stress is associated with alterations of humoral and cellular immune mechanisms in both laboratory animals and humans. Decreased response to mitogens and antigens, reduced lymphocyte-mediated cytotoxicity, reduced delayed hypersensitivity, diminished skin graft rejection and graft-*vs.*-host reactivity, and suppressed antibody response have all been observed after stress. Other studies, however, show that stress can also augment immune function. Several of the same factors mentioned earlier were also found to play an important role in determining the effects of stress on the immune system. For example, stress chronicity, the timing of stress application relative to immune measurement, stress intensity, and stress controllability, all play a role (for reviews, see References 2 to 4).

# II. STRESS-INDUCED ANALGESIA

Stress can, by its potent neural, neurochemical, and neurohormonal effects, alter the susceptibility of animals to a wide variety of diseases, including neoplasia, possibly by altering immune function. Some recent evidence from our laboratory and others suggests that exposure to stress can also release opioid peptides from central and peripheral sites and cause analgesia. We studied the phenomenon of "stress-induced analgesia" and found that stressors differing only in small parametric details can have distinctly different neurochemical and hormonal consequences. Thus, a single stressor, inescapable foot shock of constant intensity, can elicit analgesia via different mechanisms depending on the temporal parameters of its application. Exposure to prolonged, intermittent foot shock causes analgesia that is blocked by the opiate antagonist, naloxone. In contrast, exposure to a brief, continuous foot shock causes a comparable degree of analgesia not blocked by naloxone.[7]

Subsequent work confirmed the existence of opioid and nonopioid stress analgesia systems by applying criteria other than naloxone antagonism. For example, tolerance developed with daily repetitions of the naloxone-sensitive form of stress, but not the naloxone-insensitive form; and in morphine-tolerant rats, cross-tolerance was seen with only that form of stress analgesia blocked by naloxone.[8] The opioid, but not the nonopioid, form of stress analgesia is also attenuated by hypophysectomy and especially by adrenalectomy or even adrenal demedullation or denervation.[8] Because hypophysectomy can compromise the function of the adrenal medulla as well as the adrenal cortex and because the adrenal medulla is known to contain enkephalin-like peptides and secrete them in response to stress, we concluded that opioid stress analgesia depends on the release of adrenal medullary enkephalins. Finally, in a collaborative study with Maier,[9] we showed that only the naloxone-sensitive (opioid) form of stress leads to an escape learning deficit termed "learned helplessness", normally associated with inescapable, but not escapable shock.

## III. OPIATES, OPIOID PEPTIDES, IMMUNITY, AND TUMORS

There is growing evidence implicating opiates and endogenous opioid peptides in the regulation of the immune system. Opiate addicts appear to be more susceptible to infections and, in fact, to have deficits in immune function. Mice injected with morphine for two weeks also have a suppressed reactivity to concancavalin A (Con A) stimulation, and this effect is partly blocked by the opiate antagonist, naloxone. It has recently been reported that morphine can adversely affect the outcome of the infectious disease processes in mice, possibly via its suppressive effect on phagocytic activity (for review, see Reference 1).

Opiate agonists and antagonists have also been implicated in tumor development. For example, systemic administration of opiates, endogenous opioids, and opioid antagonists have been shown to retard tumor growth,[11,12] whereas intracerebral administration of β-endorphin into the nucleus raphe magnus enhances the rate of tumor metastasis.[13-15]

The studies of stress-induced analgesia described above, make it clear that stress can be a physiological trigger for an intrinsic analgesia mediated by opioid peptides. However, these findings make it equally clear that another, nonopiod mechanism of analgesia also exists. We reasoned that our dissection of stress into parameters causing opioid and nonopioid forms of analgesia might help us account for some of the variability in the literature on stress and immunity. We have, therefore, compared the effects of the opioid and nonopioid stress paradigms on the cytotoxic activity of natural killer (NK) cells and on the development of an experimental mammary tumor in rats.

### A. EFFECTS OF MORPHINE AND OPIOID AND NONOPIOID STRESS ON NK-CELL CYTOTOXICITY

Numerous reports suggest that immune surveillance against viral infections and neoplastic disease is primarily provided by cytotoxic T lymphocytes and NK cells.[16] NK cells are a subpopulation of lymphocytes that can spontaneously recognize and kill certain target cells and that have been implicated in immune surveillance against tumors, in particular against tumor metastasis.[16] It is especially noteworthy in this context that NK activity is markedly reduced by stressors, including cold water swim, starvation, and surgery. In view of these reports, we examined the effect of our opioid and nonopioid foot shock stress paradigms on NK activity.

Fischer 344 (F344) female rats were subjected to one of the two foot shock paradigms. The two foot shock paradigms were identical in shock intensity (2.0 mA) and total "shock on" time, but differed in their temporal parameters. The intermittently applied foot shock (on 1 s out of every 5 for 10 min) caused an opioid mediated analgesia; the continuously applied foot shock (on for 2 min) caused an equally potent but nonopioid analgesia. Rats were given the opioid or nonopioid stress daily for 4 consecutive days. Another group was given daily subcutaneous (s.c.) injection of the long lasting opioid antagonist, naltrexone (10 mg/kg) 20 min before opioid foot shock stress. Nonstressed control groups received either naltrexone or saline alone for 4 d or were untreated. Three hours after the last stress session or drug treatment, all animals were anesthetized and their spleens removed and dissociated into a single-cell suspension. NK activity was measured in a standard 4-h chromium release assay using YAC-1 murine lymphoma cells as targets.

Exposure to the opioid, but not the nonopioid, form of foot shock stress resulted in significant suppression of NK activity to 74% of control values, and this suppression was prevented by naltrexone administration.[17] Injections of naltrexone or saline alone had no significant effect on NK activity compared with untreated controls.

In another experiment, rats were given morphine to determine if this opiate drug would mimic the effect of endogenous opioids released by stress. F344 female rats were given daily injections of morphine for 4 d at doses of 10, 30, and 50 mg/kg. Untreated rats served as controls. Three hours after the last injection, spleens were removed and processed, and NK activity was assessed as before. Morphine caused a dose-related suppression of NK activity.[17] Whereas 10 mg/kg had no statistically significant effect, the two higher doses markedly suppressed NK activity to a degree comparable to that caused by the opioid form of stress. The findings that only the opioid form of stress suppresses NK activity, that this suppression is blocked by naltrexone, and that a suppression of similar magnitude is caused by morphine, suggest that opioid peptides can modulate NK activity.

We have subsequently reported that a single exposure to the opioid form of foot shock stress or a single high dose of morphine (>20 mg/kg, s.c.) induces suppression of NK-cell cytotoxicity.[18] This effect is evident 3 h after treatment, returning to normal by 24 h. Morphine-induced NK suppression is evident in cells derived simultaneously from the spleen, bone marrow, and peripheral blood, suggesting that this suppression does not result from selective egress of NK cells from the spleen.[18] We have extended these findings to two other narcotic agents commonly used in general anesthesia—fentanyl and sufentanil. Both agents significantly suppressed NK-cell activity in rats for up to 12 h, and this effect was blocked by naltrexone. Pretreatment with Poly I:C, an interferon inducer, significantly attenuated the suppressive effect of fentanyl on NK activity.[19]

To further examine the involvement of opioid peptides in these phenomena, we sought to determine whether tolerance would develop to the stress-induced and morphine-induced suppression of NK activity and whether cross-tolerance would develop between them. We found that NK cytotoxic activity was significantly suppressed, and suppressed to the same degree, in animals subjected to either 4, 14, or 30 daily foot shock sessions.[20] Thus, tolerance did not occur to the opioid foot shock stress. On the other hand, tolerance did develop to morphine after 14 daily injections. NK activity was significantly suppressed in rats injected with a single dose of morphine compared with untreated controls, and no such suppression was seen in morphine-tolerant rats. Exposure to opioid stress significantly suppressed NK activity in both morphine-naive and morphine-tolerant rats compared with untreated controls, indicating that cross-tolerance did not occur between the effects of opioid stress and morphine.[20]

That tolerance develops to the suppressive effect of morphine on NK activity suggests that this effect is mediated via opioid receptors. That the opioid form of stress did not manifest such tolerance or show cross-tolerance with morphine in suppressing NK activity suggests that the immune suppressive effects of opioid stress and morphine are mediated by different mechanisms, perhaps by different opioid receptor types.

In subsequent studies, we have found that morphine injected into the lateral ventricle of the brain (20 and 40 μg) suppressed NK activity to the same degree as a 1000-fold higher systemic dose, and this effect was blocked by naltrexone.[21] In contrast, NK activity was unaffected by systemic administration of *N*-methyl morphine, a morphine analog that does not cross the blood brain barrier.[21] These data implicate opiate receptors in the brain in morphine-induced suppression of NK-cell cytotoxicity.

Recently, the stereospecificity of opiate-induced NK suppression was examined by comparing the effect of the opiate alkaloid, levorphanol, with its opiate-inactive stereoisomer, dextrophan. *In vivo*, only levorphanol (10 mg/kg, s.e.) significantly suppressed NK activity. *In vitro*, however, both drugs suppressed NK activity when present at very high concentrations.[22] These results imply that the *in vivo* effect of levorphanol is mediated by opiate receptors, whereas the *in vitro* effect is not. Furthermore, these results support our earlier hypothesis that the opiate alkaloids, morphine and levorphanol, suppress NK activity by binding to brain opiate receptors, rather than acting directly upon NK cells. We next sought to determine the type of opiate receptor mediating the *in vivo* effect of opiates by using β-funaltrexamine (β-FNA), an irreversible antagonist of the μ-opiate receptor. β-FNA injected directly into the lateral cerebral ventricle significantly blocked morphine-induced NK suppression.[23] Taken together, these data further support our hypothesis that morphine's effect on NK cells is mediated by opioid receptors in the brain.

Several physiological mechanisms could account for our immunological findings. As previously mentioned, morphine and opioid peptides released by stress might suppress immune function by acting directly on opiate receptors present on various cells of the immune system.[13-15] Alternatively, the effects of morphine and opioids on NK activity could involve modulation of the autonomic nervous system, which has recently been shown to innervate immune organs such as the spleen, thymus, and lymph nodes.[24] Finally, opioids and morphine might affect NK cells by their effects on hormone release. For example, adrenocorticotropic hormone (ACTH) and adrenocorticoids are well known to suppress various aspects of immune function, including NK activity;[25] and opioids and morphine have been implicated in mediating the release of these hormones by activating the hypothalamic-pituitary-adrenal axis.[26]

We have recently studied the role of the adrenal gland in mediating the NK-suppressive effect of morphine. F344 rats were adrenalectomized and implanted with corticosterone pellets. These pellets maintain a constant, physiological level of corticosterone, independent of stress or drug treatment. Two weeks after surgery, rats were injected with morphine (30 mg/kg, s.c.), and their NK activity was studied 3 h later. Adrenalectomy blocked the suppressive effect of morphine, indicating that the adrenal gland is involved in mediating this effect.[27] To further examine the role of the adrenal cortex vs. the adrenal medulla, we studied the effect of morphine in adrenal demedullated rats. Morphine suppressed NK activity to a comparable level in both demedullated and control rats, suggesting that the effect of morphine on NK cells is mediated by hormones released from the adrenal cortex (Shavit, Martin, Gale, and Leibeskind, unpublished, 1986).

## B. EFFECTS OF MORPHINE AND OPIOID AND NONOPIOID STRESS ON TUMOR GROWTH

Using similar experimental manipulations, we also tested the effects of foot shock stress and of morphine on tumor growth. Groups of F344 female rats were administered either the opioid or nonopioid form of stress daily for 4 consecutive days before tumor injection. A third, nonshocked group injected with tumor cells at the same time served as controls. The rat mammary adenocarcinoma (MAT 13762B) was used in these experiments.

We found that rats subjected to the opioid, but not the nonopioid, form of foot shock stress prior to tumor implantation had reduced median survival time and percent survival compared to nonstressed controls.[28] The results of subsequent experiments lend support to the hypothesis that opioids are involved in the effect of this stress on tumor growth. We found that the tumor-enhancing effect of the opioid stress was blocked by naltrexone and was mimicked by morphine (50 mg/kg) administered for 4 d before tumor implantation.[29,30] The fact that tolerance developed to the tumor-enhancing effect of morphine when the drug was given for 14 d also supports this hypothesis. Consistent with the findings of our NK experiments, the effects of stress manifested neither tolerance after 14 daily stress exposures nor cross-tolerance in morphine-tolerant rats.[29,30] Thus, opioid peptides released by foot shock stress appear to be involved in the decreased resistance of rats challenged with an experimental tumor, just as they appear to be involved in the stress-induced suppression of NK cytotoxicity.

There are several possible mechanisms by which opioids and morphine might influence the development of tumors. First, some tumor cell lines have opioid receptors, making it possible that opioids released by stress directly affect tumor cells. Unfortunately, it has not yet been determined whether opioid receptors occur on MAT 13762B cells. Second, opioids and morphine may interact with tumor development via hormonal mechanisms. Many tumors are sensitive to circulating hormones, and their growth may be influenced by stress- or morphine-induced changes in hormone levels. Opioid peptides are thought to regulate the release of several of the hormones secreted during stress, including prolactin and adrenal corticosteroids.[26] Many experimental breast tumors respond to elevated levels of prolactin with increased growth, including the solid form of MAT 13762B.[31] Alternatively, adrenal corticoids can be powerful immune suppressants in rodents and in this way may contribute to the enhanced growth or metastasis of tumors following stress. Specifically, it has been shown that MAT 13762B tumor induces a specific immune response,[32] and hence may be sensitive to steroid-induced impairment of immune function. It is also possible that other physiological changes caused by opioids (e.g., cardiovascular or thermoregulatory) affect the degree of tumor vascularization or the incidence and pattern of metastases. Finally, opioids may influence tumor growth by directly modulating the immune system. Our findings concerning NK suppression induced by both the opioid form of stress and by morphine administration are particularly relevant in this regard.

Collectively, the results of our experiments suggest that exposure to opioids *in vivo* causes suppression of NK function and enhancement of tumor growth. This is in agreement with *in vivo* studies by others showing facilitatory effects of opioids on tumor development and adverse effects on several indices of immune competence. By contrast, using *in vitro* methods others have shown that opioid peptides can either enhance or suppress the responsiveness of T lymphocytes.[13-15] Studying the *in vitro* effects of opioids on NK-cell activity has shown that several opioid peptides enhance the cytotoxic activity of NK cells.[11,12,33,34]

Whether opiates and opioid peptides act directly or indirectly upon immune cells remains to be determined. Certainly there is evidence for opiate receptors on monocytes, granulocytes, lymphocytes, mast cells, and terminal complexes of human complement.[11-15] In addition, Blalock and his colleagues have shown structural similarities between interferon, ACTH, and endorphins. These investigators have also shown that intracerebral administration of interferon induces a naloxone-sensitive analgesia and catatonia; and that lymphocytes can produce ACTH and endorphin-like peptides along with interferon.[35] Thus, there seems to be a close relation between opioid peptides and the immune system.

## IV. MODULATION OF IMMUNITY BY OTHER STRESS-RELATED NEUROPEPTIDES AND PITUITARY HORMONES

In the last decade, several neuropeptides and hypophysial hormones were shown to modulate immune function. Many of these substances are known to be released during stress and many mediate some of the effects of stress on the immune system. In the following section, the effects of several of these stress-related hormones on immunity will be reviewed.

## A. VASOPRESSIN

Several lines of evidence suggest that vasopressin is involved in the hypothalamic modulation of the stress response; vasopressin is secreted by stressful stimuli,[36-38] potentiates the release of ACTH by corticotropin-releasing factor (CRF),[39,40] and thus modulates the activation of the hypothalamic-pituitary-adrenal (HPA) axis during stress.[41] Brattleboro (DI) rats, which are homozygous for diabetes insipidus and lack vasopressin, exhibit altered hormonal, physiological, and behavioral responses to stress; both an increase and a decrease in the response to stress were shown. For example, DI rats show more stress-induced stomach ulcers,[42] but reduced hypophysial-adrenal axis activation by stress[43] and no swimming stress-induced analgesia.[44]

Several lines of evidence support a role for argenine-vasopressin (AVP) in immunoregulation. AVP levels in the brain and plasma are altered by stimuli that are known to activate the immune system. For example, brain levels of immunoreactive AVP are elevated after experimental induction of swelling and inflammation of arthritic joints in the rat,[45] and plasma AVP increases after administration to rats of bacterial endotoxin, widely used as an experimental model of infection.[46] These alterations in AVP levels may result from an immune-nervous system interaction, as hypothalamic magnocellular neurons, possibly vasopressinergic, have been shown to accumulate circulating endogenous and exogenous immunoglobulins.[47] Vasopressin was found to affect directly several components of the immune system; for example, AVP was found to promote the production of $\gamma$-interferon in mitogen-stimulated lymphocytes, thus acting as a substitute for interleukin-2;[48] this effect may be mediated by AVP receptors on T-cells.[49] Arginine vasopressin also stimulates prostaglandin synthesis by human mononuclear phagocytes.[50] Finally, AVP immunoreactivity is found in lymphoid tissues such as the thymus and lymph nodes.[51,52]

We have recently studied the possibility that vasopressin is involved in regulation of the immune system by comparing the cytotoxic activity of NK cells in DI and Long Evans (LE) rats, the strain from which DI rats were derived. Additionally, we compared the effects of swimming stress, morphine administration, and vasopressin replacement on NK-cell activity in these two strains. In all experiments, NK-cell activity in DI rats was significantly higher than in LE rats, ranging between 149 and 189% above that of LE rats.[53] Both swimming stress and morphine administration significantly suppressed NK activity in DI and LE rats, but there was no difference in the level of suppression between the two strains. Vasopressin replacement, which was effective in normalizing water intake in DI rats, did not have a significant effect on their NK-cell activity. DI rats exhibited lower plasma corticosterone levels, which were not elevated by vasopressin replacement. These results suggest that vasopressin does not directly modulate NK-cell activity, but that other differences between DI and LE rats, secondary to the vasopressin deficiency (e.g., lower corticosterone levels, higher pituitary growth hormone [GH], and higher plasma prolactin [PRL] levels), are responsible for the elevated baseline NK-cell activity. Neither vasopressin nor other hormones affected by vasopressin deficiency seem to be involved in the acute modulating effects of stress and morphine on NK cells.

## B. CORTICOTROPIN-RELEASING FACTOR (CRF) AND ADRENOCORTICOTROPIC HORMONE (ACTH)

In 1936, Selye reported that diverse noxious agents cause an enlargement of the adrenal cortex as a consequence of the "stress syndrome". Later, many investigators observed that a variety of stressful events cause the release of ACTH from the anterior pituitary, which in turn stimulates synthesis and release of corticosteroids from the adrenal cortex.[54] ACTH release was found to be regulated mainly by the hypothalamic CRF, which is released following stressful stimuli and mediates the activation of the HPA axis during stress.[55,56]

Almost any aspect of the immune system can be modulated to some degree by glucocorticosteroids.[57] Although most of the effects of CRF and ACTH on immunity are mediated by the release of glucocorticosteroids, both peptides may act directly on immune function. Both high- and low-affinity ACTH receptors were described on lymphocytes.[58] Furthermore, the ACTH receptor on mouse and human mononuclear cells was found to be similar if not identical to the receptor found in the adrenal cortex.[59]

Exogenous administration of ACTH *in vivo* was shown to suppress several immune functions tested. For example, ACTH depresses antibody formation,[60] produces atrophy of lymphoid tissue,[61] suppresses inflammation, and prolongs the survival of skin grafts.[62,63] Studies *in vitro* showed that ACTH inhibits antibody response to T-cell dependent and independent antigens and suppresses the mitogen induction of $\gamma$-interferon by spleen cells.[58,64] Recent evidence suggests that CRF plays a modulatory role in NK cytotoxicity.[65] Intraventricular, but not systemic or *in vitro,* administration of CRF suppressed NK-cell activity. This suppression was antagonized by central, but not systemic, preadministration of CRF antagonist.[65]

## C. GROWTH HORMONE (GH)

In rats, a decrease in plasma GH follows both acute and prolonged stress.[66,67] This decrease was found to be mediated by CRF.[68] On the other hand, studies in rhesus and squirrel monkeys have demonstrated an increase in plasma GH levels during stress.[69,70] Humans also display elevated GH levels during psychological stress, particularly the more anxious and neurotic subjects.[71,72]

Certain lymphoid cell lines and normal mouse thymocytes have been shown to express specific and high avidity receptors for GH.[73,74] However, other studies (e.g., Reference 75) were unable to demonstrate clearly detectable receptors on any lymphoid tissue tested.

GH was shown to slightly enhance antibody levels after immunization of rats with Pasteurella pestis extracts. Additionally, GH was found to antagonize the immunosuppressive effect of ACTH in that paradigm.[60] In aged animals, injections of GH or implantation of GH- and prolactin-secreting pituitary adenoma cells could completely reverse thymic atrophy, restore T-cell dependent immune function, and enhance NK-cell cytotoxicity.[76,77] In man, nanogram concentrations of GH potentiated colony formation by normal T-cell and induced lymphoproliferation.[78,79] On the other hand, GH treatment in children with GH deficiency was reported to decrease percent B-cells, T-helper/suppressor ratio, and mitogen responses.[80]

Hypophysectomy has been shown to suppress many immunological functions. Several of these functions could be restored by GH replacement. In mice, GH treatment reversed the depression of immune reactivity to antigens *in vitro,*[81] and elevated the hypophysectomy-induced reduction in NK-cell activity.[82] In rats, GH treatment to hypophysectomized animals opposed cortisol-induced leukopenia,[83] and restored normal production of precipitating antibodies and skin rejection response.[84,85]

## D. PROLACTIN (PRL)

Many types of physical or emotional stress induce secretion of prolactin. For example, ether anesthesia,[86] cold stress,[87] immobilization stress,[88] foot shock,[89] and surgical stress[88] have all been reported to elevate plasma PRL levels. This prolactin release may be mediated by prolactin-releasing hormones[86] and regulated by other factors, such as opioid peptides.[90]

Although several studies failed to find PRL receptors on lymphoid tissue, other studies report the existence of PRL receptors on both lymphocytes and monocytes.[91,92] Daily PRL injections during immunization against sheep red blood cells (SRBC) increase antibody production.[93] Under certain conditions, PRL was also found to dramatically increase the proliferation response of mouse splenocytes to Con A.[94] Selective inhibition of PRL by the dopamine agonist drug, bromocriptine (BRC), inhibited the development of contact sensitivity reaction to dinitrochlorobenzene (DNCB), the antibody formation against SRBC, and the development of adjuvant arthritis.[95,96] BRC also inhibited the *in vivo* tumoricidal activation of peritoneal macrophages, probably by interfering with the production of macrophage activating factors by lymphocytes.[94] All of the BRC effects could be reversed by PRL replacement.

PRL was also found to restore several impaired immune functions in hypophysectomized animals, including antibody response to SRBC, contact sensitivity to DNCB, and adjuvant arthritis response.[75,96] This restoration was obtained at a minimum effective dose that was approximately ten times lower than the dose required for induction of lactation, and it was antagonized by simultaneously administered ACTH.[75,95,97]

## V. CONCLUSIONS

Our data indicate that a particular form of foot shock stress can suppress immune function in rats and decrease their resistance to tumor challenge. These effects appear to be mediated by opioid peptides released by stress, and they can be mimicked by high doses of morphine given systemically or by a vastly smaller dose delivered intracerebrally. Our recent findings indicate that the immune-suppressive effect of morphine is mediated by opiate receptors in the brain, possibly by activating the hypothalamic-pituitary-adrenal axis.

Although knowledge of the precise involvement of the immune system in surveillance against tumors is limited, there is evidence to suggest that NK cells play a pivotal role.[16] In this regard, we found that the effects of the opioid form of foot shock stress and morphine on NK cytotoxic activity closely parallel their effects on the survival of rats challenged with tumor cells. In fact, every manipulation that caused an opioid-mediated suppression of NK cytotoxicity in rats also decreased their resistance to tumor challenge. Although these data are suggestive, it is premature to conclude that NK suppressive effects of stress are responsible for the decreased survival after tumor challenge.

These studies also make it clear that Selye's monolithic concept of stress is no longer tenable.[98] The data reveal that stress is not a unitary phenomenon: seemingly minor differences in stress parameters (e.g., the same total amount of foot shock applied intermittently vs. continuously) determine whether or not stress will be immune-suppressive and tumor-enhancing, just as such parametric variations were seen to determine the neurochemical basis of stress analgesia. Moreover, others have shown that exposure to inescapable but not escapable shock suppresses T-cell responsiveness to mitogens and enhances tumor development in rats.[99-101] Corroborating this evidence for the importance of the controllability of stress, we have recently completed a study in collaboration with Maier's group from Colorado, showing that inescapable but not escapable tail shock suppresses NK activity in rats,[102] and preliminary data suggest that this effect is blocked by naloxone. It seems reasonable to conclude that the experience of helplessness is important for the immune-suppressive and tumor-enhancing effects of stress, and that opioid peptides are involved in mediating these immunologic and oncologic effects.

More generally, the findings presented in this chapter support the view that stress, by altering brain and plasma levels of certain neurotransmitters, neuropeptides, and hormones plays a role in immunoregulation. The findings also support the notion that the nervous system, by significantly modulating immune function, exercises some measure of control over the inception and development of certain disease processes. Such findings fit well into the emerging field of behavioral neuroimmunology and reinforce continuing efforts to elucidate the neural and neurohumoral mechanisms by which the environment can affect the organism's immune system and susceptibility to disease.

## ACKNOWLEDGMENTS

Yehuda Shavit was supported by the John D. and Catherine T. MacArthur Foundation. Raz Yirmiya was supported by the William S. Bettingin Endowment Fund for Psychoneuroimmunology. This research was supported in part by NIH grant NS07628.


## REFERENCES


1. **Tobach, E. and Bloch, H.,** Effect of stress by crowding prior to and following tuberculous infection, *Am. J. Physiol.*, 187, 399, 1958.
2. **Monjan, A. A.,** Stress and immunologic competence: studies in animals, in *Psychoneuroimmunology*, Ader, R., Ed., Academic Press, New York, 1981, 195.
3. **Solomon, G. F. and Amkraut, A. A.,** Psychoneuroendocrinological effects on the immune response, *Annu. Rev. Microbiol.*, 35, 155, 1981.
4. **Ader, R., Ed.,** *Psychoneuroimmunology*, Academic Press, New York, 1981.
5. **Sklar, L. S. and Anisman, H.,** Stress and cancer, *Psychol. Bull.*, 89, 369, 1981.
6. **Riley, V.,** Psychoneuroendocrine influences on immunocompetence and neoplasia, *Science*, 212, 1100, 1981.
7. **Lewis, J. W., Cannon, J. T., and Liebeskind, J. C.,** Opioid and nonopioid mechanisms of stress analgesia, *Science*, 208, 623, 1980.
8. **Terman, G. W., Shavit, Y., Lewis, J. W., Cannon, J. T., and Liebeskind, J. C.,** Intrinsic mechanisms of pain inhibition: activation by stress, *Science*, 226, 1270, 1984.
9. **Maier, S. F., Sherman, J. E., Lewis, J. W., Terman, G. W., and Liebeskind, J. C.,** The opioid-nonopioid nature of stress-induced analgesia and learned helplessness, *J. Exp. Psychol. (Anim. Behav.)*, 9, 80, 1983.
10. **Greening, A. P.,** Opiates, opioid peptides, and immunity, *Lancet*, 1, 774, 1984.
11. **Zagon, I. S. and McLaughlin, P. J.,** Endogenous opioid systems, stress, and cancer, in *Enkaphalins and Endorphins: Stress and the Immune System*, Plotnikoff, N. P., Faith, R. E., Murgo, A. J., and Good, R. A., Eds., Plenum Press, New York, 1986, 81.
12. **Plotnikoff, N. P. and Miller, G. C.,** Enkephalins as immunomodulators, *Int. J. Immunopharmacol.*, 5, 437, 1983.
13. **Weber, R. J. and Pert, C. B.,** Opiatergic modulation of the immune system, in *Central and Peripheral Endorphins: Basic and Clinical Aspects*, Müler, E. E. and Genazzani, A. R., Eds., Raven Press, New York, 1984, 35.
14. **Teschemacher, H. and Scweigerer, L.,** Opioid peptides: do they have immunological significance?, *TIPS*, 6, 368, 1985.
15. **Plotnikoff, N. P., Faith, R. E., Murgo, A. J., and Good, R. A., Eds.,** *Enkephalins and Endorphins: Stress and the Immune System*, Plenum Press, New York, 1986.
16. **Herberman, R. B. and Ortaldo, J. R.,** Natural killer cells: their role in defenses against disease, *Science*, 214, 24, 1981.
17. **Shavit, Y., Lewis, J. W., Terman, G. W., Gale, R. P., and Liebeskind, J. C.,** Opioid peptides mediate the suppresive effect of stress on natural killer cell cytotoxicity, *Science*, 223, 188, 1984.
18. **Shavit, Y., Martin, F. C., Yirmiya, R., Ben-Eliyahu, S., Terman, G. W., Weiner, H., Gale, R. P., and Liebeskind, J. C.,** Effects of a single administration of morphine or footshock stress on natural killer cell cytotoxicity, *Brain Behav. Immun.*, 1, 318, 1987.
19. **Beilin, B., Martin, F. C., Shavit, Y., Gale, R. P., and Liebeskind, J. C.,** Suppression of natural killer cell activity by high-dose narcotic anesthesia in rats, *Brain Behav. Immun.*, 1989, in press.
20. **Shavit, Y., Terman, G. W., Lewis, J. W., Zane, C. J., Gale, R. P., and Liebeskind, J. C.,** Effects of footshock stress and morphine on natural killer lymphocytes in rats: studies of tolerance and cross-tolerance, *Brain Res.*, 372, 382, 1986.
21. **Shavit, Y., Depaulis, A., Martin, F. C., Terman, G. W., Pechnick, R. N., Zane, C. J., Gale, R. P., and Liebeskind, J. C.,** Involvement of brain receptors in the immune-suppressive effect of morphine, *Proc. Natl. Acad. Sci. U.S.A.*, 83, 7114, 1986.
22. **Martin, F. C., Shavit, Y., Terman, G. W., Pechnick, R. N., Oh, C., and Liebeskind, J. C.,** Stereospecificity of opiate immunosuppression in rats, *Soc. Neurosci. Abstr.*, 11, 86, 1985.

23. **Sohn, J.-H., Shavit, Y., Morgan, M. M., Martin, F. C., Melendez, A. R. J., Gale, R. P., and Liebeskind, J. C.**, Brain μ-opioid receptors are involved in morphine-induced suppression of immune function, *Soc. Neurosci. Abstr.*, 12, 339, 1986.
24. **Felten, D. L., Felten, S. Y., Carlson, S. L., Olschowka, J. A., and Livnat, S.**, Noradrenergic and peptidergic innervation of lymphoid tissue, *J. Immunol.*, 135, 755s, 1985.
25. **Hochman, P. S. and Cudkowicz, G.**, Suppression of natural cytotoxicity by spleen cells of hydrocortisone-treated mice, *J. Immunol.*, 123, 968, 1979.
26. **Morley, J. E.**, Neuroendocrine effects of endogenous opioid peptides in human subjects: a review, *Psychoneuroendocrinology*, 8, 361, 1983.
27. **Shavit, Y., Martin, F. C., Angarita, G. M. L., Gale, R. P., and Liebeskind, J. C.**, Morphine-induced suppression of natural killer (NK) cells is mediated by the adrenal gland, *Soc. Neurosci. Abstr.*, 12, 339, 1986.
28. **Shavit, Y., Lewis, J. E., Terman, G. W., Gale, R. P., and Liebeskind, J. C.**, Endogenous opioids may mediate the effects of stress on tumor growth and immune function, *Proc. West. Pharmacol. Soc.*, 26, 53, 1983.
29. **Lewis, J. W., Shavit, Y., Terman, G. W., Nelson, L. R., Gale, R. P., and Liebeskind, J. C.**, Apparent involvement of opioid peptides in stress-induced enhancement of tumor growth, *Peptides*, 4, 635, 1983.
30. **Lewis, J. W., Shavit, Y., Terman, G. W., Gale, R. P., and Liebeskind, J. C.**, Stress and morphine affect survival of rats challenged with a mammary ascites tumor (MAT 13762B), *Nat. Immun. Cell Growth Regul.*, 3, 43, 1983.
31. **Bogden, A. E., Taylor, D. J., Kuo, E. Y. H., Mason, M. M., and Speropoulos, A.**, The effect of perphenazine-induced serum prolactin response on estrogen primed mammary tumor-host systems, 13762 and R-35 mammary adeno-carcinomas, *Cancer Res.*, 34, 3018, 1974.
32. **Kreider, J. W., Bartllet, G. L., Purnell, D. M., and Webb, S.**, Immunotherapy of an established rat mammary adenocarcinoma (13762A) with intratumor injection of *Corynebacterium parvum*, *Cancer Res.*, 38, 689, 1978.
33. **Mathews, P. M., Froelich, C. J., Sibbit, W. L., Jr., and Bankhurst, A. D.**, Enhancement of natural cytotoxicity by β-endorphine, *J. Immunol.*, 130, 1658, 1983.
34. **Froelich, C. J. and Bankhurst, A. D.**, The effect of β-endorphine on natural killer cytotoxicity and antibody dependent cellular cytotoxicity, *Life Sci.*, 35, 26, 1984.
35. **Blalock, J. E.**, The relationship between neuroendocrine hormones and lymphokines, *Lymphokines*, 9, 1, 1984.
36. **Mirski, I. A., Stein, M., and Paulisch, G.**, The secretion of an antidiuretic substance into the circulation of rats exposed to noxious stimuli, *Endocrinology*, 54, 49, 1954.
37. **Felsl, I., Gottsmann, M., Eversmann, T., Jehle, W., and Uhlich, E.**, Influence of various stress situations on vasopressin secretion in man, *Acta Endocrinol.*, 215, 122, 1978.
38. **Knepel, W., Nutto, D., and Hertting, G.**, Effect of neonatal treatment with monosodium glutamate on vasopressin release during footshock stress in the rat, *Life Sci.*, 33, 1703, 1983.
39. **Yates, F. E., Russell, S. M., Hedge, G. A., McCann, S. M., and Dhariwal, A. P. S.**, Potentiation by vasopressin of corticotropin release induced by corticotropin-releasing factor, *Endocrinology*, 88, 3, 1971.
40. **Gillies, G. E., Linton, E. A., and Lowry, P. J.**, Corticotropin releasing activity of the new CRF potentiated several times by vasopressin, *Nature (London)*, 299, 355, 1982.
41. **Gibbs, D. M.**, Vasopressin and oxytocin: hypothalamic modulators of the stress response: a review, *Psychoneuroendocrinology*, 11, 131, 1986.
42. **Wideman, C. H. and Murphy, H. M.**, Effects of vasopressin deficiency, age, and stress on stomach ulcer induction in rats, *Peptides*, 6 (Suppl.), 63, 1985.
43. **McCann, S. M., Antunes-Rodrigues, J., Nallar, R., and Valtin, H.**, Pituitary-adrenal function in the absence of vasopressin, *Endocrinology*, 79, 1058, 1966.
44. **Bodnar, R. J., Zimmerman, E. A., Nilaver, G., Mansour, A., Thomas, L. W., Kelly, D. D., and Glusman, M.**, Dissociation of cold-water swim and morphine analgesia in brattleboro rats with diabetes insipidus, *Life Sci.*, 26, 1581, 1980.
45. **Millan, M. J., Millan, M. H., Colpaert, F. C., and Herz, A.**, Chronic arthritis in the rat: differential changes in discrete brain pools of vasopressin as compared to oxytocin, *Neurosci. Lett.*, 54, 33, 1985.
46. **Kastin, N. W.**, Characteristics of body temperature, vasopressin, and oxytocin responses to endotoxin in the rat, *Can. J. Physiol. Pharmacol.*, 64, 1575, 1986.
47. **Meeker, M. L., Meeker, R. B., and Hayward, J. N.**, Accumulation of circulating endogenous and exogenous immunoglobulins by hypothalamic magnocellular neurons, *Brain Res.*, 423, 45, 1987.
48. **Johnson, H. M., Farrar, W. L., and Torres, B. A.**, Vasopressin replacement of interleukin 2 requirement in gamma interferon production: lymphokine activity of a neuroendocrine hormone, *J. Immunol.*, 129, 983, 1982.
49. **Johnson, H. M. and Torres, B. A.**, Regulation of lymphokine production by arginine vasopressin and oxytocin: modulation of lymphocyte function by neurohypophysial hormones, *J. Immunol.*, 135, 773s, 1985.

50. **Locher, R. Vetter, W., and Block, L. H.**, Interactions between 8-L-arginine vasopressin and prostaglandin-E2 in human mononuclear phagocytes, *J. Clin. Invest.*, 71, 884, 1983.
51. **Markwick, A. J., Lolait, S. J., and Funder, J. W.**, Immunoreactive arginine vasopressin in the rat thymus, *Endocrinology*, 119, 1690, 1986.
52. **Aravich, P. F., Silverman, W. F., Sladek, C. D., Felton, S. Y., Felton, D. L., and Sladek, J. R., Jr.**, Aging and water deprivation alter vasopressin content in lymphoid tissue and gut, *Soc. Neurosci. Abstr.*, 13, 1369, 1987.
53. **Yirmiya, R., Shavit, Y., Ben-Eliyahu, S., Martin, F. C., Weiner, H., and Liebeskind, J. C.**, Natural killer cell activity in vasopressin-deficient rats (Brattleboro strain), *Brain Res.*, 479, 16, 1989.
54. **Axelrod, J. and Reisine, T. D.**, Stress hormones: their interaction and regulation, *Science*, 224, 452, 1984.
55. **Britton, D. R., Koob, G., Rivier, J., and Vale, W.**, Stress-induced secretion of corticotropin releasing factor immunoreactivity in rat cerebrospinal fluid, *Soc. Neurosci. Abstr.*, 10, 94, 1984.
56. **Rivier, C., Rivier, J., and Vale, W.**, Inhibition of adrenocorticotropic hormone secretion in the rat by immunoneutralization of corticotropin-releasing factor, *Science*, 218, 377, 1982.
57. **Fauci, A. S.**, Mechanisms of the immunosuppressive and anti-inflammatory effects of glucocorticoids, *J. Immunopharmacol.*, 1, 1, 1978.
58. **Johnson, H. M., Smith, E. M., Torres, B. A., and Blalock, J. E.**, Regulation of the *in vitro* antibody response by neuroendocrine hormones, *Proc. Natl. Acad. Sci. U.S.A.*, 79, 4171, 1982.
59. **Blalock, J. E., Bost, K. L., and Smith, E. M.**, Neuroendocrine peptide hormones and their receptors in the immune system, *J. Neuroimmun.*, 10, 31, 1985.
60. **Hayashida, T. and Li, C. H.**, Influence of adrenocorticotropic and growth hormone on antibody formation, *J. Exp. Med.*, 105, 93, 1957.
61. **Dougherty, T. F.**, Effects of hormones on lymphatic tissue, *Physiol. Rev.*, 32, 279, 1952.
62. **Osgood, C. K. and Favour, C. B.**, Effect of adrenocorticotropic hormone on inflammation due to tuberculin hypersensitivity and turpentine and on circulating antibody levels, *J. Exp. Med.*, 94, 415, 1951.
63. **Medawar, P. P. and Sparrow, E. M.**, Effect of adrenocortical hormones, adrenocorticotropic hormone and pregnancy on skin transplantation, *J. Endocrinol.*, 14, 240, 1956.
64. **Johnson, H. M., Torres, B. A., Smith, E. M., Dion, L. D., and Blalock, J. E.**, Regulation of lymphokine (gamma-interferon) production by corticotropin, *J. Immunol.*, 132, 246, 1984.
65. **Irwin, M. R., Vale, W., and Britton, K. T.**, Central corticotropin-releasing factor suppresses natural killer cytotoxicity, *Brain Behav. Immun.*, 1, 81, 1987.
66. **Brown, G. M. and Martin, J. B.**, Corticosterone, prolactin, growth hormone response to handling and new environment in the rat, *Psychosom. Med.*, 36, 241, 1974.
67. **Lenox, F. H., Kant, G. J., Sessions, G. R., Pennington, L. L., Mougey, E. H., and Meyerhoff, J. L.**, Specific hormonal and neurochemical responses to different stressors, *Neuroendocrinology*, 21, 300, 1980.
68. **Rivier, C. and Vale, W.**, Involvement of corticotropin-releasing factor and somatostatin in stress-induced inhibition of growth hormone secretion in the rat, *Endocrinology*, 117, 2478, 1985.
69. **Mason, J. W., Wool, M. S., Wherry, F. E., Pennington, L. L., Brady, J. V., and Beer, B.**, Plasma growth hormone response to avoidance sessions in the monkey, *Psychosom. Med.*, 30, 760, 1968.
70. **Brown, G. M., Schalch, D. S., and Reichlin, S.**, Patterns of growth hormone and cortisol responses to psychological stress in the squirrel monkey, *Endocrinology*, 88, 956, 1971.
71. **Miyabo, S., Asato, T., and Mizushima, N.**, Psychological correlates of stress-induced cortisol and growth hormone releases in neurotic patients, *Psychosom. Med.*, 41, 515, 1979.
72. **Schalach, D. S.**, The influence of physical stress and exercise on growth hormone and insulin secretion in man, *J. Lab. Clin. Med.*, 69, 256, 1967.
73. **Lesniak, M. A., Roth, J., Gordon, P., and Gavin, J. R., III**, Human growth hormone radioreceptor assay using cultured human lymphocytes, *Nature (London)*, 241, 20, 1973.
74. **Arrenbbrecht, S.**, Specific binding of growth hormone to thymocytes, *Nature (London)*, 252, 255, 1974.
75. **Nagy, E., Berczi, I., and Friesen, H. G.**, Regulation of immunity in rats by lactogenic and growth hormones, *Acta Endocrinol.*, 102, 351, 1983.
76. **Kelly, K. W., Brief, S., Westly, H. J., Novakofski, J., Bechtel, P. J., Simon, J., and Walker, E. B.**, GH3 pituitary adenoma cells can reverse thymic aging in rats, *Proc. Natl. Acad. Sci. U.S.A.*, 83, 5663, 1986.
77. **Davila, D. R., Brief, S., Simon, J., Hammer, R. E., Brinster, R. L., and Kelly, K. W.**, Role of growth hormone in regulating T-dependent immune events in aged, nude, and transgenic rodents, *J. Neurosci. Res.*, 18, 108, 1987.
78. **Mercola, K. E., Cline, M. J., and Golde, D. W.**, Growth hormone stimulation of normal and leukemic human T-lymphocyte proliferation *in vitro*, *Blood*, 58, 337, 1981.
79. **Kiess, W., Holtmann, H., Butenandt, O., and Eife, R.**, Modulation of lymphoproliferation by human growth hormone, *Eur. J. Pediatr.*, 140, 47, 1983.

80. **Rappaport, R., Oleske, J., Ahdieh, H., Solomon, S., Delfaus, C., and Denny, T.,** Suppression of immune function in growth hormone-deficient children during treatment with human growth hormone, *J. Pediatr.*, 109, 434, 1986.
81. **Gisler, R. and Schenkel-Hulliger, L.,** Hormonal regulation of the immune response. II. Influence of pituitary and adrenal activity on immune responsiveness *in vitro*, *Cell. Immunol.*, 2, 646, 1971.
82. **Saxena, Q. B., Saxena, R. K., and Adler, W. H.,** Regulation of natural killer cell activity *in vivo*. III. Effect of hypophysectomy and growth hormone treatment on the natural killer activity of the mouse spleen cell population, *Int. Arch. Allergy Appl. Immunol.*, 67, 169, 1982.
83. **Chatterton, R. T., Jr., Murray, C. L., and Hellman, L.,** Endocrine effects on leukocytopoiesis in the rat. I. Evidence for growth hormone secretion as the leukopoietic stimulus following acute cortisol-induced lymphopenia, *Endocrinology*, 92, 775, 1973.
84. **Comsa, J., Schwarz, J. A., and Neu, H.,** Interaction between thymic hormone and hypophysial growth hormone on production of precipitating antibodies in the rat, *Immunol. Commun.*, 3, 11, 1974.
85. **Comsa, J., Leonhardt, H., and Schwarz, J. A.,** Influence of the thymus-corticotropin-growth hormone interaction on the rejection of skin allografts in the rat, *Ann. N.Y. Acad. Sci.*, 249, 387, 1975.
86. **Shin, S. H.,** Prolactin secretion in acute stress is controlled by prolactin releasing factor, *Life Sci.*, 25, 1829, 1979.
87. **Jobin, M., Feland, L., Cote, J., and Labrie, R.,** Effect of exposure to cold on hypothalamic TRH and plasma levels of TSH and prolactin in the rat, *Endocrinology*, 92, 1256, 1975.
88. **Riegel, G. D. and Meites, J.,** The effects of stress on serum prolactin in the female rat, *Proc. Soc. Exp. Biol. Med.*, 152, 441, 1976.
89. **Kant, G. J., Leu, J. R., Anderson, S. M., and Mougey, E. H.,** Effect of chronic stress on plasma corticosterone, ACTH and prolactin, *Physiol. Behav.*, 40, 775, 1987.
90. **Rossier, J., French, E., Rivier, C., Shibasaki, T., Guillemin, R., and Bloom, F. E.,** Stress-induced release of prolactin: blockade by dexamethasone and naloxone may indicate beta-endorphine mediation, *Proc. Natl. Acad. Sci. U.S.A.*, 77, 666, 1980.
91. **Russell, D. H., Matrigian, L., Kibler, R., Larson, D., and Magun, B.,** Prolactin receptors on human lymphocytes and their modulation by cyclosporin, *Biochem. Biophys. Res. Commun.*, 121, 899, 1984.
92. **Bellussi, G., Muccioli, G. Ghe, C., and Di Carlo, R.,** Prolactin binding sites in human erythyrocytes and lymphocytes, *Life Sci.*, 41, 951, 1987.
93. **Spangelo, B. L., Hall, N. R., NcGillis, J. P., and Goldstien, A. L.,** Evidence for an interaction between prolactin and the primary immune response, *Fed. Proc.*, 43, 1610, 1984.
94. **Brenton, E., Hartmann, D., Gibreath, M., Holaday, J., and Meltzer, S.,** Inhibition of macrophages *in vivo* activation by pharmacologic blockade of prolactin release, in *Leukocytes and Host Defense*, Oppenheim, J. J. and Jacobs, D. M., Eds., Alan R. Liss, New York, 1986, 213.
95. **Nagy, E., Berczi, I., Wren, G. E., Asa, S. L., and Kovacs, K.,** Immuno-modulation by bromocriptine, *Immunopharmacology*, 6, 231, 1983.
96. **Berczi, I. and Nagy, E.,** A possible role of prolactin in adjuvant arthritis, *Arthritis Rheum.*, 25, 591, 1982.
97. **Berczi, I., Nagy, E., Asa, S. L., and Kovacs, K.,** Pituitary hormones and contact sensitivity in rats, *Allergy*, 38, 325, 1983.
98. **Selye, H.,** *The Stress of Life*, McGraw-Hill, New York, 1956.
99. **Laudenslager, M. L., Ryan, S. M., Drugan, R. C., Hyson, R. L., and Maier, S. F.,** Coping and immunosuppression: inescapable but not escapable shock suppresses lymphocytes proliferation, *Science*, 221, 568, 1983.
100. **Sklar, L. S. and Anisman, H.,** Stress and coping factors influence tumor growth, *Science*, 205, 513, 1979.
101. **Visintainer, M. A., Volpicelli, J. R., and Seligman, M. E. P.,** Tumor rejection in rats after inescapable or escapable shock, *Science*, 216, 437, 1982.
102. **Shavit, Y., Ryan, S. M., Lewis, J. W., Laudenslager, M. L., Terman, G. W., Maier, S. F., Gale, R. P., and Liebeskind, J. C.,** Inescapable but not escapable stress alters immune function, *Physiologist*, 26, A-64, 1983.

Chapter 13

# MODULATION OF PHAGOCYTE ACTIVITY BY SUBSTANCE P AND NEUROTENSIN

**Zvi Bar-Shavit and Rachel Goldman**

## TABLE OF CONTENTS

# I. INTRODUCTION

Research in recent years has brought forth the notion that neural, endocrine, immunological, and inflammatory mechanisms are functionally integrated within a major regulatory axis controlling homeostasis and adaptation of the organism to its external and internal environment. The immune system and the neuroendocrine system affect each other via molecules and receptors shared by both systems. Potent effects of various neuropeptides on mast cells, lymphocytes, macrophages, and granulocytes have been documented, and this field of research is only in its beginning. To date, a wide variety of neurotransmitters have been shown to bind and affect the function of phagocytes and their precursors.

At the time we first became interested in the possible interaction of substance P and neurotensin with phagocytes, there was still no indication that these cells could be targets for peptide neurotransmitters. We were characterizing the interaction of macrophages and neutrophils with the phagocytosis-enhancing peptide tuftsin described by Najjar and his coworkers.[1] Structure-function relationship studies carried out by several groups including our work (though not always consistent) showed that most changes in tuftsin structure led to partial or total loss of activity and even produced peptides with antagonistic activity, and that a proline residue between two basic amino acids is essential for expression of tuftsin activity.[2,3] A literature search for similar sequences showed that substance P[4] and neurotensin[4] were good candidates for a tuftsin-like activity (Table 1), and subsequent studies showed that the prediction was correct.

This chapter will be devoted to the effects of these neuropeptides on neutrophil and macrophage activities and to the possible role of these neuropeptides in inflammatory processes.

# II. SUBSTANCE P

The undecapeptide substance P (SP) (Table 1) was discovered almost 60 years ago, and since then it has been one of the most thoroughly studied neuropeptides. SP has a wide distribution both in the central nervous system and in the periphery.[5] Consistent with its wide distribution, SP possesses numerous biological activities. In the periphery, SP is recognized for its hypotensive, vasodilatory, and smooth muscle contracting properties, whereas in the central nervous system, SP is suggested to play a role in exciting many neurons, mainly those involved in sensory neurotransmission.[4] Structure-function studies of SP indicate that a whole range of biological activities of the neuropeptide depend on conservation of the sequence of its C-terminal six or seven amino acids.[6] One SP activity (cAMP increase in cultured neuroblastoma) was found to depend on the amino terminal 4 residues [SP (1 to 4)].[7]

Studying the interaction of SP with phagocytes, we have found that SP enhances the phagocytic capability of murine-resident macrophages (Table 2), thioglycollate-elicited macrophages, and human neutrophils in a dose dependent manner.[8] Enhanced phagocytosis was observed with both an Fc receptor-mediated phagocytosis (IgG coated erythrocytes) and a glucan receptor-mediated phagocytosis (heat killed yeast).[9,10] The enhancing activity resided in the amino terminal 4 amino acid sequence, while the C-terminal heptapeptide was devoid of enhancing activity. The dose response curves of both SP and SP (1 to 4) were similar to that obtained with tuftsin,[11-13] and 50% of the augmenting effect was obtained at 15 n*M*. Binding competition studies with $^3$H-tuftsin showed that both SP and SP (1 to 4), but not the C-terminal sequence peptides, shared the same binding site of tuftsin and exhibited the same affinity for the receptor.[8]

Exploring the mechanism of signal transduction following tuftsin binding to murine macrophages and human neutrophils we found that increased $^{45}Ca^{2+}$ extrusion from pre-

## TABLE 1
## The Sequence of Tuftsin, Substance P (SP), Neurotensin (NT), and Some of their Fragments[a] — Peptides Sharing the Capacity to Augment the Phagocytic Capability of Pbagocytic Cells

| | |
|---|---|
| Tuftsin[b] | Thr–Lys–Pro–Arg |
| SP (1 to 4) | Arg–Pro–Lys–Pro |
| SP[4] | Arg–Pro–Lys–Pro–Gln–Gln–Phe–Phe–Gly–Leu–Met($NH_2$) |
| NT (8 to 13) | Arg–Arg–Pro–Tyr–Ile–Leu |
| NT (6 to 13) | Lys–Pro–Arg–Arg–Pro–Tyr–Ile–Leu |
| NT (1 to 10) | pGlu–Leu–Tyr–Glu–Asn–Lys–Pro–Arg–Arg–Pro |
| NT[4] | pGlu–Leu–Tyr–Glu–Asn–Lys–Pro–Arg–Arg–Pro–Tyr–Ile–Leu |

[a] All amino acids are in the L-configuration.
[b] Superscripts denote references for the sequence.

## TABLE 2
## Effects of SP and Partial Sequences of SP on Phagocytosis by Resident Mouse Peritoneal Macrophages

| | Phagocytosis of[a] | |
|---|---|---|
| | EIgG | HK-yeast |
| Peptide | (% increased phagocytosis over control) | |
| SP | 54 | 77 |
| SP (1 to 4) | 48 | 86 |
| SP (4 to 11) | −3 | 9 |
| SP (5 to 11) | 6 | 20 |
| Tuftsin | 37 | 58 |

[a] Macrophage monolayers were incubated with or without (control) 100 n*M* of a specified peptide, for 15 min at 37°C. The particles (IgG-coated sheep red blood cells (EIgG) or heat killed yeast cells (HK-yeast) were added for additional 45 min. Adapted from Reference 8.

labeled cells was one of the early events.[14] Using a $Ca^{2+}$ sensitive electrode, we could demonstrate that the interaction of SP (1 to 4) with human neutrophils also leads to increased $Ca^{2+}$ extrusion from the cells.[12] The concentration of SP (1 to 4), at which maximum $Ca^{2+}$ release was attained, was also the concentration at which optimal stimulation of phagocytosis was observed.

Several groups have tested the possible role of SP as a chemotaxin. The neuropeptide was found to induce chemotaxis in rabbit and rat neutrophils.[15,16] This activity was attributed to binding to the receptor of the classical chemotaxin *N*-formyl-met-leu-phe (FMLP). SP also stimulated chemotaxis of human monocytes.[17] In this case, however, the possibility of interaction with FMLP receptors was excluded. The $ED_{50}$ in the latter study was approximately 0.1 p*M*, while for chemotaxis of rabbit and rat neutrophils high concentrations ($10^{-5}$ *M*) of SP were required. Thus, the effect on granulocyte chemotaxis may not be physiologically significant.

SP ($10^{-5}$ *M*) was found to be a weak activator of human monocyte cytotoxicity against tumor cells and did not significantly enhance that induced by interferon-γ.[18] SP could however reverse the inhibitory effects that ACTH and somatostatin had on interferon-γ-mediated increased macrophage cytotoxicity. Using mouse peritoneal macrophages, we were unable to demonstrate potentiation of antitumor activity by SP ($10^{-10}$ to $10^{-6}$ *M*) (Bar-Shavit, Z. and Goldman, R., unpublished results, 1980). SP was shown to substantially increase the proportion of monocytes/macrophages able to secrete superoxide anions ($O_2^-$), suggesting an enhanced antimicrobial activity.[18] SP was also shown to modulate some functions of albumin-elicited guinea pig macrophages. In this system a 24 h interaction with SP led to a down regulation of 5′-nucleotidase activity and induced release of lysosomal enzymes.[19] Short periods of up to 1 h sufficed for stimulation of synthesis and release of metabolites of arachidonic acid, such as leukotriene C4 (LTC4), prostaglandin $E_2$ and thromboxane $B_2$.[19] SP was also shown to evoke the release of $O_2^-$ and $H_2O_2$ from *C. parvum*-activated macrophages.[20]

Direct binding studies of SP to guinea pig macrophages revealed a homogeneous class of receptors with a kDa of 14 to 24 n*M*.[19] However, in these studies SP (1 to 4) was practically unable to compete with radiolabeled SP for its binding ($ED_{50}$ $1.7 \times 10^{-4}$ *M*). The C-terminal heptapeptide was also a poor inhibitor of SP binding ($2.1 \times 10^{-5}$ *M*), and thus the authors suggest the need for both ends of the molecule for binding. The reason for this discrepancy from the results obtained with mouse macrophages is not clear. It is possible that in the guinea pig macrophages, the receptor site for SP is different from that of the mouse and still differs form that of mouse synaptosomes in that the C-terminal peptides is not sufficient for the interaction.

## III. NEUROTENSIN

Neurotensin (NT) is a tridecapeptide (Table 1) which was first isolated from bovine hypothalami and subsequently, from a variety of tissues of other species.[4] In the central nervous system, NT participates in the regulation of a variety of physiological and behavioral processes. In the periphery, NT activities include hypotension, inhibition of gastric acid secretion, stimulation of ileum contraction, vasodilation, increase in vascular permeability, and mast cell degranulation.[4] The three latter phenomena are early components of the inflammatory response, which are followed by the recruitment and activation of granulocytes and macrophages.

NT and several partial sequences (Table 1) were tested for their effects on phagocyte functions. Thioglycollate-elicited mouse peritoneal macrophages, resident peritoneal macrophages, and mouse bone marrow-derived *in vitro* differentiated macrophages (BMM) responded to increasing concentrations of NT in a biphasic manner.[13,21] In thioglycollate-elicited macrophages incubation with $10^{-14}$ to $10^{-9}$ *M* NT led to a dose-dependent augmentation of phagocytosis of $^3$H-zymosan particles.[13,22] Further increase in NT concentration resulted in a reduction of the stimulation, reaching the basal response (in the absence of NT) at $>10^{-7}$ *M* NT. The optimal stimulation was observed at 1 n*M* of NT and amounted to an increase of 120% over the control. The Kp (the concentration of NT at which 50% augmentation of phagocytosis was attained) derived from the ascending part of the bell-shaped dose response curve was at 0.01 n*M*, and 50% inhibition of this activity (the descending part) was attained at 10 n*M*. Partial sequences of NT, NT (8 to 13), NT (6 to 13), NT (1 to 10) showed a normal dose response curve with Kp of 18 n*M*, 18 n*M*, and 16 n*M*, respectively, and a plateau in the enhancing effect from $10^{-7}$ to $10^{-5}$ *M*. The dose response curves for the three peptides almost coincided with those of SP (Kp 26 n*M*) and tuftsin (Kp 22 n*M*) carried out in the same experiment.[13] NT (11 to 13) had no effect on the phagocytic capability of macrophages. The interaction of NT with resident peritoneal macrophages and

**TABLE 3**
**Comparison of the Interaction of Neurotensin with Bone Marrow-Derived Macrophages and Thioglycollate Elicited Peritoneal Macrophages**

| Measurement | Bone marrow macrophages | | Peritoneal macrophages | |
|---|---|---|---|---|
| | High-affinity sites | Low-affinity sites | High-affinity sites | Low-affinity sites |
| $K_D$ | 0.4 n*M* | 11 n*M* | 0.9 n*M* | 28 n*M* |
| Binding sites/cell | 15,800 | 178,000 | 4,800 | 33,500 |
| $K_P$ | 0.016 n*M* (↑) | 3.1 n*M* (↓) | 0.01 n*M* (↑) | 10 n*M* (↓) |

*Note:* Thioglycollate-elicited macrophages were cultured at $7 \times 10^5$/well, 71 μg protein/$10^6$ cells. Bone marrow derived macrophages had 38 μg protein/well on the day of assay, (estimated to be $2.4 \times 10^5$ cells/well). $K_P$ equals the concentration of peptide that produces 50% of the maximal enhancement of phagocytosis (↑) or 50% of reduction of the maximal enhancement (↓).

Adapted from References 13 and 21.

with BMM gave essentially similar biphasic dose response curves of phagocytosis. The effect of NT on human neutrophil phagocytosis of starch particles was also tested.[23] The results of five experiments with neutrophils from different donors were varied, but in all cases an augmentation of phagocytosis (40 to 300%) was obtained. The dose response curves were biphasic with an optimum in the range of $10^{-11}$ to $10^{-8}$ *M*. Unlike the findings with macrophages, a second concentration range of augmented phagocytosis, two orders of magnitude higher than the respective optimal values of the first peak of augmented phagocytic activity, was consistantly observed. The latter phenomenon can stem from either a population heterogeneity or the expression of two different binding sites (Goldman, R., Bar-Shavit, Z., and Romeo, D., unpublished results, 1980).

Some insight into the bewildering biphasic effect of NT on phagocytosis (as opposed to the normal dose response curves obtained with the active NT partial sequences, SP, and tuftsin) was gained from direct binding studies of $^3$H-NT. $^3$H-NT bound specifically to both thioglycollate-elicited macrophages and BMM. Scatchard analysis suggested the existence of two classes of binding sites, high affinity sites, and low affinity sites (Table 3).[13] Binding competition studies revealed that NT, NT (8 to 13), NT (6 to 13), SP, SP (1 to 4), and tuftsin inhibited the binding of $^3$H-NT to thioglycollate-elicited macrophages with a $K_I$ of 12, 13, 13, 20, 17, and 16 n*M*, respectively. Only NT inhibited 100% of the binding of $^3$H-NT. The rest of the peptides appeared to be as effective as neurotensin in inhibiting about 85 to 90% of the binding (implying competition for the low affinity sites). Of interest was the observation that NT (1 to 10) was almost noninhibitory, i.e., 20 to 30% inhibition of $^3$H-NT binding at $10^{-4}$ *M* of the peptide. The following conclusions were derived from the binding and phagocytosis studies:

1. The augmentation of phagocytosis by NT is the result of occupation of the high affinity receptors. A rather low occupation is required for the cellular effect (Table 3).
2. Occupation of the low affinity sites by NT results in an inhibition of the enhanced phagocytosis.
3. The low affinity sites of NT are probably identical with the sites that recognize NT (8 to 13), NT (6 to 13), SP, SP (1 to 4), and tuftsin. NT (1 to 4) is probably bound to the high affinity sites.
4. NT serves as an antagonist to activation through the low affinity sites, i.e., at $10^{-7}$

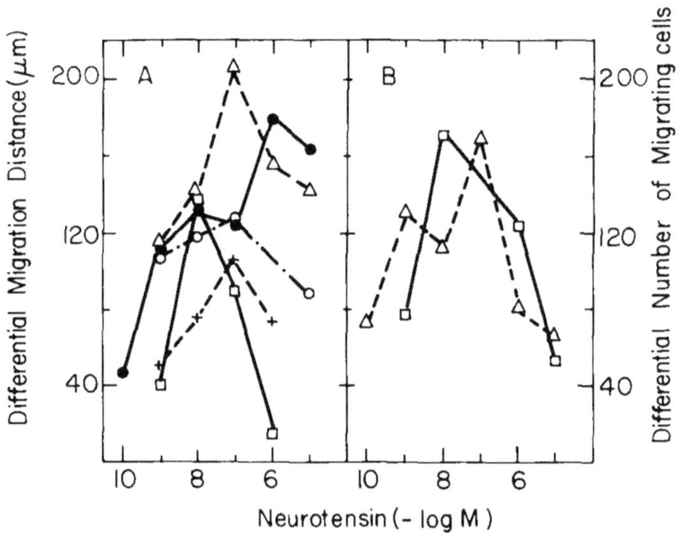


FIGURE 1. Activation by NT of directional migration of human neutrophils. Points are means of three to four determinations run on different slides (SEM not exceeding 5 to 15%) with five neutrophil preparations. A modified method of chemotaxis under agarose[24] (Romeo, D., Bar-Shavit, Z., and Goldman, R., unpublished, 1980).

*M* NT inhibits also the stimulatory effect of $10^{-7}$ *M* SP. NT partial sequences augment phagocytosis by interacting with the low affinity sites, except NT (1 to 10) which interacts with the high affinity sites, but shows the same dose response curve for enhanced phagocytosis as the low affinity site agonists (probably has reduced affinity to the high affinity sites).

5. It is possible that interaction with the high affinity sites requires certain portions of the amino terminal of NT which are shared by NT and NT (1 to 10).

In our hands, SP failed to affect the locomotion of human neutrophils. NT was however effective in enhancing the locomotion of these cells.[23] Figure 1 shows the biphasic effect that NT has on oriented neutrophil locomotion. With different cell preparations maximal oriented migration was seen at $10^{-8}$ to $10^{-6}$ *M* of initial NT concentration. With NT as attractant, neutrophils migrate lower distances than with 10% yeast-activated serum and a lower number of neutrophils respond to NT than to activated serum (unpublished, and Reference 23). In the method used for evaluation of neutrophil migration, the attractant molecules diffuse from wells, where they are placed in the specified initial concentrations into the agarose gel and where they generate a chemical gradient. Thus, the initial NT concentrations sensed by neutrophils are much lower than indicated and the $ED_{50}$ is most likely close to that assessed for phagocytosis-enhancing activity. The dose response curve of the locomotion stimulating activity of NT is biphasic, a characteristic of a number of chemotaxins[25] This has been ascribed to a process of desensitization which may be linked to down regulation of chemoattractant receptors and/or (as will be discussed below) to alterations of intracellular $Ca^{2+}$ steady states.

An increase in the intracellular concentration of free $Ca^{2+}$ is often observed in leukocytes triggered by chemoattractants, and phagocytosis.[26,27] $Ca^{2+}$ can activate several $Ca^{2+}$-dependent proteins whose functions are connected with cell migration and phagocytosis, as well as the plasma membrane $Ca^{2+}$ extrusion system.[28] The extent of a shift of $Ca^{2+}$ from membrane sites to cytosol can be directly evaluated from the rate and the amount of $Ca^{2+}$

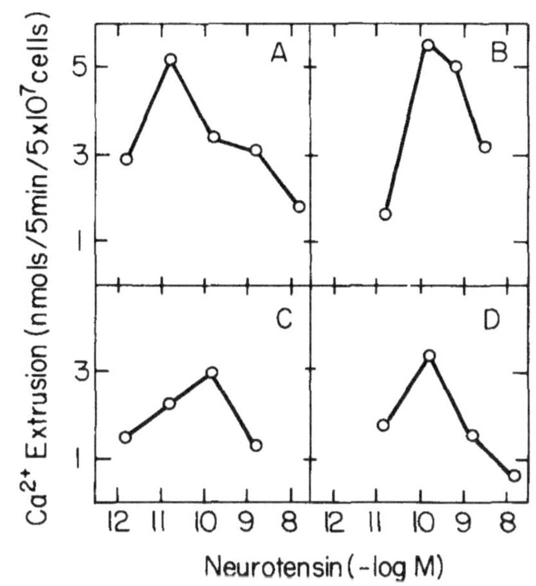


FIGURE 2. $Ca^{2+}$ extrusion from human neutrophils exposed to different concentrations of NT. Experiments were carried out with preparations from four individuals (A to D). $Ca^{2+}$ net fluxes were measured by a very sensitive selective electrode/amplifier/recorder system[29] (Romeo, D., Bar-Shavit, Z., and Goldman, R., unpublished, 1980).

extruded from the cell.[29] It was mentioned previously that both tuftsin[14] and SP (1 to 4)[12] increase $Ca^{2+}$ extrusion from macrophages and neutrophils. In a more extensive study, we have shown that NT also activates $Ca^{2+}$ extrusion from human neutrophils with a maximum cation release at $10^{-10}$ *M* (Figure 2). $Ca^{2+}$ extrusion induced by NT proceeds much more slowly than that elicited by the ionophore A23187 and a phorbol diester.[12,23] Beyond the optimal concentration of about $10^{-10}$ *M*, NT elicits a lower extrusion of $Ca^{2+}$ from neutrophils. The mechanism by which $Ca^{2+}$ extrusion from neutrophils is inhibited at increased NT concentrations is not known. It is possible however that the biphasic dose response of $Ca^{2+}$ extrusion reflects changes in intracellular free $Ca^{2+}$ and relates to the biphasic effects seen on locomotion and phagocytosis.

NT has recently also been shown to potentiate the cytolytic capability of mouse peritoneal macrophages activated with interferon-γ.[30] In this system SP had no enhancing effect on the cytolytic activity induced by interferon-γ. This is in accord with the lack of potentiation of interferon-γ-induced cytotoxicity by human monocytes.[18]

## IV. CONCLUDING REMARKS

Above we have shown that phagocytes (macrophages and neutrophils) express specific binding sites for SP and NT. At least two different binding sites are responsible for the effects of these neuropeptides. The high-affinity binding sites for NT mediate enhancement of phagocytosis and oriented migration and their occupation evokes an enhanced $Ca^{2+}$ extrusion from the cells. SP binding to its receptor likewise induces an enhanced phagocytic response (and possibly chemotaxis) and $Ca^{2+}$ extrusion. In addition, several parameters that indicate cellular activation and perturbation have been described, such as increased tumoricidal activity (induced by SP and NT) and SP-induced, increased superoxide ion release; down regulation of 5′-nucleotidase; release of lysosomal enzymes; and increased release of prostaglandin $E_2$, thromboxane $B_2$, and $LTC_4$. SP also increases the expression of the Fc IgG and Fc IgE receptors on human eosinophils and increases their antibody dependent cellular cytotoxicity (ADCC).[31]

The SP binding sites seem to be shared by tuftsin and to recognize the N-terminal tetrapeptide SP (1 to 4). These same sites bind also NT with a similar affinity (the low-affinity NT binding sites), and their occupation with NT leads to inhibition of responses mediated by these sites. Fragments of NT which lack the N-terminal portion NT (8 to 13) and NT (6 to 13) share the same site with SP and tuftsin and enhance phagocytosis with the same $K_p$.

What is the biological relevance of the phenomena described? There is no definite answer to this question, but the components of phagocyte response and availability of neuropeptides for interaction with phagocytes at various stages of inflammatory reactions can be readily visualized.

Both SP and NT were shown to interact with mast cells and to be strong vasodilators. Both affect the mobilization of granulocytes and enhance phagocytosis by macrophages and granulocytes. The SP affects eosinophil functions.[31] The physiological levels of NT in the blood are in the range that we have shown to be effective *in vitro* and are further increased following fat intake.[32] SP is released from sensory nerves following injury, axon reflex, and antidromic stimulation, and thus is thought to be involved in neurogenic inflammation.[33] Studies with experimental arthritis suggested a direct role for SP in the inflammatory process, i.e., the concentration of SP increases in peripheral nerves that have branches innervating inflamed joints, and infusion of SP into the knee increases the severity of the arthritis.[34] Extension of these studies indicated that the increased inflammatory response is probably due to increased release of collagenase from synovial cells.[35] Macrophages are a source for regulators of synovial cells collagenase release (IL-1 and TNF), and thus, SP (and perhaps NT) may indirectly affect synovial cells by macrophage activation).[36]

It is also of interest that SP significantly stimulates the proliferation of T lymphocytes, an activity which resides in the carboxyterminal octapeptide SP (4 to 11).[37]

It is well recognized now that the autonomic nervous system directly innervates the parenchymal tissue of the thymus, spleen, lymph nodes, and bone marrow, providing an important neural link to the immune system.[38] T and B lymphocyte functions have been shown to be modulated by various neuropeptides[39] and other neurotransmitters as were macrophages and granulocytes and their precursors. The list receptors for neurotransmitters on these cells is ever increasing. There is thus a strong basis for the notion of regulation or modulation of the immune system via neuronal communication with the brain, and neuropeptide-phagocyte communication may be one specific component of this intricate network.

## REFERENCES

1. **Najjar, V. A.,** Biological and biochemical characteristics of the tetrapeptide tuftsin, Thr-Lys-Pro-Arg, in *Macrophages and Lymphocytes*, Escobar, M. R. and Friedman, H., Eds., Plenum Press, New York, 1980, 131.
2. **Fridkin, M. and Gottlieb, P.,** Tuftsin, Thr-Lys-Pro-Arg. Anatomy of an immunologically active peptide, *Mol. Cell. Biochem.*, 41, 73, 1981.
3. **Siemion, I. Z. and Knopinska, D.,** Tuftsin analogues and their biological activity, *Mol. Cell. Biochem.*, 41, 99, 1981.
4. **Leeman, S. E., Mroz, E. A., and Carraway, R. E.,** Substance P and neurotensin, in *Peptides in Neurobiology*, Gainer, H., Ed., Plenum Press, New York, 1977, 99.
5. **Hokfelt, T., Kellerth, J. O., Nilsson, G., and Pernow, B.,** Substance P — localization in the central nervous system and primary sensory neurons, *Science*, 190, 889, 1975.

6. **Bury, R. W. and Mashford, M. L.**, Biological activity of C-terminal partial sequences of substance P, *J. Med. Chem.*, 19, 854, 1976.
7. **Narumi, S. and Maki, Y.**, Stimulatory effects of substance P on neurite extension and cyclic AMP levels in cultured neuroblastoma, *J. Neurochem.*, 30, 1321, 1978.
8. **Bar-Shavit, Z., Goldman, R., Stabinsky, Y., Gottlieb, P., Fridkin, M., Teichberg, V. I., and Blumberg, S.**, Enhancement of phagocytosis — a newly found activity of substance P, residing in its N-terminal tetrapeptide sequence, *Biochem. Biophys. Res. Commun.*, 94, 1445, 1980.
9. **Goldman, R.**, Characteristics of the β-glucan receptor of murine macrophages, *Exp. Cell. Res.*, 174, 481, 1988.
10. **Kadish, J. L., Choi, C. C., and Czop, J. K.**, Phagocytosis of unopsonized zymosan particles by trypsin-sensitive and β-glucan-inhibitable receptors on bone marrow-derived murine macrophages, *Immunol. Res.*, 5, 129, 1986.
11. **Bar-Shavit, Z., Stabinsky, Y., Fridkin, M., and Goldman, R.**, Tuftsin-macrophage interaction: specific binding and augmentation of phagocytosis, *J. Cell. Physiol.*, 100, 55, 1979.
12. **Goldman, R. and Bar-Shavit, Z.**, On the mechanism of the augmentation of the phagocytic capability of phagocytic cells by tuftsin, substance P, neurotensin and kentsin and the interrelationship between their receptors, in *Antineoplastic, Immunogenic and other Effects of the Tetrapeptide Tuftsin: A Natural Macrophage Activator*, Ann. N.Y. Acad. Sci., 419, 1983, 143.
13. **Bar-Shavit, Z., Terry, S., Blumberg, S., and Goldman, R.**, Neurotensin-macrophage interaction: specific binding and augmentation of phagocytosis, *Neuropeptides*, 2, 325, 1982.
14. **Stabinsky, Y., Bar-Shavit, Z., Fridkin, M., and Goldman, R.**, On the mechanism of action of the phagocytosis-stimulating peptide tuftsin, *Mol. Cell. Biochem.*, 30, 71, 1980.
15. **Marasco, W. A., Showell, H. J., and Becker, E. L.**, Substance P binds to the formylpeptide receptor on the rabbit neutrophil, *Biochem. Biophys. Res. Commun.*, 99, 1065, 1981.
16. **Helme, R. D., Eglezos, A., and Hosking, C. S.**, Substance P induces chemotaxis of neutrophils in neural and capsaicin-treated rats, *Immunol. Cell Biol.*, 65, 267, 1987.
17. **Ruff, M., Wahl, S. M., and Pert, C. B.**, Substance P mediated chemotaxis of human monocytes, *Peptides*, 6 (Suppl. 2), 107, 1985.
18. **Peck, R.**, Neuropeptides modulating macrophage function, in *Neuroimmune Interactions: Proc. 2nd Int. Workshop on Neuroimmunomodulation*, Ann. N.Y. Acad. Sci., 496, 1987, 264.
19. **Hartung, H. R., Wolters, K., and Toyka, K. V.**, Substance P: binding properties and studies on cellular responses in guinea pig macrophages, *J. Immunol.*, 136, 3856, 1986.
20. **Hartung, H. P. and Toyka, K. V.**, Activation of macrophages by substance P: induction of oxidative burst and thromboxane release, *Eur. J. Pharmacol.*, 89, 301, 1983.
21. **Goldman, R., Bar-Shavit, Z., Shezen, E., Terry, S., and Blumberg, S.**, Enhancement of phagocytosis by neurotensin, a newly found biological activity of the neuropeptide, in *Macrophages and Natural Killer Cells: Regulation and Function*, Plenum Press, New York, 1982, 133.
22. **Bar-Shavit, Z. and Goldman, R.**, Substance P and neurotensin, *Methods Enzymol.*, 132, 326, 1986.
23. **Goldman, R., Bar-Shavit, Z., and Romeo, D.**, Neurotensin modulates human neutrophil locomotion and phagocytic capability, *FEBS Lett.*, 159, 63, 1983.
24. **Nelson, R. D., Quie, P. G., and Simmons, R. L.**, Chemotaxis under agarose: a new and simple method for measuring chemotaxis and spontaneous migration of human polymorphonuclear leukocytes and monocytes, *J. Immunol.*, 115, 1650, 1975.
25. **Zigmond, S. H.**, Chemotaxis by polymorphonuclear leukocytes, *J. Cell Biol.*, 77, 269, 1978.
26. **Hoffstein, S. T.**, Ultrastructural demonstration of calcium loss from local regions of the plasma membrane of surface-stimulated human granulocytes, *J. Immunol.*, 123, 1395, 1979.
27. **Petroski, R. J., Naccache, P. H., Becker, E. L., and Sha'afi, R. I.**, Effect of chemotactic factors on calcium levels of rabbit neutrophils, *Am. J. Physiol.*, 237, C43, 1979.
28. **Mottola, C., Dolzani, L. and Romeo, D.**, The peripheral $Ca^{2+}$ pump activity of macrophages and neutrophils, *Cell Calcium*, 1, 371, 1980.
29. **Mottola, C. and Romeo, D.**, Calcium movement and membrane potential changes in the early phase of neutrophil activation by phorbol myristate acetate: a study with ion selective electrodes, *J. Cell Biol.*, 93, 129, 1982.
30. **Koff, W. C. and Dunegan, M. A.**, Modulation of macrophage-mediated tumoricidal activity by neuropeptides and neurohormones, *J. Immunol.*, 135, 350, 1985.
31. **De Simone, C., Ferrari, M., Ferrarelli, G., Rumi, C., Pugnaloni, L., and Sorice, F.**, The effects of substance P on human eosinophil receptors and functions, in *Neuroimmune Interactions: Proc. Sec. Int. Workshop on Neuroimmunomodulation*, Ann. N.Y. Acad. Sci., 496, 1987, 226.
32. **Miller, R. J.**, Neurotensin as a gastrointestinal hormone, *Med. Biol.*, 59, 65, 1981.
33. **Lembeck, F. and Gamse, R.**, Substance P in peripheral sensory processes, in *Substance P in the Nervous System, Ciba Found. Symp.*, Porter, R. and O'Connor, M., Eds., Pitman, London, 91, 35, 1982.

34. **Levine, D. J., Glark, R., Devor, M., Helms, C., Moskowitz, M. A., and Basbaum, A, I.,** Intraneuronal substance P contributes to the severity of experimental arthritis, *Science*, 226, 547, 1984.
35. **Lotz, M., Carson, D. A., and Vaughan, J. H.,** Substance P activation of rheumatoid synoviocytes: neural pathway in pathogenesis of arthritis, *Science*, 235, 893, 1987.
36. **Marx, J. L.,** Cytokines are two-edged swords in disease, *Science*, 239, 257, 1988.
37. **Payan, D. G., Brewster, D. R., and Goetzl, E. J.,** Spcific stimulation of human T lymphocytes by substance P, *J. Immunol.*, 131, 1613, 1983.
38. **Morley, J. E., Kay, N. E., Solomon, G. F., and Plotnikoff, N. P.,** Neuropeptides: conductors of the immune orchestra, *Life Sci.*, 41, 527, 1987.
39. **Payan, D. G., McGillis, J. P., Renold, F. K., Mitsuhashi, M., and Goetzel, E. J.,** Neuropeptide modulation of leukocyte function, in *Neuroimmune Interactions: Proc. Sec. Int. Workshop on Neuroimmunomodulation, Ann. N.Y. Acad. Sci.*, 496, 1987, 182.

Chapter 14

# THE ROLE OF SUBSTANCE P, SOMATOSTATIN AND VASOACTIVE INTESTINAL PEPTIDE IN MODULATION OF MUCOSAL IMMUNITY

**M. Sue O'Dorisio**

## TABLE OF CONTENTS

# I. INTRODUCTION

Phenomenological studies have suggested the existence of a neuro-endocrine-immune axis. Communication between cells of such diverse tissues requires common messengers and transducing pathways. Neuropeptides, hormones, and lymphokines were so named because they were thought to be unique to their respective tissues; each class of peptide was proposed to function only within a well-delineated system. Many of these molecules are synthesized in diverse tissues; peptides originally discovered in brain are now known to be synthesized in endocrine cells as well as in immune effector cells. Similarly, immunopeptides are also synthesized in cells of the nervous system. Table 1 illustrates how arbitrary the distinction between hormone, neuropeptide, and lymphokine has become.

The transducing mechanism for neuropeptides, peptide hormones, and lymphokines involves specific receptors which, upon ligand binding, induce the production of second messengers. Just as the messengers are shared in the neuro-endocrine-immune axis, so too, the receptors which specifically recognize these peptide messengers and transduce the cellular message are also shared by nervous, endocrine, and immune systems. Among the first clues to the existence of this neuro-endocrine-immune axis was the discovery of receptors for neuropeptides and peptide hormones on immune effector cells. These observations were followed shortly by discoveries of receptors for immunopeptides on neuronal cells (Table 2).

Neuropeptide modulation of the immune response constitutes one communication pathway between nervous, endocrine, and immune systems. This review will focus on three peptides: substance P, vasoactive intestinal peptide, and somatostatin, which appear to modulate immune function in gut-associated lymphoid tissue.

# II. SUBSTANCE P

Substance P belongs to the tachykinin family of peptides, only a few of which are synthesized in mammalian tissue;[30] these peptides possess a common C-terminal sequence Phe-X-Gly-Leu-Met-$NH_2$ with the specificity of each tachykinin thought to reside in the N-terminal portion of the molecule (Table 3). Substance P and substance K appear to be synthesized from a common preprotachykinin mRNA by alternative RNA splicing.[31] The dopamine antagonist, haloperidol, inhibits synthesis of this mRNA in rat basal ganglia.[32]

Substance P is found in high concentrations in both brain and intestine. In central nervous tissue, it is found in basal ganglia and hypothalamus.[5] In the peripheral nervous system, it can be localized in descending neurons, often colocalized with serotonin, in interneurons, and in primary sensory afferents.[33]

Neuropeptide modulation of the immune response has been well documented for substance P. This neuropeptide appears to have a stimulatory effect on the immune response, both *in vivo* and *in vitro*. *In vivo*, substance P-containing afferents are found in high concentrations in joints which are most susceptible to arthritis. This finding led to the hypothesis of neurogenic inflammation.[34] Injection of substance P into joints with a low propensity for development of arthritis produced inflammation and arthritic symptoms.[35] The mechanism by which substance P mediates these inflammatory changes appears to involve stimulation of several classes of immune effector cells.

*In vitro*, substance P has been shown to stimulate proliferation of human peripheral blood T lymphocytes in response to Con A and PHA.[36] The effects of substance P were dose dependent; thymidine uptake was optimal at 0.01 $\mu M$ substance P, while the peak uptake of leucine was seen at 0.1 $\mu M$ peptide. Substance P also stimulates proliferation of murine lymphocytes in response to Con A.[37]

Proliferative responses were enhanced by 60% in lymphocytes harvested from spleen,

## TABLE 1
## Peptide Messengers in the Neuro-Endo-Immune Axis

| Messenger | Sites of synthesis: Neuro | Endo | Immuno | Ref. |
|---|---|---|---|---|
| VIP | Hypothalamus | Gonads | Neutrophil | 1, 2 |
| Somatostatin | Frontal cortex | Pancreas | Basophil | 3, 4 |
| Substance P | Spinal cord | Pituitary | — | 5, 6 |
| β-Endorphin | Hypothalamus | Lymphocyte | Pituitary | 7, 8 |
| ACTH | — | Pituitary | Lymphocyte | 10—12 |
| Interleukin-1 | Astrocytes | — | Macrophage | 13, 14 |

## TABLE 2
## Target Cells in the Neuro-Endocrine-Immune Axis

| Messenger | Target tissue: Neuronal | Endocrine | Immune | Ref. |
|---|---|---|---|---|
| Interleukin-1 | — | Astroglia | Lymphocyte | 15, 16 |
| Somatostatin | Cortical neurons | Pancreatic acinar cell | T & B lymphocytes | 17—19 |
| Prolactin | Cortical neurons | Ovary | Lymphocytes | 20—22 |
| Substance P | Cortical neurons | Gall bladder smooth muscle | T & B lymphocytes ? Mast cell | 23—25 |
| VIP | Cortical neurons | Intestinal epithelial cell | T lymphocytes B lymphocytes | 26—29 |

## TABLE 3
## Substance P Receptor in Nervous, Endocrine, and Immune Tissues

| Tissue | Receptor Characteristics: $K_D$ | Subtype | Affinity | Ref. |
|---|---|---|---|---|
| Frontal cortex | 12 n$M$ | P | SP > Phy > $SP_{4-11}$ > E | 94 |
| Frontal cortex | 0.9 n$M$[a] | E | E > K > $SP_{6-11}$ > Phy > SP | 24 |
| IM-9 lymphoblast | 0.65 n$M$ | P | SP = $SP_{4-11}$ > E > K | 38, 41 |
| Lymphocyte | 0.5 n$M$ | P | SP = $SP_{4-11}$ > E > SK | 39, 40 |
| Parotid cell | — | P | SP >>> E | 95 |
| Mast cell | ? | ? | SP > Phy > K | 26, 45 |

*Note:* P type prefers substance P; E type prefers eledoisin.
Substance P (SP): Arg-Pro-Lys-Pro-Gln-Phe-Phe-Gly-Leu-Met-$NH_2$
Eledoisin (E): pGlu-Pro-Ser-Lys-Asp-Ala-Phe-Ile-Gly-Leu-Met-$NH_2$
Kassinin (K): Asp-Val-Pro-Lys-Ser-Asp-Gln-Phe-Val-Gly-Leu-Met-$NH_2$
Physalaemin (Phy): pGlu-Ala-Asp-Pro-Asn-Lys-Phe-Tyr-Gly-Leu-Met-$NH_2$
Substance K (SK): His-Lys-Thr-Asp-Ser-Phe-Val-Gly-Leu-Met-$NH_2$

[a] $K_D$ for eledoisin

mesenteric lymph nodes, and Peyer's patches. These findings suggested the presence of substance P receptors on T lymphocytes and that hypothesis has been confirmed. Using classical binding techniques, Payan and colleagues,[38] demonstrated a high-affinity binding site for substance P on the IM-9 lymphoblastic cell line ($k_D$ = 0.65 n$M$) and estimated the number of receptors per cell at 22,600.

Substance P also enhanced the synthesis of immunoglobulins in spleen, mesenteric lymph nodes, and Peyer's patches.[37] In the presence of 0.01 $\mu M$ substance P, IgA synthesis was increased threefold in Peyer's patches, 70% in spleen cells, and 40% in cells isolated from mesenteric lymph nodes. IgM synthesis was significantly increased only in spleen cells, while substance P did not enhance the synthesis of IgG in any of these tissues. These results might be explained by a direct effect of substance P on B-cells or IgA-secreting plasma cells; alternatively, the effects of substance P might be mediated via its effects on T-helper cells.

Using a fluorescent analog of substance P, Payan and colleagues were able to identify large subpopulations of both T and B lymphocytes isolated from spleen and Peyer's patches which specifically bound substance P.[39] The substance P receptor appeared to be expressed on one third of splenic T-cells and on nearly half of splenic B-cells. Peyer's patch lymphocytes expressed the highest number of binding sites with 53% and 65% of T- and B-cells, respectively, being labeled with fluorescent substance P. Thus, the substance P receptor bearing cells appeared to home onto the mucosal immune system. This migration pattern may explain why these same investigators[40] found significant binding of substance P to only a small set of human peripheral blood T lymphocytes.

Substance P receptors have been characterized according to their affinity for various natural and synthetic tachykinins. Those with high affinity for substance P are P type, while E-type receptors prefer eledoisin.[23] Kinetic binding studies suggest homologous P-type receptors on lymphocytes and in brain cortex (Table 3); both tissues bind substance P with high affinity ($k_D$ = 0.5 and 12 $nM$ in lymphocytes and brain, respectively). In both tissues, the C-terminal amino acids effectively compete with substance P for the binding site, while the N-terminal tetrapeptide of substance P does not appear to be crucial for ligand binding.[23,38] Payan and colleagues have further characterized two lymphocyte proteins of 58,000 and 33,000 $M_r$, which were specifically cross-linked to substance P.[41] They have also produced a monoclonal antibody to the lymphocyte substance P receptor and found that it precipitated the 33,000 $M_r$ protein which was identified by covalent cross-linking techniques.[42]

No second messenger has been identified in lymphocytes following stimulation with substance P. In other tissues such as salivary gland, the proposed mechanism of action is hydrolysis of inositol phospholipids.[43,44] Whether substance P induces formation of diacylglycerol and activation of protein kinase C, or induces $Ca^{++}$ release via $IP_3$ generation in salivary gland, nervous tissue, or lymphocytes has not been determined.

Substance P has been called the primary mediator of neurogenic inflammation.[45] Antidromic stimulation of sensory neurons results in vasodilation and increased vascular permeability thus causing the flare and plasma extravasation typical of neurogenic inflammation. This response is diminished in rats treated with capsaicin.[46] Production of the flare and of plasma extravasation also involves histamine; substance P appears to be capable of releasing histamine from mast cells and thus enhancing the inflammatory response. Interestingly, the mast cell receptor for substance P appears to require the amino terminal sequence.[25] The molecular structure of this receptor remains to be delineated.

## III. SOMATOSTATIN

Somatostatin 14 is a tetradecapeptide which is synthesized in both neuronal and endocrine cells.[3] The peptide is synthesized as a high molecular weight prepropeptide; posttranslational processing results in the known tissue forms, somatostatin 28 and somatostatin 14. Somatostatin is found in high concentrations in both brain and intestine. The intestinal sources include neurons in the lamina propria and gut endocrine cells. Another source of somatostatin in intestinal tissue may be the basophil; Goetzl et al.[4] have isolated several somatostatin-like peptides from a rat basophilic leukemia cell line. These peptides show great similarity

**TABLE 4**
**Somatostatin Receptors in Neuro, Endocrine, and Immune Tissues**

| Tissue | Receptor characteristics $K_D$ | Subtype[a] | Transducer | Ref. |
|---|---|---|---|---|
| Adrenal medulla | 0.15 n*M* | A | $G_i$ | 49 |
| Adrenal cortex | 0.42 n*M* | A | $G_i$ | 49 |
| Pancreatic acini | 0.32 n*M* | A | $G_i$ | 53 |
| Gastric parietal cell | 3.2 n*M* | A | $G_i$ | 51 |
| Pituitary | 8.6 n*M* | A | $G_i$ | 50 |
| Frontal cortex | 0.52 n*M* | A | $G_i$ | 17 |
| Frontal cortex | 1 μ*M* | B | ? | 17 |
| Lymphocyte | ? | ? | ? | 19, 55 |
| IM-9 lymphoblast | 16 n*M* | ? | ? | 54 |

[a] Subtypes distinguished by affinity for somatostatin 14 or the synthetic analog, SMS 201—995.[17]

to somatostatin 14 in both amino acid sequence and biologic activity, but do not appear to be products of the somatostatin gene.

Somatostatin was originally discovered in hypothalamus, where it was shown to inhibit release of growth hormone.[47] Subsequent studies have demonstrated that this peptide is a general inhibitor of hormone release; it inhibits release of insulin, thyroid hormones, gastrin, secretin, motilin, enteroglucagon, and vasoactive intestinal peptide (VIP).[47] Somatostatin appears to have inhibitory actions in the immune system as well; it inhibits lymphocyte proliferation by a mechanism which appears to block both RNA and DNA synthesis.[48] Similarly, somatostatin has been shown to inhibit DNA synthesis *in vitro* in cultures of intestinal tissue.[3] Conceivably, somatostatin effects on DNA synthesis may be secondary to its inhibition of release of growth factors and/or lymphokines.

Somatostatin has been shown to inhibit the release of IgA by plasma cells isolated from spleen and Peyer's patches.[37] The inhibition is seen at concentrations as low as 0.1 n*M* somatostatin. Whether somatostatin-mediated inhibition of peptide hormone release and immunoglobulin release occur by the same mechanism is an open question.

One approach to this question is to characterize the molecular structure of somatostatin receptors in endocrine and immune cells and to compare their transducing mechanisms. High-affinity binding sites have been identified in adrenal cortex,[49] pituitary,[50] gastric parietal cells,[51] brain,[17,52] and pancreas.[53] As shown in Table 4, the affinity varies from a $k_D$ of 0.15 n*M* in adrenal medulla[49] to 16 n*M* in a murine plasmacytoma.[54]

Two classes of high-affinity receptors have been identified in the brain, using a synthetic analog of somatostatin 14.[17,52] The class A receptor subtype with high affinity for somatostatin and its analog is found in brain, pancreas, and pituitary, while the class B subtype appears to be found only in brain (Table 4). The class A subtype has been characterized in pancreas as a 92,000 $M_r$ glycoprotein.[19]

Somatostatin receptors were first identified on lymphocytes by Bhathena and colleagues.[55] Although the lymphocyte receptor for somatostatin has not been characterized as to subtype, Scicchitano et al.[19] have developed a fluorescent analog of somatostatin and characterized the distribution of receptors on both T and B lymphocytes isolated from spleen and Peyer's patches. As was true for substance P, somatostatin receptors are found on subpopulations of both T- and B-cells. In spleen, 35% of T and 37% of B lymphocytes appear to express the somatostatin receptor. An even larger proportion of Peyer's patch lymphocytes are somatostatin receptor-positive (49 and 52% for T- and B-cells, respectively; see Table 5).

**TABLE 5**
**Murine Lymphocytes Expressing Somatostatin or Substance P Receptors**

| Cell type | % Receptor + cells | |
|---|---|---|
| | Substance P | Somatostatin |
| Splenic T | 34 ± 3 | 35 ± 2 |
| Splenic B | 47 ± 1 | 37 ± 2 |
| Peyer's patch T | 53 ± 3 | 49 ± 2 |
| Peyer's patch B | 65 ± 7 | 52 ± 4 |

*Note:* Receptors identified by cytofluorographic analysis using fluorescein-substance P[39] or fluorescein-somatostatin.[19]

The mechanism of action of somatostatin in any tissue may, of course, depend on the subtype of receptors expressed in that tissue. In intestinal epithelial cells,[56] gastric parietal cells,[51] anterior pituitary cells,[57] and human lymphocytes,[58] somatostatin inhibits agonist-induced activation of adenylate cyclase. The precise mechanism has been further characterized in both pituitary $GH_3$ cells and in S49 lymphoma cells.[59-61] Agonists such as VIP activate a stimulatory guanine nucleotide binding protein, $G_s$, which, in turn, stimulates the catalytic unit of adenylate cyclase. When VIP is added to pituitary $GH_3$ cells, a VIP receptor-$G_s$-cyclase complex is formed which induces cAMP accumulation. In S49 lymphoma cells, forskolin bypasses the receptor requirement and activates the catalytic unit of adenylate cyclase directly. In both cell lines, the effect of somatostatin appears to be mediated by an inhibitory guanine nucleotide binding protein, $G_i$, which forms a $G_i$-cyclase complex and inhibits cAMP accumulation.

In both cell lines, pretreatment with pertussis toxin prevents somatostatin-mediated inhibition of adenylate cyclase activity. Pertussis toxin is believed to inactivate $G_i$, thus leaving VIP free to induce the receptor-$G_s$-cyclase complex in $GH_3$ cells and allowing forskolin to activate the adenylate cyclase catalytic unit directly in S49 lymphoma cells.[59-61]

In some cells, including the pituitary $GH_3$ cell, the inhibitory effects of somatostatin can occur by a cAMP-independent mechanism, but whether this mechanism involves $Ca^{++}$ or phospholipid pathways is controversial,[59,60,62-64] and whether the cAMP-dependent and the cAMP-independent pathways are mediated via different subtypes of the somatostatin receptor remains to be investigated.

# IV. VASOACTIVE INTESTINAL PEPTIDE

This 28-amino acid peptide was originally described by Said as a vasodilatory substance in rabbit lung[65] and subsequently isolated from porcine intestine by Said and Mutt.[66] The peptide was later localized to central and peripheral nervous tissue where it was found to be a neurotransmitter, often coreleased with acetylcholine.[67,68] The clinical significance of VIP was first recognized when investigators observed that VIP-secreting tumors caused watery diarrhea.[69] The peptide is now known to be a major regulator of water and elctrolyte secretion in the gut.[70] VIP regulates exocrine secretion in pancreatic acinar cells[71] and secretion of prolactin from pituitary.[59,60]

The role of VIP in modulation of the immune response is less clear. VIP was unexpectedly found in human neutrophils;[2] this observation led O'Dorisio and colleagues to hypothesize that other immune effector cells might express VIP receptors, thus allowing for a physiological response to neuropeptide released from neutrophils.[72]

Other investigators have subsequently begun to probe the question of neuropeptide modulation of the immune response; one physiologic role of VIP in immune modulation appears to be regulation of T-cell proliferation in response to mitogens. Incubation of lymphocytes in the presence of VIP has been shown to inhibit their ability to proliferate in response to Con A or to phytohemagglutinin.[73,74] VIP-inhibited lymphocyte proliferation in response to Con A by 73% when added simultaneously with Con A; the effectiveness was reduced to 46% inhibition if added 24 h after addition of mitogen. The VIP-mediated inhibition was reversible when VIP added at time 0 was removed up to 6 h later, but was irreversible if VIP was removed after 24 h.[75] The ability of Con A-stimulated lymphocytes to secrete IL-2 is also inhibited by VIP, while the Con A-mediated generation of suppressor activity is not affected by VIP.[75] The inhibitory effect of VIP was dependent on the cell concentration,[74] being most effective when 0.5 to 1.0 $\times$ $10^5$ cells per microtiter well were used, and when high concentrations (0.1 to 1.0 $\mu M$) of VIP were added.[74,75] These observations suggest either that full occupancy of VIP receptors is required for inhibition, or alternatively, may reflect the susceptibility of VIP to degradation by proteases in serum (mitogenic assays were performed in 8 to 10% fetal calf serum).

Studies from two laboratories have confirmed that VIP does not affect the response of lymphocytes to lipopolysaccharide.[73,74] This suggested that B-cells did not express receptors for VIP. On the other hand, VIP inhibits synthesis and/or secretion of immunoglobulins in culture,[37] an effect which may be directly on plasma cells or alternatively on T-helper cells. The fact that VIP inhibits synthesis of IgA more significantly than synthesis of IgM or IgG suggests that lymphocytes in mucosal immune tissues may be regulated by this peptide. Thus, peripheral blood B-cells may not express VIP receptors. In support of this concept, recent studies have demonstrated specific, high affinity VIP receptors on a human pre-B-cell-line and on an IgA secreting plasma cell line.[26]

VIP has been shown to modulate lymphocyte trafficking in the gut.[76] Murine lymphocytes were incubated *in vitro* with 1 $\mu M$ VIP for 3 h; under these conditions the number of high-affinity VIP receptors was down-regulated by 50%. When these desensitized lymphocytes were reinjected into mice, the number of lymphocytes which migrated to Peyer's patches was significantly reduced. Migration to the spleen was not affected. The mechanism by which VIP modulates lymphocyte trafficking is not yet understood, but presumably requires VIP receptors on a subpopulation of circulating lymphocytes.

Functional VIP receptors were first identified on human peripheral blood lymphocytes by demonstration of VIP-mediated activation of adenylate cyclase in lymphocyte membranes.[58] Guerrero[77] and Danek[29] subsequently identified high-affinity binding sites for VIP on human peripheral blood lymphocytes using classical binding techniques. This finding has since been confirmed by several investigators using murine lymphocytes.[73,74] The binding kinetics are remarkably similar in all of these studies. Affinity constants for the high-affinity binding site are in the nanomolar range for both human and murine lymphocytes. The nature of the low-affinity site observed by Guerrero[77] remains to be explained. The authors suggest that monocytes bind VIP. VIP does not activate adenylate cyclase in monocytes,[58] but monocytes do chemotax in response to VIP.[78] The low-affinity site observed in mononuclear cells of peripheral blood conceivably is a monocyte receptor which, although not linked to adenylate cyclase, may be a chemotactic receptor. The estimated number of high-affinity binding sites per cell varies from 1500 in unseparated mononuclear cells to 5000 binding sites per cell in enriched T-cell populations.[73,77] The estimated number of high-affinity binding sites on lymphocytes is lower than has been observed on intestinal epithelial cells, colon carcinoma cells, or pancreatic acinar cells (Table 6). The human lymphoblastic cell line, Molt 4b, has an estimated 15,000 binding sites per cell.[79] This suggests the possibility that only selected lymphocyte subpopulations express the VIP receptor.

As noted above, functional studies suggest that in peripheral blood, only T lymphocytes

**TABLE 6**
**Comparison of VIP Receptor in Nervous, Endocrine, and Immune Tissues**

| Tissue | $K_D$ (n*M*) | $B_{max}$ (n*M*) (sites/cell) | $M_r$ (Da) | Ref. |
|---|---|---|---|---|
| F cortex | | 0.6 n*M* | 46K | 27, 82 |
| T lymphocyte | 0.5 | 0.025 n*M* (1,700) | 47K[a] | 29 |
| Molt lymphoblast | 17 | 0.24 n*M* (15,000) | 47K | 79, 81 |
| Pituitary $GH_3$ | 17 n*M* | — | 47K | 84 |
| Intestinal epithelial | 1.6 n*M* | — (140,000) | 73K | 28, 90 |
| HT29 colon carcinoma | 0.3 | — (69,000) | 67K[b] | 86, 91 |
| Liver | 3 n*M* | — | 75K[b] | 87, 92 |
| Lung | 4.4 n*M* | — | 66K[b] | 85, 93 |
| Pancreatic acinar | 3 n*M* | — | 47K | |

[a] O'Dorisio, M. S., unpublished observations, 1989.
[b] Converted to 48K by removal of carbohydrate.

possess the receptor, since VIP inhibited T-cell, but not B-cell proliferation in response to mitogens. Recent studies by O'Dorisio and colleagues[26,80] demonstrated both high-affinity binding and VIP-mediated activation of adenylate cyclase in a human pre-B-cell line and in leukemic B-cell blasts, but high-affinity binding to peripheral blood B-cells has not been directly demonstrated.

Using bifunctional reagents to covalently cross-link radiolabeled VIP to intact cells or to a plasma membrane fraction, the VIP receptor on Molt 4b lymphoblasts has been characterized as a 47,000 $M_r$ protein without disulfide bonds.[81] Using this same cross-linking technique with a variety of bifunctional cross-linking reagents, the ligand binding protein appears to be identical in lymphoblasts, frontal cortex, pancreas, and pituitary $GH_3$ cells.[27,81-84] A homologous receptor protein is found in lung, colon carcinoma cells, and intestinal epithelial cells.[28,85,86] In each of these tissues (Table 6), a higher molecular weight form of the receptor is observed which can be converted to a 47,000 to 50,000 $M_r$ band by treatment with glycohydrolases, suggesting that the ligand binding component is a 47K protein and that various amounts of carbohydrate account for the molecular weight differences in these tissues.

Characterization of the VIP receptor in liver using photoaffinity labeled VIP rather than a bifunctional cross-linking reagent, also demonstrated a 47,000 $M_r$ ligand binding protein and a higher molecular weight form which could be reduced to 47,000, again supporting the concept of the receptor protein being a single, 47K polypeptide chain.[87] Whether the protein structure of the receptor is identical in all of these tissues will not be definitively answered until purification and sequencing of the protein or identification and sequencing of the gene is accomplished.

The VIP receptor on lymphoblasts is linked to the adenylate cyclase catalytic unit via interaction with a stimulatory guanine nucleotide binding protein, $G_s$.[58,88] The ability of VIP to stimulate adenylate cyclase is dose dependent, being observed at 0.1 n*M* to 1 μ*M* concentrations.[58] The effects of VIP and guanine nucleotide (GppNHp) are synergistic, supporting the concept that the VIP-receptor complex interacts with the $G_s$-adenylate cyclase complex. The effects of VIP and prostaglandin $E_2$ are simply additive, indicating that VIP and $PGE_2$ act through separate and distinct receptors. VIP mediated activation of lymphoblast adenylate cyclase results in phosphorylation of a 41,000 $M_r$ protein.[88] The identical protein is phosphorylated in the presence of the diterpene, forskolin, indicating that this 41K protein is a physiological substrate for cAMP dependent protein kinase.

# V. PEPTIDE-PEPTIDE INTERACTIONS

Both *in vivo* and *in vitro* experiments suggest that substance P, somatostatin and VIP modulate immune function; thus delineation of possible peptide-peptide interactions may lead to a better understanding of immune regulation by these peptides. VIP induces the synthesis of somatostatin mRNA in cerebral cortex by a cAMP dependent mechanism.[47] On the other hand, somatostatin has been shown to down regulate peptide synthesis in VIP secreting tumors.[89]

Thus, a feedback loop appears to operate in nervous tissue, whereby high levels of VIP lead to synthesis of somatostatin which, in turn, down regulates VIP synthesis. Whether expression of genes for somatostatin and VIP synthesis is reciprocally regulated in immune tissue promises to be a fruitful area for further investigation.

Substance P clearly enhances the immune response; it stimulates both lymphocyte proliferation and immunoglobulin synthesis. Since VIP has inhibitory effects on lymphocyte proliferation and on IgA synthesis, VIP and substance P may act in concert to regulate these immune functions. The delineation of the physiologic importance of these possible peptide-peptide interactions will depend on development of techniques to look at individual immune effector cells, determination of whether individual cells have receptors for more than one peptide, and delineation of the mechanism of action of each peptide in immune effector cells. Neuropeptide modulation of the immune response is a science in its infancy, but one which promises to yield a greater understanding of the mucosal immune system with relevance to treatment of infections, autoimmunity, and immunodeficiencies.

# REFERENCES


1. **Khalil, T., Alinder, G., and Rayford, P. L.,** Vasoactive intestinal peptide, in *Gastrointestinal Endocrinology,* Thompson, J. C., Greeley, G. H., Jr., Rayford, P. L., and Townsend, C. M., Jr., McGraw-Hill, New York, 1987, chap. 19.
2. **O'Dorisio, M. S., O'Dorisio, T. M., Cataland, S., and Balcerzack, S. P.,** Vasoactive intestinal polypeptide as a biochemical marker in polymorphonuclear leukocytes, *J. Lab. Clin. Med.,* 96, 666, 1980.
3. **Newman, J. B., Lluis, F., and Townsend, C. M.,** Somatostatin, in *Gastrointestinal Endocrinology,* Thompson, J. C., Greeley, G. H., Jr., Rayford, P. L., and Townsend, C. M., Jr., Ed., McGraw-Hill, New York, 1987, chap. 21.
4. **Goetzl, E. J., Chernov-Rogan, T., Cooke, M. P., Renold, F., and Payan, D. G.,** Endogenous somatostatin-like peptides of rat basophilic leukemia cells, *J. Immunol.,* 135, 2707, 1985.
5. **Walker, J. P. and Thompson, J. C.,** Substance P, in *Gastrointestinal Endocrinology,* Thompson, J. C., Greeley, G. H. Jr., Rayford, P. L., and Townsend, C. M., Jr., McGraw-Hill, New York, 1987, chap. 24.
6. **Morel, G., Chayvialle, J. A., Kerdelhue, B., and Dubois, P. M.,** Ultrastructural evidence for endogenous substance P-like immunoreactivity in the rat pituitary gland, *Neuroendocrinology,* 35, 86, 1982.
7. **Watson, S. J., Akil, H., Richard, C., and Barchas, J. D.,** Evidence for two separate opiate peptide neuronal systems, *Nature (London),* 275, 226, 1978.
8. **Li, C. H. and Chung, D.,** Isolation and structure of a peptide with opiate activity from camel pituitary gland, *Proc. Natl. Acad. Sci. U.S.A.,* 73, 1145, 1976.
9. **Smith, E. M., Harbour-McMenamin, D., and Blalock, J. E.,** Lymphocyte production of endorphins and endorphin-mediated immunoregulatory activity, *J. Immunol.,* 135, 779s, 1985.
10. **Eberwine, J. H., Jonassen, J. A., Evinger, M. J., and Roberts, J. L.,** Complex transcriptional regulation by glucocorticoids and corticotropin-releasing hormone of proopiomelanocortin gene expression in rat pituitary cultures, *DNA,* 6, 483, 1987.
11. **Westly, H. J., Kleiss, A. J., Kelley, K. W., Wong, P. K., and Yuen, P.,** Newcastle disease virus-infected splenocytes express the proopiomelanocortin gene, *J. Exp. Med.,* 163, 1589, 1986.
12. **Smith, E. M., Morrill, A. C., Meyer, W. J., III, and Blalock, E.,** Corticotropin releasing factor induction of leukocyte-derived immunoreactive ACTH and endorphins, *Nature (London),* 321, 881, 1986.
13. **Fontana, A., Weber, E., Grob, P. J., Lim, R., and Miller, J. F.,** Dual effect of glia maturation factor on astrocytes: differentiation and release of interleukin-1-like factors, *J. Neuroimmunol.,* 5, 261, 1983.

14. **Auron, P. E., Warner, S. J., Webb, A. C., Cannon, J. G., Bernheim, H. A., McAdam, K. J., Rosenwasser, L. J., LoPreste, G., Mucci, S. F., and Dinarello, C. A.,** Studies on the molecular nature of human interleukin 1, *J. Immunol.*, 138, 1447, 1987.
15. **Giulian, D. and Lachman, L. B.,** Interleukin-1 stimulation of astroglial proliferation after brain injury, *Science*, 228, 497, 1985.
16. **Bernton, E. W., Beach, J. E., Holaday, J. W., Smallridge, R. C., and Fein, H. G.,** Release of multiple hormones by a direct action of interleukin-1 on pituitary cells, *Science*, 238, 519, 1987.
17. **Tran, V. T., Beal, M. F., and Martin, J. B.,** Two types of somatostatin receptors differentiated by cyclic somatostatin analogs, *Science*, 228, 492, 1985.
18. **Susini, C., Bailey, A., Szecowka, J., and Williams, J. A.,** Characterization of covalently cross-linked pancreatic somatostatin receptors, *J. Biol. Chem.*, 261, 16738, 1986.
19. **Scicchitano, R., Dazin, P., Bienenstock, J., Payan, D. G., and Stanisz, A. M.,** Distribution of somatostatin receptors on murine spleen and peyer's patch T and B lymphocytes, *Brain Behav. Immun.*, 1, 173, 1987.
20. **Posner, B. I., Kelly, P. A., Shiu, R. P., and Friesen, H.,** Studies of insulin growth hormone and prolactin binding: tissue distribution, species variation and characterization, *Endocrinology*, 95, 521, 1974.
21. **Mitani, M. and Dufau, M.,** Purification and characterization of prolactin receptors from rat ovary, *J. Biol. Chem.*, 261, 1309, 1986.
22. **Russel, D. H., Kibler, R., Matrisan, L., Larson, D. F., Poulos, B., and Magun, B. E.,** Prolactin receptors on human T and B lymphocytes: antagonism of prolactin binding by cyclosporine, *J. Immunol.*, 134, 3027, 1985.
23. **Cascieri, M. A., Chicchi, G., and Liang, T.,** Demonstration of two distinct tachykinin receptors in rat brain cortex, *J. Biol. Chem.*, 260, 1501, 1985.
24. **Rosell, S., Bjorkroth, U., Chang, D., et al.,** Effects of Substance P and analogs on isolated guinea pig ileum, in *Substance P*, von Euler, U. S. and Pernow, B., Eds., Raven Press, New York, 1977, 83.
25. **Piotiowski, W., Devoy, M. A. B., Jordan, C. C., and Foreman, J. C.,** The substance P receptor on rat mast cells and in human skin, *Agents Actions*, 14, 420, 1984.
26. **O'Dorisio, M. S., Shannon, B. T., Fleshman, D. J., and Campolite, L. B.,** Identification of high affinity receptors for VIP on human lymphocytes of B lineage, *J. Immunol.*, 142, 3533, 1989.
27. **Couvineau, A., Gammeltoft, S., and Laburthe, M.,** Molecular characteristics and peptide specificity of vasoactive intestinal peptide receptors from rat cerebral cortex, *J. Neurochem.*, 47, 1469, 1986.
28. **Laburthe, M., Breant, B., and Rouyer-Fessand, C.,** Molecular identification of receptors for VIP in rat intestinal epithelium by covalent crosslinking, *Eur. J. Biochem.*, 139, 181, 1984.
29. **Danek, A., O'Dorisio, M. S., O'Dorisio, T. M., and George, J. M.,** Specific binding of vasoactive intestinal polypeptide to human T lymphocytes. *J. Immunol.*, 131, 1173, 1983.
30. **Nawa, H., Hirose, T., Takashima, H., Inayama, S., and Nakanishi, S.,** Nucleotide sequences of cloned cDNA's for two types of bovine brain substance P precursor, *Nature (London)*, 306, 32, 1983.
31. **Nawa, H., Kotani, H., and Nakanishi, S.,** Tissue-specific generation of two preprotachykinin mRNA's from one gene by alternative RNA splicing, *Nature (London)*, 312, 729, 1984.
32. **Bannon, M. J., Lee, J., Giraud, P., Young, A., Affolter, H., and Bonner, T. I.,** Dopamine antagonist haloperidol decreases substance P, substance K, and preprotachykinin mRNA's in rat striatonigral neurons, *J. Biol. Chem.*, 261, 6640, 1986.
33. **Hokfelt, T., Johannson, O., Ljungdahl, A., Lundberg, J. M., and Schultzberg, M.,** Peptidergic neurones, *Nature (London)*, 284, 515, 1980.
34. **Levine, J. D., Moskowitz, M. A., and Basbaum, A. I.,** The contribution of neurogenic inflammation in experimental arthritis, *J. Immunol.*, 135, 843s, 1985.
35. **Levine, J. D., Clark, R., Devon, M., Helms, C., Moskowitz, M. A., and Basbaum, A. I.,** Intraneuronal substance P contributes to the severity of experimental arthritis, *Science*, 226, 547, 1984.
36. **Payan, D. G., Brewster, D. R., and Goetzl, E. H.,** Specific stimulation of human T lymphocytes by substance P, *J. Immunol.*, 131, 1613, 1983.
37. **Stanisz, A. M., Befus, D., and Bienenstock, J.,** Differential effects of vasoactive intestinal peptide, substance P and somatostatin on immunoglobulin synthesis and proliferations by lymphocytes from peyer's patches, mesenteric lymph nodes and spleen, *J. Immunol.*, 136, 152, 1986.
38. **Payan, D. G., Brewster, D. R., and Goetzl, E. J.,** Stereospecific receptors for substance P on cultured human IM-9 lymphoblasts, *J. Immunol.*, 133, 3260, 1984.
39. **Stanisz, A. M., Scicchitano, R., Dazin, P., Bienenstock, J., and Payan, D. G.,** Distribution of substance P receptors on murine spleen and peyer's patch T and B cells, *J. Immunol.*, 139, 749, 1987.
40. **Payan, D. G., Brewster, D. R., Missirian-Bastian, A., and Goetzl, E. J.,** Substance P recognition by a subset of human T lymphocytes, *J. Clin. Invest.*, 74, 1532, 1984.
41. **Payan, D. G., McGillis, J. P., and Organist, M. L.,** Binding characteristics and affinity labeling of protein constituents of the human IM-9 lymphoblast receptor for substance P, *J. Biol. Chem.*, 261, 14321, 1986.

42. **Organist, M. L., Harvey, J., McGillis, J. P., Mitsuhashi, M., Melera, P., and Payan, D. G.,** Characterization of a monoclonal antibody against the lymphoblast substance P receptor, *J. Immunol.*, 139, 3050, 1987.
43. **Watson, S. P. and Downes, C. P.,** Substance P induced hydrolysis of inositol phospholipids in guinea-pig ileum and rat hypothalamus, *Eur. J. Pharmacol.*, 93, 245, 1983.
44. **Hanley, M. R., Lee, C. M., Jones, L. M., and Michell, R. H.,** Similar effects of substance P and related peptides on salivation and on phosphatidylinositol turnover in rat salivary glands, *Mol. Pharmacol.*, 18, 78, 1980.
45. **Foreman, J. C. and Jordan, C. C.,** Neurogenic inflammation, *Trends Pharmacol. Sci.*, 5, 116, 1984.
46. **Helme, R. D., Eglezos, A., Dandie, G. W., Andrews, P. V., and Boyd, R. L.,** The effect of substance P on the regional lymph node antibody response to antigenic stimulation in capsaicin-pretreated rats, *J. Immunol.*, 139, 3470, 1987.
47. **Reichlin, S.,** Somatostatin: historical aspects, *Scand. J. Gastroenterol.*, 21, 1, 1986.
48. **Payan, D. G., Hess, C. A., and Goetzl, E. J.,** Inhibition by somatostatin of the proliferation of T lymphocytes and molt-4 lymphoblasts, *Cell. Immunol.*, 84, 433, 1984.
49. **Maurer, R. and Reubi, J. C.,** Somatostatin receptors in the adrenal, *Mol. Cell Endocrinol.*, 45, 81, 1986.
50. **Aguilera, G. and Parker, D. S.,** Pituitary somatostatin receptors, *J. Biol. Chem.*, 257, 1134, 1982.
51. **Park, J., Chiba, T., and Yamada, T.,** Mechanisms for direct inhibition of canine gastric parietal cells by somatostatin, *J. Biol. Chem.*, 262, 14190, 1987.
52. **Reubi, J. C.,** Evidence for two somatostatin-14 receptor types in rat brain cortex, *Neurosci. Lett.*, 49, 259, 1984.
53. **Sakamoto, C., Goldfine, I. D., and Williams, J. A.,** The somatostatin receptor on isolated pancreatic acinar cell plasma membranes, *J. Biol. Chem.*, 259, 9623, 1984.
54. **Stead, R. H., Bienenstock, J., and Stanisz, A. M.,** Neuropeptide regulation of mucosal immunity, *Immunol. Rev.*, 100, 333, 1987.
55. **Bathena, S. J., Louie, J., Schecter, G. P. P., Redman, R. S., Wahl, L., and Recant, L.,** Identification of human mononuclear leukocytes bearing receptors for somatostatin and glucagon, *Diabetes*, 30, 127, 1981.
56. **Carter, R. F., Bitar, K. N., Zfass, A. M., and Makhlouf, G. M.,** Inhibition of VIP-stimulated intestinal secretion and cyclic AMP production by somatostatin in the rat, *Gastroenterology*, 74, 726, 1978.
57. **Spada, A., Vallar, L., and Giannattasio, G.,** Presence of an adenylate cyclase dually regulated by somatostatin and human pancreatic growth hormone (GH)-releasing ractor in GH-secreting cells, *Endocrinology*, 115, 1203, 1984.
58. **O'Dorisio, M. S., Hermina, N., Balcerzak, S. P., and O'Dorisio, T. M.,** Vasoactive intestinal polypeptide stimulation of adenylate cyclase in purified human leukocytes, *J. Immunol.*, 127, 2551, 1981.
59. **Koch, B. D., Dorflinger, L. J., and Schonbrunn, A.,** Pertussis toxin blocks both cyclic AMP-mediated and Cyclic AMP-independent actions of somatostatin, *J. Biol. Chem.*, 260, 13138, 1985.
60. **Yajima, Y., Akita, Y., and Saito, T.,** Pertussis toxin blocks the inhibitory effects of somatostatin on cAMP-dependent vasoactive intestinal peptide and cAMP-independent thyrotropin releasing hormone-stimulated prolactin of $GH_3$ cells, *J. Biol. Chem.*, 261, 2684, 1986.
61. **Jakobs, K. H., Aktories, K., and Schultz, G.,** A nucleotide regulatory site for somatostatin inhibition of adenylate cyclase in S49 lymphoma cells, *Nature (London)*, 303, 177, 1983.
62. **Martin, T. F., Lucas, D. O., Bajjalieh, S. M., and Kowalchyk, J. A.,** Thyrotropin-releasing hormone activates a $Ca^{2+}$-dependent polyphosphoinositide phosphodiesterase in permeable $GH_3$ cells, *J. Biol. Chem.*, 261, 2918, 1986.
63. **Aub, D. L., Frey, E. A., Sekura, R. D., and Cote, T. E.,** Coupling of the thyrotropin-releasing hormone receptor to phospholipase C by a GTP-binding protein distinct from the inhibitory or stimulatory GTP-binding protein, *J. Biol. Chem.*, 261, 9333, 1986.
64. **Ronning, S. A. and Martin, T. F.,** Characterization of phorbol ester- and diacylglycerol-stimulated secretion in permeable $GH_3$ pituitary cells, *J. Biol. Chem.*, 261, 7840, 1986.
65. **Said, S. I.,** Vasodilator action of VIP: introduction and general considerations, in *Vasoactive Intestinal Peptide: Advances in Peptide Hormone Research Series*, Said, S. I., Ed., Raven Press, New York, 1982, 145.
66. **Said, S. I. and Mutt, V.,** Isolation from porcine-intestinal wall of a vasoactive octacosapeptide related to secretin and to glucagon, *Eur. J. Biochem.*, 28, 199, 1972.
67. **Hokfelt, T., Schultzbert, M., Lundberg, J. M., Fuxe, K., Mutt, V., Fahrenkrug, J., and Said, S. I.,** Distribution of vasoactive intestinal polypeptide in the central and peripheral nervous systems as revealed by immunochemistry, in *Vasoactive Intestinal Peptide: Advances in Peptide Hormone Research Series*, Said, S. I., Ed., Raven Press, New York, 1982, 65.
68. **Lundberg, J. M., Anggard, A., Fahrenkrug, J., and Hokfelt, T.,** Vasoactive intestinal polypeptide in cholinergic neurons of exocrine glands, in *Vasoactive Intestinal Peptide: Advances in Peptide Hormone Research Series*, Said, S. I., Ed., Raven Press, New York, 1982, 373.

69. **Said, S. I. and Faloona, G. R.**, Elevated plasma and tissue levels of vasoactive intestinal polypeptide in the watery-diarrhea syndrome due to pancreatic, bronchogenic and other tumors, *N. Engl. J. Med.*, 293, 155, 1975.
70. **Schwartz, C. J., Kimberg, D. V., Sheerin, H. E., Field, M., and Said, S. I.**, Vasoactive intestinal peptide stimulation of adenylate cyclase and active electrolyte secretion in intestinal mucosa, *J. Clin. Invest.*, 51, 536, 1974.
71. **Zhou, Z., Gardner, J. D., and Jensen, R. T.**, Receptors for vasoactive intestinal peptide and secretin on guinea pig pancreatic acini, *Peptides*, 8, 633, 1987.
72. **O'Dorisio, M. S., Wood, C. L., and O'Dorisio, T. M.**, Vasoactive intestinal polypeptide as a neuromodulator of lymphocyte function: a review, *J. Immunol.*, 135, 7925, 1985.
73. **Ottaway, C. A. and Greenberg, G. R.**, Interaction of VIP with mouse lymphocytes: specific binding and the modulation of mitogen responses, *J. Immunol.*, 132, 417, 1984.
74. **Krco, T. J., Gores, A., and Go, V. L. W.**, Gastrointestinal regulatory peptides modulate in vitro immune reactions of mouse lymphoid cells, *Clin. Immunol. Immunopathol.*, 39, 308, 1986.
75. **Ottaway, C. A.**, Selective effects of VIP on the mitogenic response of murine T cells, *Immunology*, 62, 291, 1987.
76. **Ottaway, C. A.**, In vitro alteration of receptors for vasoactive intestinal peptide changes in the in vivo localization of mouse T cells, *J. Exp. Med.*, 160, 1054, 1984.
77. **Guerrero, J. M., Prieto, J. C., Elorza, F. L., Ramierez, R., and Goberna, R.**, Interaction of vasoactive intestinal peptide with human blood mononuclear cells, *Mol. Cell Endocrinol.*, 21, 151, 1981.
78. **Sacerdote, P., Ruff, M. R., and Pert, C. B.**, Vasoactive intestinal peptide 1-12: a ligand for the CD4 (T4)/human immunodeficiency virus receptor, *J. Neurosci. Res.*, 18, 102, 1987.
79. **Beed, E. A., O'Dorisio, M. S., O'Dorisio, T. M., and Gaginella, T. S.**, Demonstration of a functional receptor for vasoactive intestinal polypeptide on molt 4b T lymphoblasts, *Regul. Pept.*, 6, 1, 1983.
80. **Shannon, B. T., O'Dorisio, M. S., Mulne, A. F., Grossman, N. J., Zwick, D., and Ruymann, F. B.**, Neuropeptide receptor on a subset of leukemias positive for common ALL antigen (CALLA), *Pediatr. Res.*, 20, 287A, 1986.
81. **Wood, C. L. and O'Dorisio, M. S.**, Covalent cross-linking of vasoactive intestinal polypeptide to its receptor on intact human lymphoblasts, *J. Biol. Chem.*, 260, 1243, 1985.
82. **O'Dorisio, M. S., Vasiloff, J., Campolito, L. B., Beattie, M. S., and Bresnahan, J. C.**, Characterization of the VIP receptor in frontal cortex, *Neurosci. Res. Commun.*, in press.
83. **McArthur, K. E., O'Dorisio, M. S., Wood, C. L., Zhou, Z., Gardner, J. D., and Jensen, R. T.**, Characterization of receptors for VIP on pancreatic acinar cell plasma membranes using covalent cross-linking, *Am. J. Physiol.*, 252, 8404, 1987.
84. **Wood, C. L., O'Dorisio, M. S., Vassalo, L. M., Malarkey, W. B., and O'Dorisio, T. M.**, Characterization of VIP receptors in pituitary $GH_3$ cells, *Regul. Pept.*, 12, 237, 1985.
85. **Paul, S. and Said, S. I.**, Characterization of receptors for vasoactive intestinal peptide solubilized from the lung, *J. Biol. Chem.*, 262, 158, 1987.
86. **Battari, A. E., Luis, J., Martin, J., Fantani, J., Muller, J., Maryaldi, J., and Pichon, J.**, The vasoactive intestinal peptide receptor on intact human colonic adenocarcinoma cells (HT29-D4), *Biochem. J.*, 242, 000, 1987.
87. **Robichon, A., Kuks, P. F., and Besson, J.**, Characterization of VIP receptors by a photoaffinity label, *J. Biol. Chem.*, 262, 11539, 1987.
88. **O'Dorisio, M. S., Wenger, G. D., Wood, C. L., and Vassalo, L. M.**, cAMP dependent protein kinase in molt 4b lymphoblasts: identification by photoaffinity labeling and activation by VIP and PHI, *J. Immunol.*, 134, 4078, 1985.
89. **Vinik, A. I., Tsai, S., Moattari, A. R., Cheung, P., Eckhauser, F. E., and Cho, K.**, Somatostatin analogue (SMS 201-995) in the management of gastroenteropancreatic tumors and diarrhea syndromes, *Am. J. Med.*, 81, 23, 1986.
90. **Prieto, J. C., Laburthe, M., and Rosselin, G.**, Interaction intestinal peptide with isolated intestinal epithelial cells from rat, *J. Biochem.*, 96, 229, 1979.
91. **Luis, J., Muller, J., Abadie, B., Martin, J., Marvaldi, J., and Pichon, J.**, Cycle of the vasoactive intestinal peptide and its binding site in a human adenocarcinoma cell line (HT 29), *Eur. J. Biochem.*, 156, 631, 1986.
92. **Nguyen, T. D., Williams, J. A., and Gray, G. M.**, Vasoactive intestinal peptide receptor on liver plasma membranes: characterization as a glycoprotein, *Biochemistry*, 25, 361, 1986.
93. **Dickinson, K. E., Schachter, M., Miles, C. M., Coy, D. H., and Sever, P. S.**, Characterization of vasoactive intestinal peptide (VIP) receptors in mammalian lung, *Peptides*, 7, 791, 1986.
94. **Cascieri, M. A. and Liang, T.**, Characterization of the substance P receptor in rat brain cortex membranes and the inhibition of radioligand binding by guanine nucleotides, *J. Biol. Chem.*, 258, 5158, 1985.
95. **Liang, T. and Cascieri, M. A.**, Substance P receptor on parotid cell membranes, *J. Neurosci.*, 1, 1133, 1981.

Chapter 15

# HORMONAL INFLUENCE ON THE SECRETORY IMMUNE SYSTEM OF THE EYE

**David A. Sullivan**

## TABLE OF CONTENTS

# I. INTRODUCTION

During the past decade, it has become increasingly clear that the endocrine and nervous systems exert a tremendous, regulatory impact on immunological expression.[1-5] Hormones and neuropeptides, including androgens,[6] estrogens,[7,8] progestins,[7,8] glucocorticoids,[9] adrenocorticotrophic hormone,[10] growth hormone,[11] prolactin,[12] α-melanocyte stimulating hormone,[13] insulin,[14] oxytocin,[15] vasopressin,[15] somatostatin,[16-18] substance P,[16-19] vasoactive intestinal peptide,[16,17,20] thyroxine,[21] acetylcholine,[22] catecholamines,[22] thyroid-stimulating hormone,[21] α-, β- and γ-endorphins,[23] methionine enkephalin and leucine enkephalin,[24,25] secretin, and possibly pancreozymin-cholecystokinin,[26,27] all bind to and/or modulate lymphocytes and monocytes. The consequence of this neuroendocrine-immune interaction results in alterations in lymphoid cell function both *in vivo* (e.g., mucosal defense, hypersensitivity responses, phagocytosis, natural killer cell activity, graft survival, reticuloendothelial processing, thymic function, and autoimmunity) and *in vitro* (e.g., lymphocyte proliferation, lymphokine and immunoglobulin production, antigen presentation, Fc receptor expression, and mixed-lymphocyte reactivity).[1-6,28] The complexity of these interrelationships is further accentuated by the demonstrations that lymphocytes synthesize and secrete a variety of hormones[29] and that antigenic perturbations of the immune system give rise to lymphocytic signals that control the endocrine and nervous systems.[29,30] Thus, an elaborate regulatory triangle exists among the immune, endocrine, and nervous systems; this extensive association, which acts to maintain a homeostatic *milieu interieur,* has been termed neuroimmunoendocrinology.[5,12]

The extent and diversity of hormone-immune interactions is well illustrated by the endocrine and neural regulation of the secretory immune system. Hormones and neuropeptides appear to modulate most immunological aspects of this system,[6,31,32] which is designed to defend mucosal surfaces against antigenic challenge.[33-36] The consequence of hormone action is that secretory immunity may be enhanced, antagonized, or suppressed, depending upon the target mucosal tissue. In the present article, which focuses upon the secretory immune system of the eye, the processes involved in the endocrine control of mucosal immunity are addressed.

# II. SECRETORY IMMUNE SYSTEM

## A. CHARACTERISTICS AND FUNCTION

The secretory immune system serves as a first line of defense to protect mucosal surfaces against microbial invasion.[33-36] Effector actions of this system are exerted primarily through secretory immunoglobulin A (sIgA),[33-36] which has been demonstrated to prevent the adhesion, colonization, and function of bacteria;[37-41] inhibit the attachment and internalization of viruses;[42,43] abolish potential infestation of parasites;[44-46] and block toxin-induced damage[47-49] in mucosal epithelial cells. Deficiency of the secretory immune system may significantly augment the susceptibility to infectious and allergic disease.[50-53] In fact, complete absence of secretory IgA in critical tissues may possibly impair survival.[54,139]

Secretory IgA is the predominant antibody in external secretions and consists of polymeric IgA coupled to the polypeptides J chain and secretory component (SC).[33-35] Immunoglobulin A and J chain are produced by plasma cells in the interstitial space of the mucosal lamina propria, whereas SC is synthesized by overlying epithelial cells.[55] Immunoglobulin A is constructed intracellularly in monomeric form, apparently induced to polymerize by J chain and a sulfhydryl oxidase catalyst, and released as a polymer complexed to J chain.[35,36] The production of polymeric IgA is uniquely regulated by the microenvironment,[57,58] endocrine system,[6] and class-specific lymphokines from T cells.[59-61] These lymphokines are secreted by an elaborate T cell circuit, which controls the differentiation, maturation, and possible recruitment of IgA-positive lymphocytes in mucosal sites.[60-63] T cells, in turn, appear to be modulated by hormones[64-67] and epithelial cell signals.[61,68-71] The lymphocytic output of IgA exceeds that of all other immunoglobulin classes throughout the body.[34] This comparison testifies to the very high density and activity of IgA-containing cells in mucosal sites[34,35] and underscores the importance of IgA in immunological defense.

The transport of pIgA into external secretions is regulated by SC.[33,72,73] This glycoprotein directs pIgA movement against a concentration gradient into secretions of the eye,[78,79] nose,[80] mouth,[81,82] lung,[83-85] mammary gland,[86-88] liver,[33,72,73] gall bladder,[89] stomach,[90] intestine,[91,92] bladder,[93] uterus,[94,95] and cervix.[96] Secretory component, which bears striking homology to the variable region of the immunoglobulin k chain,[97-98] undergoes synthesis and cotranslational core-glycosylation on the rough endoplasmic reticulum.[33,99,100] The translational product, which is a precursor receptor composed of ectoplasmic, membrane-spanning, and cytoplasmic domains,[97] acquires terminal glycosylation in the Golgi and is then incorporated into the basolateral margin of the epithelial cell.[33,72,73] At this site, the SC precursor binds with high affinity and specificity to J chain positive pIgA[101,102] and the receptor-ligand complex is endocytosed into coated pits.[103-105] Unidirectional transcytosis of sIgA occurs in distinct and segregated intracytoplasmic vesicles, which bypass the lysosomal compartment and are sorted to the apical membrane.[106-111] This vesicular transport depends upon microfilaments and microtubules[112,113] and is completed with the proteolytic cleavage of the SC's membrane-anchoring domain and the exocytosis of SC's ectoplasmic segment bound to pIgA;[114,115] the residual components of the SC precursor are either internally degraded or released.[114-116] The constitutive secretion of SC does not depend upon the presence of IgA,[114,117,118] and free SC varies both in concentration[95] and form[115,119] in different external secretions. The structure of free SC appears to be determined by alternate gene splicing,[120] oligosaccharide processing,[114] and the carboxyterminal site of precursor cleavage.[121] In addition to its role in pIgA transfer, SC has been demonstrated to eliminate pIgA-immune complexes from the serum,[122-125] interfere with IgA binding to Fc receptors,[126] diminish lymphocyte proliferation,[127] and curtail immunoglobulin production[128] *in vitro*. Under the influence of hormones,[6] lymphokines,[129] development,[77,130-133] or certain mucosal injuries,[90,131,134-137] the synthesis and expression of SC may be increased, with possible alteration in cellular distribution. Deficiency or absence of SC has been correlated with immune dysfunction, possibly resulting in, or being associated with, atalectasis,[131] inflammation,[85,138] anaphylaxis,[139] and non-Hodgkin's lymphoma.[140]

Antigenic exposure to mucosal tissues induces an influx of IgA plasma cells, specific sIgA antibodies in local secretions and an anamnestic IgA response after secondary challenge.[141-146] The extent and duration of the IgA reaction are significantly enhanced by antigen-induced retention and proliferation of IgA-positive and helper T cells;[146-150] this lymphocytic infiltration may also be modulated by vascular,[63,151] endocrine,[152-156] and epithelial[157] influences. In addition to the local response, IgA-bearing and T lymphoblasts migrate to distant mucosal sites, thereby providing defense in unprotected areas against potential antigenic challenge.[158-160] This extensive interconnection between secretory tissues has been termed the common mucosal immune system.[161] In external secretions, sIgA antibodies bind to plasmid,[39] protease,[40] toxin,[48,49] and microbial surface antigens,[43,44] resulting in agglutination, decreased adhesion, or lessened virulence[37,39,40,41,44] and permitting elimination through nonspecific factors, such as hydrokinetics and enzymatic destruction.[40] Immunoglobulin A may also serve as an immunological barrier within mucosal tissues by Fc association with polymorphonuclear leukocytes, macrophages, T cells, and NK cells: these interactions have been shown to stimulate phagocytosis and antibody-mediated cellular cytotoxicity.[35,162,163] Thus, IgA plays a unique, but crucial role in immunological defense.

## B. ENDOCRINE-IMMUNE INTERACTIONS

In the secretory immune system, hormones and neuropeptides influence essentially all parameters of immune activity. This neuroendocrine control, which is elicited through circulating hormones or direct tissue innervation, serves to (a) modulate the inherent migratory capability of lymphocytes;[164] (b) regulate the migration of IgA- and IgG-positive and T lymphoblasts to mucosal tissues,[64,153-156] (c) control the attachment and retention of circulating lymphoblasts in external sites;[32,152,165] (d) alter the proliferative capacity of mucosal lymphocytes;[17] (e) modify the synthesis and secretion of IgA and IgG antibodies;[6,17,166] (f) regulate the constitutive production, expression and output of SC by epithelial cells and hepatocytes,[6,94,167-172,366] apparently through genomic interaction;[171] (g) control the epithelial cell binding, uptake, and transport of pIgA antibodies into external secretions;[74,94,173,174] (h) enhance the pIgA concentration gradient from tissue to lumen;[75] (i) modulate the accumulation of IgA and IgG antibodies to defined antigens in mucosal fluids;[175,176] (j) alter the adherence potential of microorganisms to epithelial cells;[177-179] and (k) significantly influence mucosal defense against infectious agents.[175,177-181]

Several general principles seem to underlie the interaction between hormones and neuropeptides and the mucosal immune system. First, the neuroendocrine control of secretory immunity is site-specific (Table 1). Thus, prolactin, estrogens, and progestins regulate the secretory immune system of the mammary gland,[32] estrogens stimulate uterine immunity,[31] progestins control cervical immune expression,[96] glucocorticoids enhance hepatic immunological function[170] and substance P, vasoactive intestinal polypeptide and pancreozymin-cholecystokinin exert a positive effect on intestinal mucosal immunity.[16-19,26,27] Second, individual hormone action on the secretory immune system tends not to be generalized. Rather, depending upon the mucosal tissue, the endocrine interrelationship may strengthen, diminish, or elicit no effect on immunological activity.[6] For example, estrogens stimulate uterine,[94] antagonize vagina,[172] and are inactive on hepatic[170] secretory immune sites. Similarly, androgen action on mucosal immunity is positive in the eye,[6] negative in the mammary gland,[156] and absent in the uterus.[94] As an additional example, glucocorticoids are known to enhance hepatic,[170] ameliorate intestinal,[182] suppress uterine,[94,183] and not alter mammary[184] mucosal immune function. These observations indicate that endocrine action on mucosal tissues per se does not predict potential hormonal influence on local immunity.

Another consideration is that hormone and neuropeptide activity on the secretory immune system may or may not be direct. Hormone and neuropeptide receptors have been demonstrated in or on T and B lymphocytes and on mucosal tissue and epithelial cells.[8,23,67,185-187] However, receptor presence does not conclusively indicate direct action on immune function. Neuroendocrine binding may enable or disenable the activity of unrelated hormones.[2,3] Further, hormone effects may be supported or mediated through other endocrine circuits.[6] Definitive clarification of hormone and neuropeptide activity may be gained by structure/function correlations both *in vivo* and *in vitro*. In summary, the endocrine and nervous systems have a significant impact on antigen-dependent and independent processes of mucosal immunity. The mechanisms underlying the site- and hormone-specific regulation of the secretory immune system remain to be determined.

## III. SECRETORY IMMUNE SYSTEM OF THE EYE

### A. GENERAL ASPECTS

The lacrimal gland serves as the principal tissue in the secretory immune system of the eye.[6] This gland harbors a very high density of IgA plasma cells,[201,202] which far exceeds the inherent lymphocytic populations expressing IgG, IgM, or IgD.[34] The majority of IgA-containing cells in lacrimal tissue produce both J chain and polymeric IgA[34] and, in humans, present with an IgA1/IgA2 subclass distribution that is either unique[203] or analagous[204,205] to other mucosal sites. Coexistent with IgA-positive cells in the lacrimal gland is a diverse complement of both helper and suppressor T cells.[206-208] The primary origin of lymphocytes

## TABLE 1
## Influence of Various Hormones on the Secretory Immune System

| Tissue | Hormone treatment | Tissue[a] | | | Secretion[b] | | | Ref. |
|---|---|---|---|---|---|---|---|---|
| | | IgA | IgG | SC | IgA | IgG | SC | |
| Lacrimal gland | Testosterone | ↑ | — | ↑ | ↑ | — | ↑ | A |
| | Estradiol | | | | – | | – | B |
| | Progesterone | | | | – | | – | B |
| | Cortisol | | | | – | | – | B |
| | Testosterone + estradiol | | | | ↑ | | ↑ | C |
| | Testosterone + progesterone | | | | ↑ | | ↑ | C |
| | Prolactin | | | | – | | – | D |
| | Growth hormone | | | | – | | – | D |
| | Thyroxine + dexamethasone | | | | – | | – | E |
| | α-MSH | | | | – | | – | D |
| Salivary gland | Inhibitor of thyroid hormone secretion | ↓ | ↓ | — | ↓ | ↓ | | F |
| | Testosterone | | | | – | | ± | G |
| Mammary gland | Prolactin and/or | ↑ | | ↑[c] | | | | H |
| | Progesterone and/or | ↑ | | ↑[c] | | | | H |
| | Estradiol | ↑ | | ↑[c] | | | | H |
| | Testosterone | ↓ | | ↓[c] | | | | I |
| | Glucocorticoids | | | | – | | | J |
| Lung | Testosterone | | | | – | | – | G |
| Hepatocytes | Dexamethasone | | | | | | ↑ | K |
| | Estradiol | | | | | | – | K |
| Intestine | Testosterone | | | | – | | – | G |
| | Thyroxine | | | | | | ↓ | L |
| | Corticosterone | | | | | | ↓ | L |
| | Pancreozymin-cholecystokinin | | | | | | | |
| | Secretin | | | | ↑ | — | | M |
| | α-Interferon | | | | ↑ | ↑ | | M |
| | | | | | | | ↑ | N |
| Peyer's patches | Vasoactive intestinal peptide | ↓ | — | | | | | O |
| | Substance P | ↑ | — | | | | | O |
| | Somatostatin | ↓ | — | | | | | O |
| Uterus | Estradiol | ↑ | ↑ | ↑ | ↑ | ↑ | ↑ | P |
| | Progesterone | | | | – | – | – | Q |
| | Cortisol/dexamethasone | | | | – | – | – | Q |
| | Testosterone | | | | – | – | – | Q |
| | Estradiol + dexamethasone | ↓[d] | | | ↓[d] | | | R |
| | Estradiol + progesterone | ↓[d] | | | ↓[d] | ↓[d] | ↓[d] | S |
| Cervix | Progesterone[e] | ↑ | | ↑ | | | | T |
| Vagina | Estradiol | | | | ↓ | ↓ | ↓ | U |
| | Progesterone | | | | ↓ | | ↓ | U |
| | Dexamethasone | | | | – | | – | U |
| Bladder | Testosterone | | | | – | | | G |

*Note:* References for Table 1 are as follows: A = 166—169, 188—191, 366; B = 166, 167, 169; C = 169; D = 190; E = Sullivan, D. A., unpublished data; F = 371; G = 188, 189; H = 32, 156, 174, 192; I = 156; J = 184; K = 171; L = 182; M = 26, 27; N = 129; O = 17, 19; P = 31, 75, 94, 153, 154, 170, 173, 176, 183, 193—200; Q = 94, 193, 196; R = 94, 183; S = 94, 195, 196; T = 96; U = 172. Data was obtained in part from References 6 and 166. Relative to control: — = no effect, ↑ = increase, and ↓ = decrease.

[a] The designation "tissue" refers to the density of Ig-containing cells, the content of IgA, IgG, or SC, and/or the production of these proteins *in vitro*.

[b] The term "secretion" refers to the level of IgA, IgG, or SC measured in specific mucosal secretions.

[c] Hormone administration increased or decreased epithelial cell binding and accumulation of dimeric IgA, which actions most likely reflect effects on SC.

[d] Dexamethasone or progesterone antagonized the influence of estradiol on IgA, IgG, and SC.

[e] Apparent progesterone effect, given serum hormone profile during the menstrual cycle.

in lacrimal tissue is uncertain, but may in part be derived from local cervical[209,210] and distant gut-associated[211] lymphoid aggregates. Migration of IgA-positive lymphoblasts into the lacrimal gland appears to be random,[212] but the selective retention and topographical distribution of these cells is not random.[368] The specific and enduring recruitment of IgA-containing lymphocytes, as well as their nonuniform locations, may be controlled by hormonal,[132,213,214] neural,[215] T cell, and epithelial cell[69] signals. Secretory component, which has also been identified in acinar and ductal cells of the lacrimal gland,[78,201] may regulate the binding and transcytosis of pIgA antibodies into tears. In addition to lacrimal tissue, the conjunctiva has been proposed as active contributor to ocular mucosal immunity.[216] However, though IgA-containing cells are found in the substantia propria of some species,[202,217] the conjunctival epithelium of the rat[168,201] and human[78] does not express SC and therefore, like vaginal tissue,[95,218] is unable to transport pIgA.

The predominant immunoglobulin in tears is IgA, with minimal levels of IgG and essentially no IgM.[6,219,220] Tear IgA, which is polymeric,[84,201,204,221] originates from local synthesis in the lacrimal gland[79,222] and is apparently transported by SC into lacrimal fluids against a concentration gradient.[77,219] In contrast, IgG movement from lacrimal tissue to tears is down a steep concentration gradient.[219] Tear IgA concentrations are among the highest of all mucosal secretions and are accompanied by substantial levels of free SC.[6,84,188,189]

Antigenic challenge to the ocular surface results in a marked infiltration of IgA lymphoblasts into lacrimal tissue[223] and a significant increase in tear IgA antibodies.[211,224-227] The mechanism underlying this response, though, remains to be clarified; retrograde antigen transfer up the lacrimal duct has not been demonstrated and potential conjunctival processing[216] is constrained by the molecular weight restriction of epithelial architecture.[228] It may be that infective and toxic substances are cleared through the nasolacrimal duct, thereby inducing a primary immune reaction in gut-associated lymphoid tissue. In support of this contention, oral administration of bacteria[211,229,230] leads to a specific tear IgA response, apparently through IgA lymphoblast migration from mesenteric and peripheral lymph nodes to the lacrimal gland.[211] Thus, immune protection in the eye may be conferred by both local and distant antigenic exposure.

The role of the secretory immune system of the eye is to protect the ocular surface against microbial agents and toxic compounds, thereby preserving corneal integrity and preventing severe visual impairment and possible blindness. Given the extreme importance of this immunological role, it would be of tremendous clinical value to be able to modulate and to enhance this surface defense. Such a therapeutic intervention, however, requires a thorough understanding of the regulatory mechanisms controlling the ocular secretory immune system.

## B. GENDER-RELATED IMPACT

"Males are by no means as prone to diseases of the eye from sexual causes as females."[232] This statement, included in a 1888 monograph addressing clinical ophthalmology, summarized a widespread belief among physicians that ocular health and dysfunction were dramatically influenced by gender.[232-239] During the past 40 years, scientists have extended this observation by demonstrating that gender also exerts a significant impact on the structure and function of the lacrimal gland,[240-249,304] the principal tissue involved in ocular mucosal immunity.[6] Thus, significant differences between males and females have been noted in the morphology, histochemistry, biochemistry, and genetics of lacrimal tissue in mice, rats, guinea pigs, rabbits, and/or humans.[240-249] In addition, researchers have also reported that distinct, gender-associated variations may occur in lymphocyte densities in human lacrimal glands,[250] as well as in immunoglobulin levels in human tears.[251]

To explore the extent, and possible mechanisms, of these gender-related differences in ocular mucosal immunity, a series of investigations were performed in our laboratory by

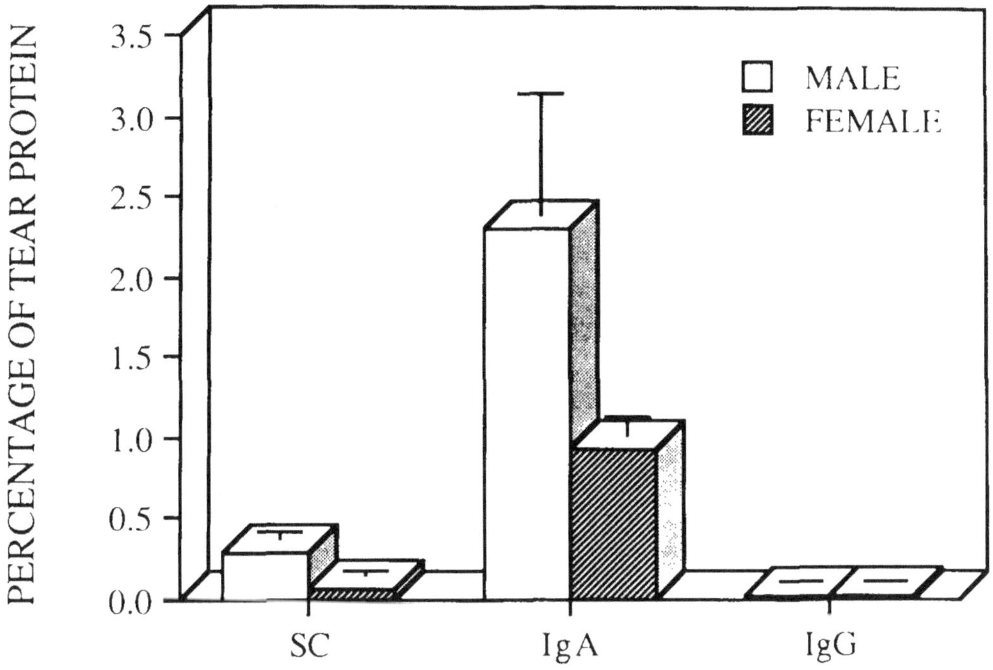


FIGURE 1. Comparison between the levels of IgA and free SC in tears of male and female rats. Data from References 166, 167, and 219.

utilizing rats as an experimental model. Analysis of tears from young adult males and females demonstrated a striking sexual dimorphism in the content of free SC and IgA.[166,167] As shown in Figure 1, free SC and IgA concentrations were two- to fivefold greater in tears of males (free SC = 51.1 ± 3.2 μg/ml; IgA = 326 ± 28 μg/ml), as compared to females (free SC = 9.9 ± 0.7 μg/ml; IgA = 156 ± 36 μg/ml). These differences could not be accounted for by fluctuations in either tear volume or total tear protein. Moreover, gender had no effect on the concentration of tear IgG (M = 5.88 ± 1.19 μg/ml; F = 8.11 ± 2.44 μg/ml)[219,249] (Figure 1). This differential expression of free SC and IgA in tears appeared to be due to a gender-associated influence on the synthesis and/or secretion of these proteins by the lacrimal gland. Overnight incubation of lacrimal tissue *in vitro* resulted in a significantly greater cycloheximide-sensitive accumulation of free SC and IgA in culture media of "male" glands, relative to those of females[166,168] (Figure 2). In addition, evaluation of lacrimal glands from age-matched, adult rats by immunofluorescence microscopy revealed that tissues from male rats contained significantly higher densities of IgA plasma cells (18.6 ± 1.0 cells/microscopic field) than those of females (11.5 ± 0.6 cells/microscopic field).[213]

Several explanations might account for these gender-related variations in the secretory immune system of the eye.

First, hormonal changes during the estrous cycle could theoretically induce considerable fluctuations in the levels of free SC and IgA in lacrimal tissue and tears, resulting in a selective reduction of these proteins during certain stages of the cycle. Asynchronous patterns of systemic and mucosal immune expression are known to occur during both the estrous and menstrual cycles.[7,252-254] In fact, cycle-associated variations have been identified in mucosal lymphoblast populations, epithelial cell production of SC and luminal immunoglobulin levels.[7,94-96,175,180,193,194,218,252-269] However, analysis of tears collected during the proestrous, estrous, diestrous day 1, and diestrous day 2 phases of estrus, as well as those from anestrous rats, demonstrated that the estrous cycle had no effect on the concentration of free SC.[167]

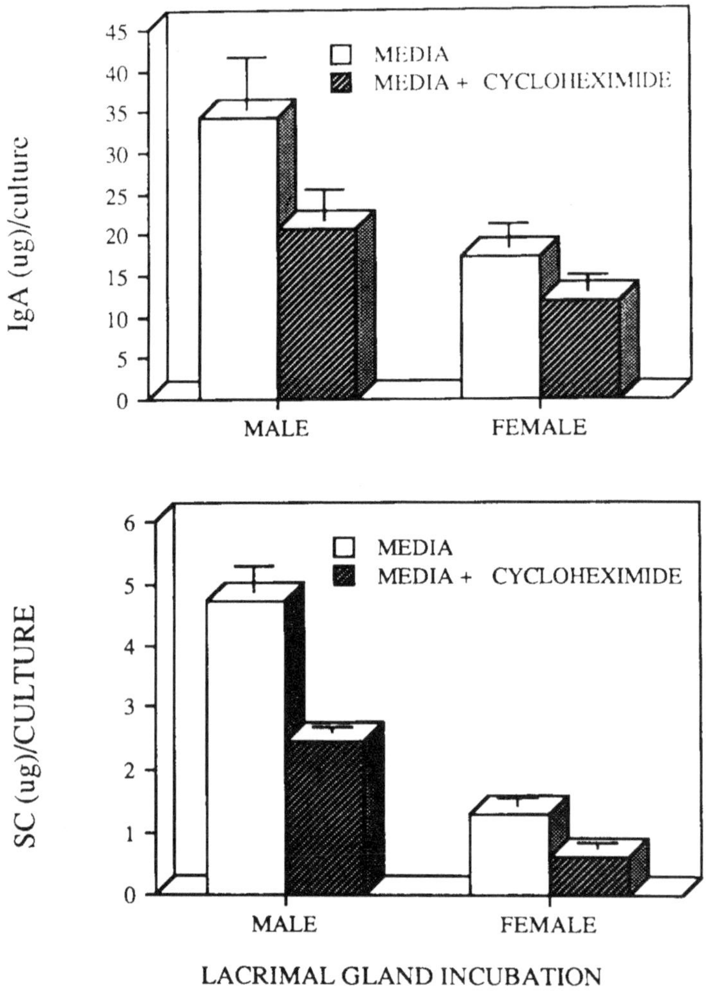


FIGURE 2. Influence of gender on the production of IgA and free SC by lacrimal glands cultured *in vitro*. Data from References 166 and 168.

Second, differences might reflect a general phenomenon common to other mucosal sites. For example, it is recognized that a sexual dimorphism exists in the development and cellular composition of the lung[270,271] and that changes in the sex steroid environment may be associated with alterations in lymphoblast homing to intestinal[165] and bronchial[272] tissue, the composition of bile,[273] and various functional activities of salivary glands.[274,275] Furthermore, researchers have shown that the immune response to oral immunization may depend upon gender.[276] To assess whether gender-related variations in the mucosal immune system of the eye might be duplicated in other secretory immune tissues, intestinal, respiratory, and salivary secretions, as well as serum, were evaluated.[213] Results demonstrated that IgA content in lung lavage fluid was significantly higher in female, as compared to male, rats (Figure 3). This difference, which was opposite to the situation in tears, was present irrespective of whether data was standardized in terms of concentration or total protein. In contrast, no gender-associated variations were found in IgA levels of salivary and intestinal secretions and serum (Figure 3). Moreover, analysis of free SC content in saliva, respiratory, and intestinal fluids did not reveal gender-dependent differences in these samples (Sullivan, D. A., unpublished data, 1989). These findings demonstrated that the gender-related profile in the ocular secretory immune system is unique to the eye.

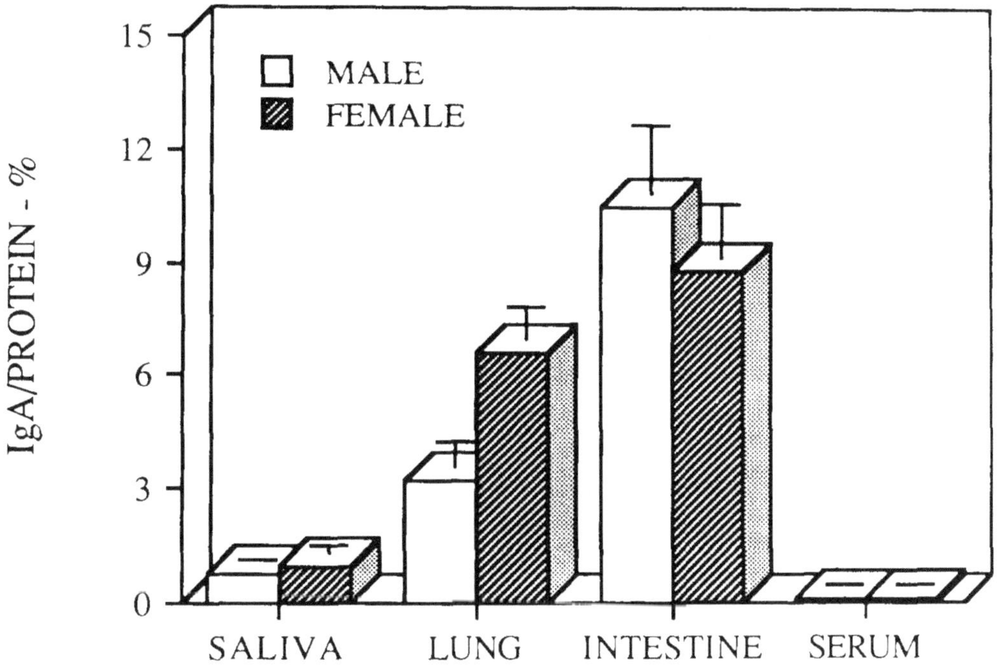


FIGURE 3. Comparison between IgA levels in mucosal secretions and serum of male and female rats. Data from Reference 213.

Third, immune differences might hypothetically represent a developmental phenomenon evident in young adults, but not necessarily present at earlier or older ages. Consequently, to examine the influence of aging on gender-associated variations in the secretory immune system of the eye, multiple immune parameters were evaluated in male and female rats at 0.2 (infant), 0.6 (pre-pubertal), 1.3 (pubertal), 3.0 (adult), 8.0 (mid-life), 14.0 (mid-life), and/or 17 (senescent) months of age. Analyses showed that the content of tear IgA underwent a precipitous increase over a 3-week time span from before (0.6 months) to during (1.3 months) puberty, rising by factors of 9- and 13-fold in females and males, respectively[77,132] (Figure 4). This developmental period also coincided with profound and parallel changes in the concentration of tear free SC[77,132] (Figure 4): prior to the onset of puberty, free SC could not be detected in 93% of tear samples, whereas after pubertal maturation, tear-free SC levels had attained adult concentrations. This tear free SC profile was analagous to the age-related pattern of free SC output by lacrimal tissue incubated *in vitro*. Following puberty, tear IgA content continued to increase in both sexes until adulthood (3 months) and then plateaued in females through 8 to 17 months of age. In contrast, tear IgA amounts in males appeared to stabilize from 3 to 8 months and then rose significantly to the highest levels at 17 months of age. This increase was also reflected in lacrimal tissue cytosol in males; IgA content underwent a six-fold elevation from 3 to 17 months. A striking finding in these studies was the persistence of a sexual dimorphism in the secretory immune system of the eye. Thus, after pubertal development, IgA and free SC levels were significantly higher in tears of males, as compared to those of females, at all ages tested up to 17 months. These gender- and age-related variations in tear IgA and free SC amounts could not be accounted for by changes in either the volume of, or total protein content in, tears.[77,132] In addition, no gender-associated differences were found in tear IgG concentrations throughout aging.[219] With regard to IgA-containing lymphocytes in lacrimal tissue, these underwent a progressive increase from infancy to adulthood and then effectively plateaued to senescence. Both the kinetic profile and magnitude of this IgA-containing cell accumulation varied according to

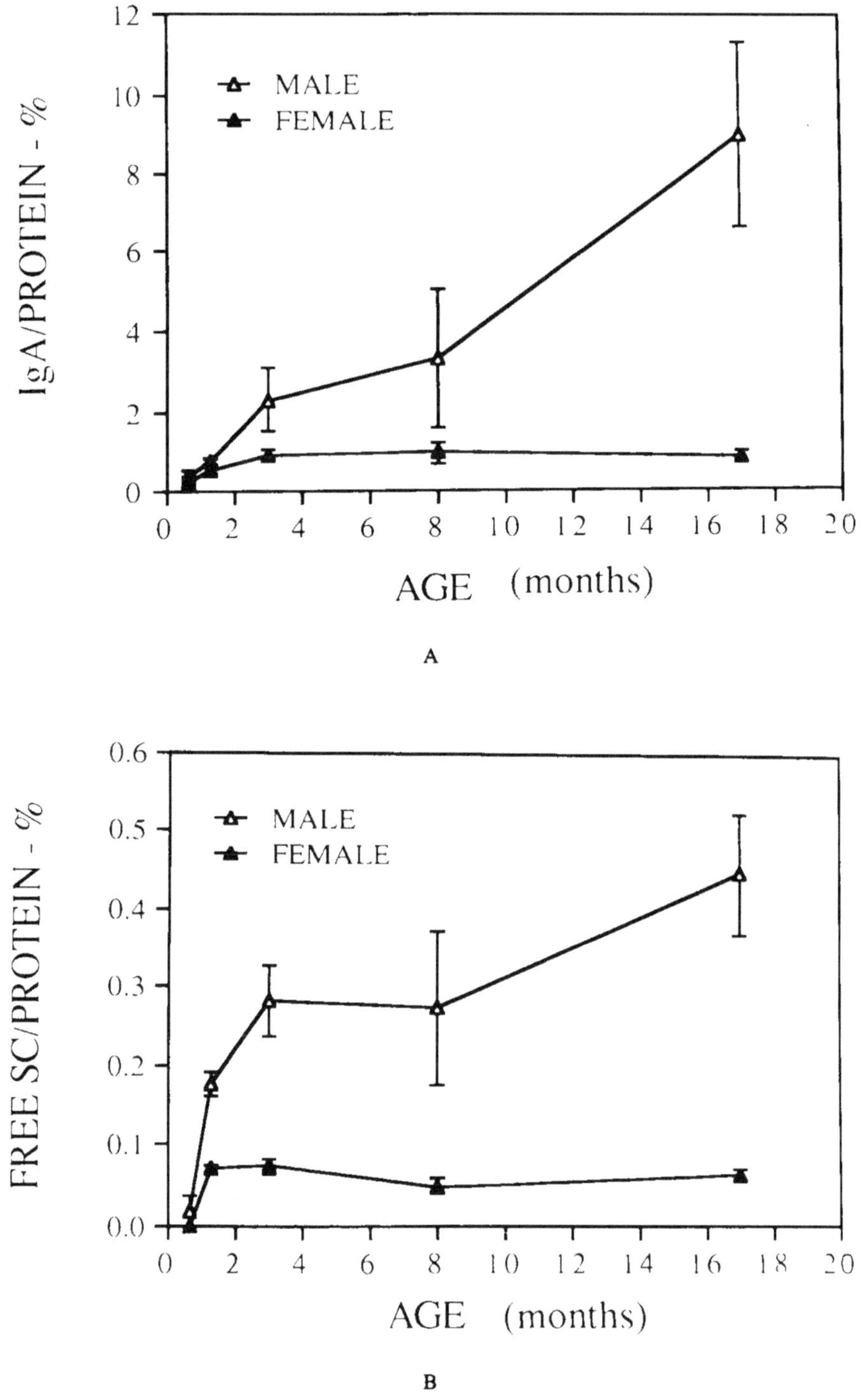


FIGURE 4. Impact of age on the tear IgA (A) and free SC (B) content in male and female rats. Data from Reference 77.

gender. In fact, the total quantity of these cells in glands of males was greater than, or equal to, that in tissues of females at all ages examined.[132,214] An intriguing observation in these experiments was that the IgA-containing cell density in lacrimal tissue of male prepubertal (0.6 months old) rats was twice that of females. This gender-related difference might be explained in part by the significantly greater content of helper T cells in male lacrimal glands

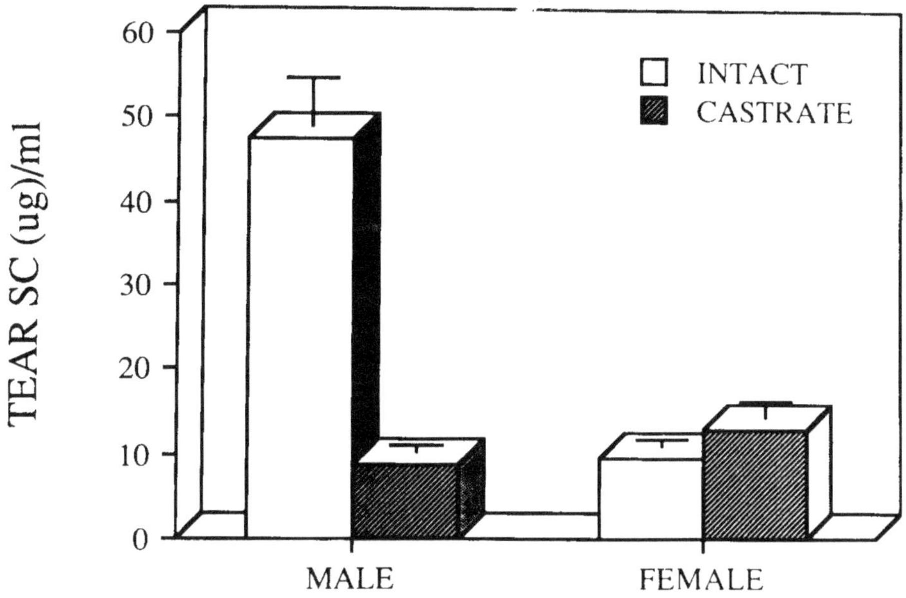


FIGURE 5. Effect of castration on free SC concentrations in tears of male and female rats. Data from Reference 167.

at this age;[206] T cells appear to modulate the recruitment, differentiation, and proliferation of IgA plasma cells in secretory immune sites.[60-63] In contrast to IgA lymphocyte dynamics, neither age nor gender influenced the distribution of IgM-containing cells in lacrimal tissue.[214] Overall, these ''aging'' studies demonstrated that gender plays a fundamental and continuing role in the organization and expression of the secretory immune system in the eye. In addition, the findings that ocular mucosal immunity is either conserved or enhanced from adulthood to senescence served as a sharp contrast to the known, age-associated reductions in immune function at other mucosal sites. Researchers have reported a direct correlation between senescence and significantly reduced concentrations of IgA in intestinal[277] and nasal[278] secretions, diminished hepatobiliary clearance of IgA antibodies,[279] abrogated salivary IgA response to antigenic challenge,[280] and a marked decline in the density of IgA-containing cells in the mesenteric lymph nodes.[281] These age-related comparisons indicate that the regulation of the ocular secretory immune system may be unlike that of other sites.[6]

## C. ANDROGEN REGULATION

### 1. Kinetics and Specificity

The mechanism(s) underlying the sexual dimorphism in the secretory immune system of the eye may be due to the selective impact of androgenic hormones. This hypothesis follows from the observation that most gender-related differences in the lacrimal gland may be attributed to androgen influence.[167-169,240-244,246,282-290] For example, castration of, or administration of the anti-androgen cyproterone acetate to, male rats results in a time-dependent alteration of lacrimal tissue morphology to that structural profile presented by females.[240,241] Treatment with androgens, but not other classes of steroid hormones, reverses the orchiectomy effect and restores the lacrimal gland characteristics of ''intact'' males.[240]

To evaluate whether sex hormones might be involved in the gender-associated variations in the ocular mucosal immune system, tears were obtained from male and female rats both before and after castration. Orchiectomy, but not sham-surgery, induced a marked and uniform decline in tear-free SC levels; within 8 d after surgery, free SC concentrations (Figure 5) were similar to those found in tears of females.[167] In contrast, ovariectomy did not alter the free SC content of tears (Figure 5). Additional experiments also demonstrated that orchiectomy, but not sham-operation, led to a significant reduction in tear IgA levels.[166] Castration of males was associated with variations in tear volume,[291,292] but these changes could not account for SC or IgA effects.[166,167] Furthermore, total protein content in tears of intact and castrated male rats were not significantly different.[166,167]

To determine whether androgen replacement might reverse the influence of orchiectomy on tear-free SC and IgA levels, a series of kinetic experiments were performed. The rationale for conducting temporal studies was that hormone-induced effects on immunity in other mucosal tissues may be either stimulatory or suppressive, depending upon whether exposure is acute (e.g., hours),[183] extended (e.g., days),[75,94,193,195] or chronic (e.g., weeks).[195]

To assess the impact of *acute* testosterone exposure, orchiectomized rats were injected once with saline or testosterone (2 mg) and tears were collected immediately before and 1.5, 4, 7, 12, and 24 h after treatment. This dosage of testosterone has been shown to elicit physiological responses in adult rats. Analysis of results showed that no rapid variations occurred in tear IgA or free SC content during the first 12 h of testosterone exposure, compared to that of saline-injected controls. From 12 to 24 h after hormone administration, the free SC concentration did undergo a significant rise in tears, but this effect was not paralleled by a concomitant increase in IgA. No hormone-associated fluctuations in total tear protein concentrations were evident during the course of this experiment. These findings indicated that abbreviated androgen action (i.e., hours) is insufficient to elicit substantive immune responses in the eye.[188]

To evaluate the effect of *extended* testosterone treatment, orchiectomized rats were administered either saline or testosterone (2 mg/day) for 4 consecutive days and tears were obtained before each injection and 24 h after the fourth treatment. This hormonal schedule resulted in a significant, time-dependent rise in both tear free SC[167] and IgA[166] concentrations (Figure 6); by 4 d of testosterone administration, free SC and IgA levels were equivalent to those measured in tears of intact male rats. Of interest, treatment of ovariectomized females with testosterone for 4 d also induced a significant, four-fold increase in the tear SC concentration.[168]

To explore the impact of *chronic* androgen exposure on the ocular secretory immune system, pellets (designed for slow release over 21 d) containing 50 mg of testosterone or placebo (cholesterol, methyl cellulose, lactose) were implanted subcutaneously in the sub-scapular region of orchiectomized rats. Tears were obtained prior to pellet implantation (day 0) and on days 4, 7, 11, 14, and 17 of hormone administration. Chronic testosterone treatment resulted in a significant and progressive elevation in the tear content of IgA and free SC[188] (Figure 7). After 4 d of androgen therapy, the magnitude of the IgA and free SC response was similar to that observed with daily testosterone injections (e.g., Figure 6). Following 17 d of hormone exposure, testosterone had dramatically increased tear IgA and free SC concentrations by 13.5- and 18.1-fold amounts, respectively, compared to those of placebo-treated controls. These hormone-related effects could not be explained by variations in either the volume of, or total protein content in, tears. Kinetic analysis of total tear IgA following androgen exposure demonstrated that mean IgA levels on day 4 were 59% of those measured on day 17. In contrast, free SC amounts of 4-d-hormone-treated rats were only 26% of mean levels observed on day 17. These findings indicated that the testosterone-induced, temporal pattern of IgA and free SC accumulation was different.[188] Experimental results also demonstrated that extended or chronic testosterone exposure had no effect on tear IgG, whether data was expressed in terms of total amount or as a percentage of total tear protein.[367] Furthermore, there was a tremendous difference between tear IgG and IgA content. Thus, by day 17 of testosterone treatment, tear concentrations of IgG and IgA were $6.70 \pm 1.11$ μg/ml, and $2{,}439 \pm 793$ μg/ml, respectively.[367] This hormone action specifically on IgA, but not IgG, appears to distinguish the secretory immune system of the eye from other

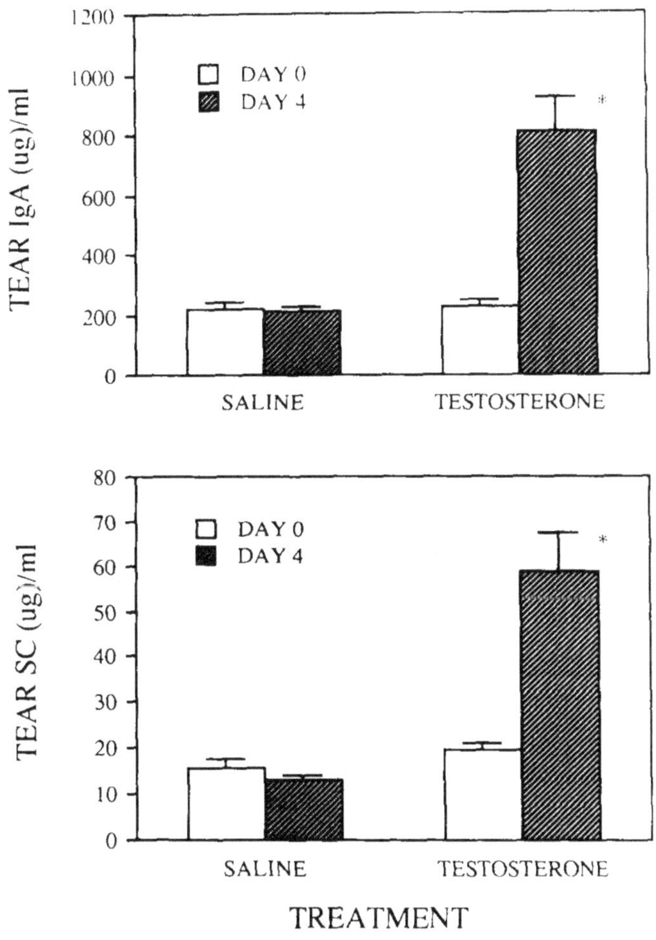


FIGURE 6. Influence of testosterone administration for 4 d on tear IgA and free SC levels in orchiectomized rats. Data from References 166 and 167.

endocrine-responsive mucosal sites, wherein both IgA and IgG are modulated.[31,172] Moreover, the positive influence of androgens on the ocular secretory immune system contrasts sharply with their known suppressive effect on mucosal immunity in the mammary gland.[32,156,174,192]

The tear free SC and IgA response to extended or chronic testosterone exposure was both dose-dependent[166,167,188] and specific for androgens.[166,167,169] Administration of physiological amounts of estradiol, progesterone, or cortisol for 4 d to orchiectomized rats had no effect on tear concentrations of free SC or IgA.[166,167] The inability of estradiol to influence these tear parameters indicated that testosterone action could not be attributed to androgen aromatization to estrogens. This conclusion was consistent with an additional finding that 5α-dihydrotestosterone, an androgen that is not aromatized,[293] elicited an ocular SC response similar to that of testosterone.[167] Furthermore, estrogen treatment alone does not appear to alter either the structure or function of the lacrimal gland[166,167,169,240,291,294] and estrogen receptors are either undetectable[294] or present in negligible quantities[295] in lacrimal tissue. This lack of estrogen action on lacrimal tissue helps explain, for example, the absence of menstrual cycle,[296] oral contraceptive,[297] and estrogen therapy[298,299] effects on tear production in humans. In contrast, androgens do modulate tear volume, depending upon the

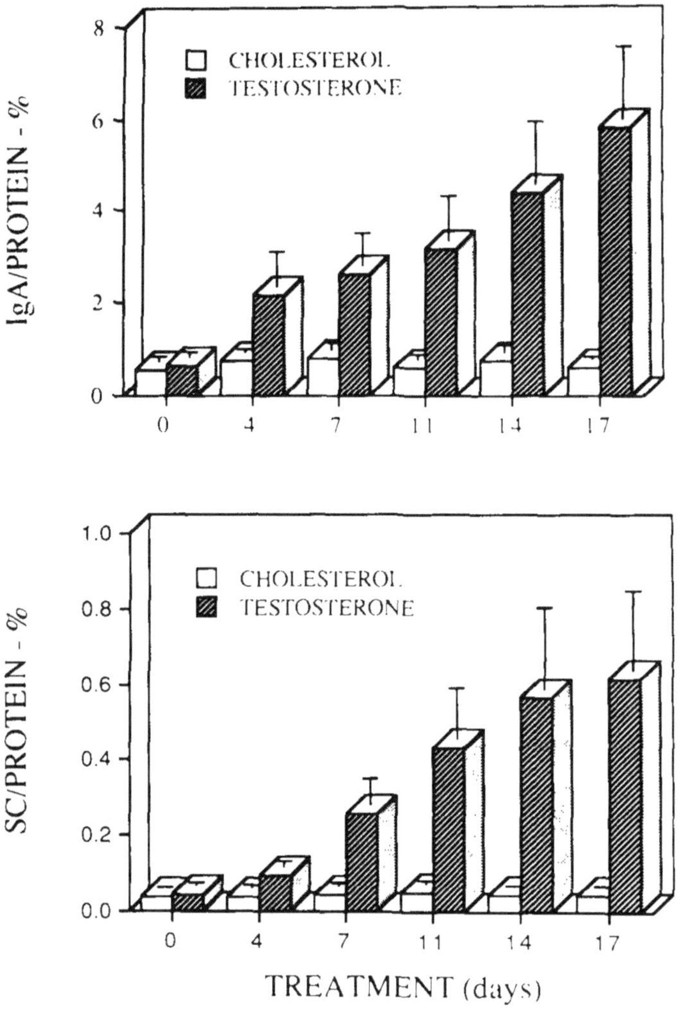


FIGURE 7. Time course of the effect of chronic testosterone exposure on IgA and free SC levels in tears of orchiectomized rats. Data from Reference 188.

experimental[291,292] or clinical[300-302] conditions, and specific, high affinity (at $10^{-9}$ *M*) androgen receptors are distributed in the rat lacrimal gland.[303] Of interest, analysis of androgen receptors showed that gender-related differences occur in receptor density as well as receptor binding capacities for DNA.[303] Whether gender modifies intracellular receptor location[344] remains to be determined.

The ocular immune potential of androgens also appeared to be very dependent upon hormone structure[188] (Table 2). Introduction of a 7α-methyl group into nortestosterone, which yields 4-estren-7α-methyl-17β-ol-3-one, seemed to enhance androgen action on tear IgA and free SC, relative to that of testosterone. Conversely, utilization of the weak androgen 5α-androstan-17β-ol resulted in almost complete abrogation of hormone response. For comparison, these steroid compounds were reported to exhibit significantly more (4-estren-7α-methyl-17β-ol-3-one) or less (5α-androstan-17β-ol) activity than testosterone in the induction of morphological alterations in the lacrimal gland.[240,285] Administration of danazol, a synthetic 17α-alkyl derivative of ethinyltestosterone,[305] to orchiectomized rats had no impact on IgA and free SC levels in tears.[188] This anabolic hormone has been shown to bind androgen

**TABLE 2**
**Influence of Various Androgenic Hormones on Tear IgA and Free SC Levels**

| Treatment | IgA (μg/ml) | | Free SC (μg/ml) | |
|---|---|---|---|---|
| | Day 0 | Day 4 | Day 0 | Day 4 |
| Saline | 201 ± 42 | 221 ± 38 | 15.1 ± 3.0 | 15.2 ± 3.0 |
| Testosterone | 193 ± 35 | 687 ± 93[b] | 17.7 ± 2.7 | 95.5 ± 7.6[c] |
| Danazol | 160 ± 25 | 165 ± 15 | 18.1 ± 2.8 | 21.6 ± 4.5 |
| 5α-androstan-17β-ol | 215 ± 47 | 375 ± 143 | 18.6 ± 1.7 | 29.0 ± 2.8[a] |
| 4-estren-7α-methyl-17β-ol-3-one | 414 ± 85 | 2396 ± 709[a] | 24.4 ± 5.3 | 165.0 ± 47[a] |

*Note:* Age-matched and orchiectomized rats (n = 5-6/ treatment group) were injected daily with saline or one of the above-mentioned androgenic compounds (2 mg/d). Tears were obtained immediately before (day 0) and 24 h after (day 4) the last injection. Numbers equal the mean ± SE. Data from Reference 188.

[a] Significantly ($p$ <0.05) greater than pretreatment and saline-injected control values.
[b] Significantly ($p$ <0.001) greater than pretreatment and saline-injected control values.
[c] Significantly ($p$ <0.0001) greater than pretreatment and saline-injected control values.

receptors in target tissues[306] and is currently used therapeutically in the treatment of hereditary angioedema,[307] idiopathic thrombocytopenic purpura,[308] and hemophilia.[309] Nevertheless, as demonstrated in the murine autoimmune model of lupus erythematosus,[310] danazol is unable to consistently replace testosterone in androgen-immune interactions.

In summary, extended and chronic, but not acute, androgen treatment stimulates a progressive rise in tear IgA and SC content of orchiectomized rats. This hormone action is both dose-dependent and specific for androgens and may underlie, in part, the sexual dimorphism observed in the secretory immune system of the eye.

### 2. Lacrimal Gland Involvement

Several possibilities might explain the androgen control of the ocular secretory immune system:

1. Androgens may stimulate the synthesis and/or secretion of SC by the lacrimal gland. Hormones are known to modulate SC production in other mucosal tissues[94,170] and enhanced lacrimal output might account for the testosterone-induced accumulation of free SC in tears. This androgen action on SC could also mediate the hormone-associated increase in tear IgA; SC regulates IgA movement into external secretions,[33,34,72] and, following steroid hormone exposure, heightened SC secretion into mucosal fluids either precedes or parallels IgA transport.[94,188]
2. Androgens might control the density of IgA-containing lymphocytes in lacrimal tissue, since the numbers of these cells are higher in glands of males, compared to females,[132,213,214] and sex steroid hormones have been demonstrated to modulate the migration of IgA-positive cells to other secretory tissues.[153,154,156]
3. Androgens may elevate IgA production, given their known ability to stimulate antibody synthesis by plasma cells.[7] It is unlikely that this hormone effect could be direct, though, because androgen receptors have not been identified in lymphocytes.[311,312] Rather, androgens might control IgA output by indirectly acting on T cells, which regulate IgA production in mucosal sites[60-63] and are susceptible to androgen control.[64-67]

To evaluate the possible androgen regulation of SC synthesis and/or secretion by the lacrimal gland, orchiectomized rats were administered testosterone (2 mg/d) or saline for 4 d, and lacrimal glands were obtained 24 h after the last injection. Tissues were cultured overnight in the presence or absence of cycloheximide, a known inhibitor of protein synthesis. Results demonstrated that lacrimal glands from testosterone-exposed animals released significantly more SC than did tissues from saline-treated controls throughout the time course of the experiment.[168] Inclusion of cycloheximide in the cultures significantly reduced the accumulation of SC in the incubation media. Of interest, testosterone administration to ovariectomized rats also stimulated a marked, fourfold increase in the SC production of cultured lacrimal tissue.[168] Whether this hormone effect is mediated through an increase in the net rate of intracellular SC synthesis or an enhancement of the percentage of acinar cells involved in SC production remains to be clarified. In this regard, immunofluorescent analysis of SC in acinar cells suggests that a heterogeneity exists in SC location, such that only a small percentage of acinar cells typically contain much SC.[78,201] Moreover, electron microscopic examination of acinar cells suggests that, based upon morphological criteria, different types of acinar cells populate the lacrimal gland.[313,316]

To determine whether an increased number of IgA-containing cells in the lacrimal gland might be involved in the androgen-induced accumulation of tear IgA, orchiectomized rats were administered testosterone, the potent androgen analog 4-estren-7α-methyl-17β-ol-3-one or vehicle for 4, 7, 14, or 17 d. In five out of eight separate experiments, testosterone increased the IgA-containing cell density by a factor of 18.1 to 41.2% over that in vehicle-treated controls. However, when densities were corrected for total tissue weight, no consistent differences were apparent between hormone and placebo-treated groups in the total numbers of IgA-positive cells. Administration of 4-estren-7α-methyl-17β-ol-3-one for 4 d, though, did elicit a significant, 49% rise in the number of IgA-containing cells in lacrimal tissue, compared to that of saline-injected controls.[367] Preliminary studies have also indicated that testosterone treatment has no influence on the IgG-containing cell population of lacrimal tissue.[367] Thus, androgen action may not play a primary role in the significantly higher density of IgA-positive lymphocytes in lacrimal glands of adult male, relative to female, rats.[132,213,214] An intriguing observation in these and other studies was the marked heterogeneity in the frequency distribution of IgA-containing cell densities in lacrimal tissue sections.[214,368] Statistical analysis of this distribution by chi square goodness of fit demonstrated that the topographical location of IgA-containing cells per section or through the gland was not random. This lymphocytic variation raises the possibility of a neural:lymphocyte interrelationship; others have shown that direct neural innervation of lymphoid tissue and small intestinal mucosa, coupled with transmitter release, has a profound impact on the immigration, distribution, and function of lymphocytes, including IgA-positive cells.[1,3,5,17,22,152] Similar neural-lymphocytic interactions may also occur in the lacrimal gland, which contains substantial amounts of neural vasoactive intestinal polypeptide and substance P, methionine-enkephalin, leucine enkephalin, and autonomic transmitters.[215,314-316] These neurotransmitters are known to influence lymphocyte localization and activity,[3,5,16,17,20,22,24,25,152] and their action might account for the distinct, nonrandom distribution of IgA plasma cells in lacrimal tissue.

To examine whether testosterone administration regulates the lacrimal gland content of IgA or IgG, as well as their concentration gradient from tissue to tears, orchiectomized rats were exposed to subcutaneous testosterone (50 mg) of placebo (50 mg) pellets for 7 d. The results of these studies[367] showed that androgen exposure stimulated a two- to three-fold rise in the IgA/protein ratios in lacrimal gland cytosol (control = 0.38 ± 0.06%; T = 0.70 ± 0.13%) and tears (control = 0.89 ± 0.30%; T = 2.93 ± 0.61%). In consequence, testosterone induced a twofold elevation in the tissue to tear transfer [(tear IgA/protein)/(cytosol IgA/protein)] of IgA (control = 2.16 ± 0.46; T = 4.37 ± 0.48); this IgA movement was against an apparent gradient. With regard to IgG, testosterone had no significant effect on IgG/protein ratios in cytosol (control = 0.35 ± 0.05%; T = 0.35 ± 0.06) or tears (control = 0.0035 ± 0.0007; T = 0.0085 ± 0.0023%). Moreover, transfer of IgG from the lacrimal to tears was down a 90-fold gradient.[367]

Overall, these studies demonstrated that androgens regulate the synthesis and/or secretion of SC by the lacrimal gland. This hormone effect appears to be paralleled by an increase in the lacrimal gland concentration of IgA, perhaps through heightened local production but not through IgA-containing cell accumulation. Androgen action is also associated with an enhanced transfer of IgA from tissue to tears against an apparent concentration gradient. In contrast, testosterone does not influence the IgG-containing cell density or IgG content in lacrimal tissue, nor the movement of IgG down a concentration gradient from tissue to tears.

#### 3. Mucosal Selectivity

The regulatory influence of testosterone on the ocular secretory immune system may be unique to the eye. This hypothesis is based upon the observation that endocrine effects on mucosal immunity tend not to be generalized; instead, specific hormones appear to enhance, curtail, or have no action on certain external sites.[6] However, given the significant role of androgens in the development and/or function of the salivary gland,[274,275] respiratory tract,[270,271] and small intestine,[318,319] it might be that additional secretory immune tissues are stimulated or suppressed by androgen exposure.

To test the tissue selectivity of possible androgen action, orchiectomized rats were administered four daily injections of testosterone (2 mg/day) or saline. Tears, saliva, respiratory fluids, intestinal secretions, urine and/or serum were collected 24 h after the last treatment.[188,189] Testosterone exposure induced a significant increase in tear IgA and SC levels in orchiectomized rats. This effect was observed irrespective of whether data were expressed in terms of concentration, total amount, as a percentage of tear protein or standardized to pretreatment values. In addition, androgen exposure enhanced the tear protein concentration, but this response could be accounted for by a reduction in tear volume; no variations were found in the total protein content. Testosterone treatment also stimulated a slight, but significant rise in salivary free SC levels and decreased the serum IgA concentration and intestinal IgA/protein ratio (Figure 8). These latter serum and intestinal responses could be attributed to changes in total protein. The serum IgA/protein ratio did not vary after hormone treatment, but the serum protein concentration was diminished. Similarly, IgA content in intestinal secretions did not differ between testosterone- and saline-treated groups, but androgen exposure resulted in a 29% increase in intestinal protein recovery. With regard to saliva, the magnitude of the free SC response to androgens was very low and could not be consistently observed in three separate experiments. Administration of testosterone had no influence on IgA levels in saliva, respiratory secretions, and urine, or on free SC content in respiratory and intestinal fluids (Figure 8). No free SC could be detected in urine. Thus, these results demonstrate that androgen regulation of the ocular secretory immune system appears to be unique to the eye.

## D. ENDOCRINE INTERACTIONS: ROLE OF THE HYPOTHALAMIC-PITUITARY AXIS

The sexual dimorphism in the mucosal immune system of the eye may not be entirely due to the effect of androgens. It has become increasingly evident in endocrine research that steroid hormone action on target tissues is often regulated by other hormones. Thus, androgen action may be significantly influenced by thyroid,[275,317,320,321] adrenal,[322,323] ovarian,[240,244] or pituitary[282,324] hormones. In fact, a number of androgen effects on peripheral tissues may be mediated entirely through another endocrine system, such as the hypothalamic-pituitary axis.[325-327] Given this information, the following series of studies were conducted.

The purpose of the initial study was to determine whether other hormones might modulate or control androgen action on the ocular secretory immune system. Experiments involved the daily administration of saline or hormones to adult, orchiectomized rats, the collection of tears 24 h after the fourth hormone injection, and the measurement of free SC, IgA, and

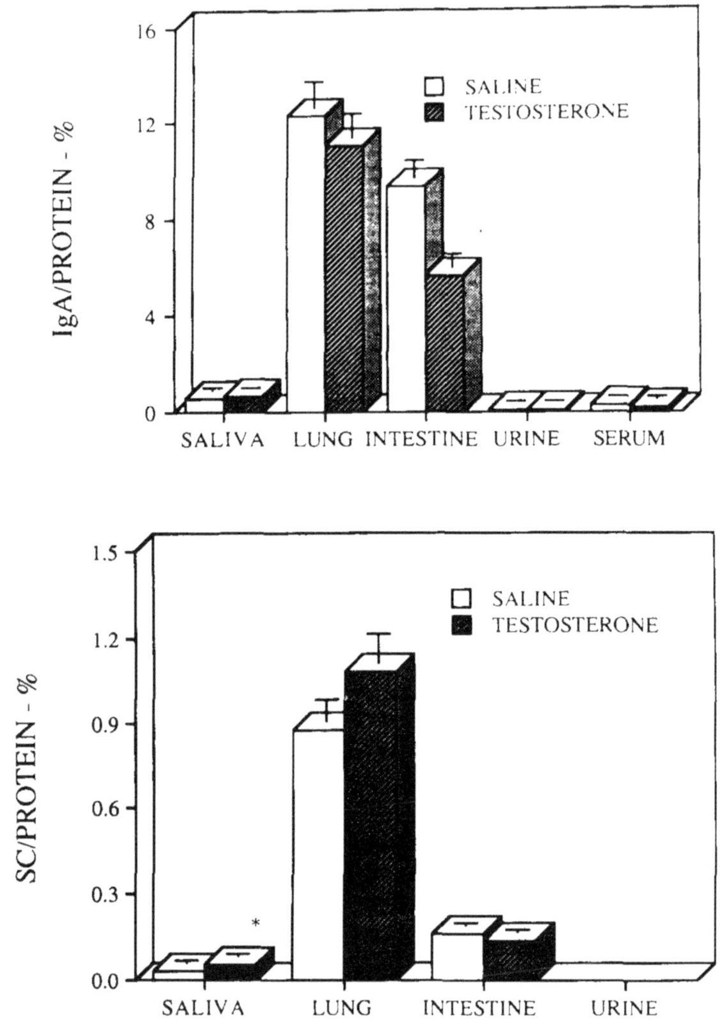


FIGURE 8. Influence of testosterone on IgA and free SC content in various mucosal secretions and serum of orchiectomized rats. Data from References 188.

total protein levels in tears. The first aim was to evaluate whether female sex steroids might antagonize androgen action on tear IgA and SC. Estrogen administration has been shown to inhibit testosterone-induced effects on lacrimal gland morphology.[240] Moreover, estrogen and/or progesterone interference could explain why tear SC levels do not increase during the proestrous phase of the estrous cycle;[167] at this stage, ovarian secretion of androgens is highest, but serum concentrations of estrogen are also maximal and are soon followed by rising levels of progesterone.[328,329] Orchiectomized rats were treated with physiological combinations of saline, testosterone, estradiol, and/or progesterone. As illustrated in Figure 9, testosterone induced a significant increase in the tear IgA and SC concentrations, as compared to those of saline-injected controls. However, this androgen effect was not inhibited by cotreatment with estradiol or progesterone, nor duplicated by the administration of these hormones alone.[169] Consequently, the ability of ovarian steroids to antagonize androgen-associated morphological action in the lacrimal gland does not extend to androgen-immune interactions in this tissue.

The second aim was to assess whether the absence of certain hormones might alter tear SC and IgA levels, or influence the ocular response to androgen exposure. The focus in these experiments were hormones from the

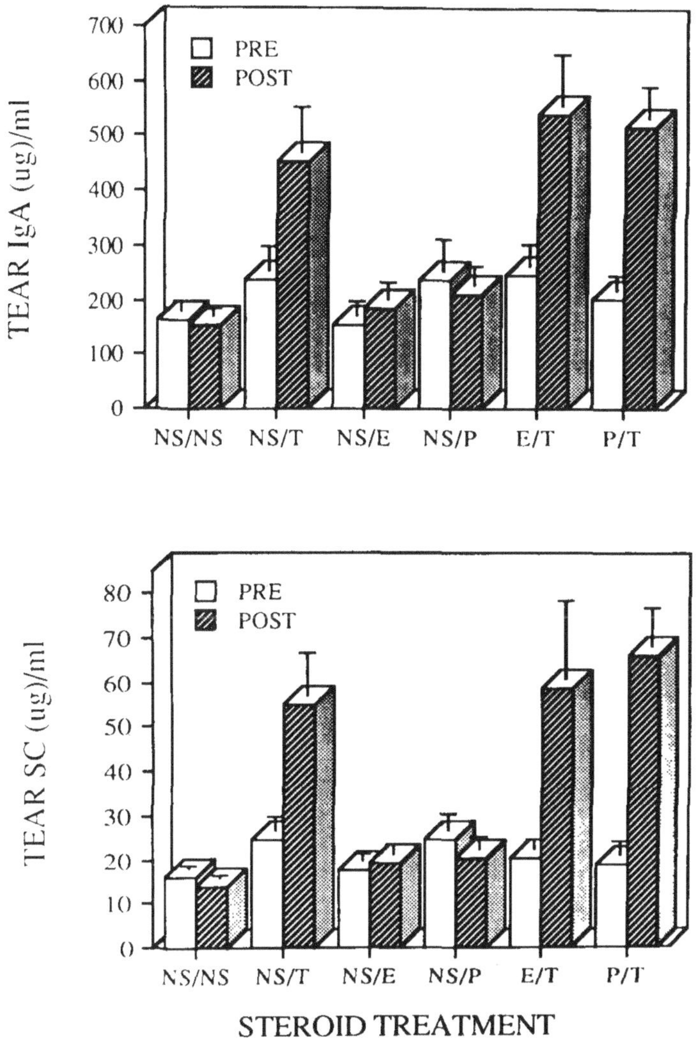


FIGURE 9. Effect of estrogen and progesterone on the testosterone-induced accumulation of IgA and free SC in tears of orchiectomized rats. Data from Reference 169.

1. Adrenal, which support androgen activity in various target tissues[321-323]
2. Thyroid, which increase cellular receptivity to androgens,[320] modulate androgen-induced cellular growth[274,275,319] and protein synthesis,[321,323] may control steroid receptor concentrations,[330] and regulate lacrimal gland growth, development, and function[318,331,332]
3. Pineal, which affect pituitary and gonadal activity and are modulated by androgens[333]
4. Pituitary, an androgen target organ,[334,340] which exert a tremendous impact on the lacrimal gland[282,287,288,335-339] and mucosal immunity.[156,174]

Rats underwent orchiectomies and specific endocrine organ ablations or appropriate sham-surgery.[169] Results demonstrated that absence of the pituitary gland (but not the thyroid, adrenal, or pineal glands) resulted in a significant decrease in tear IgA, SC, and total protein content. In addition, removal of the thyroid and adrenal glands did not prevent the testosterone-associated increase in tear SC and IgA, although thyroidectomy or adrenalectomy did diminish the magnitude of the androgen response. In contrast, hypophysectomy com-

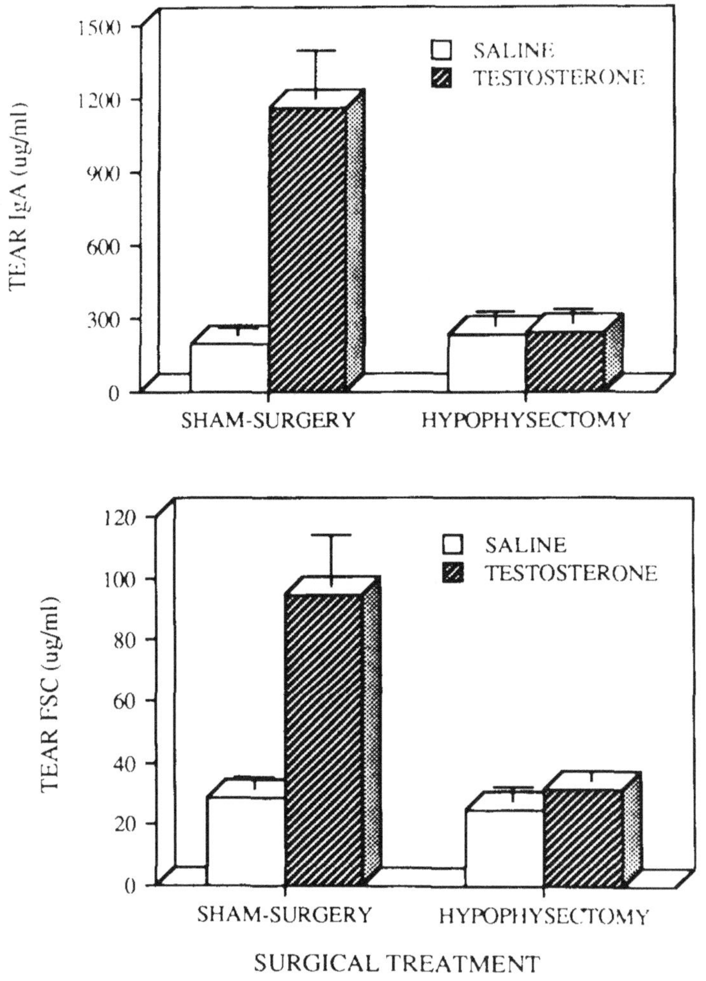


FIGURE 10. Influence of testosterone administration on tear IgA and free SC levels in orchiectomized and hypophysectomized or sham-hypophysectomized rats. Data from Reference 169.

pletely blocked the effect of testosterone on both tear SC and IgA (Figure 10). These results indicate that other hormones, but especially those from the hypothalamic-pituitary axis, may play an integral role in the androgen regulation of the rat ocular secretory immune system.[169]

The purpose of the second study was to further investigate the influence of the hypothalamic-pituitary axis on the androgen control of the secretory immune system in the eye. The experimental approach was threefold. First, studies assessed whether an intact hypothalamic-pituitary axis is required for the androgen regulation of SC and IgA levels in tears.[189,190] Age-matched, orchiectomized rats underwent hypophysectomies, pituitary transplant to under the kidney capsule or sham-surgeries. After a 2-week recovery period, animals were administered saline or testosterone (2 mg/d) for 4 d and tears were collected 24 h after the last injection. Androgen treatment induced a significant increase in the level of free SC and IgA in tears of rats with sham-hypophysectomies and sham-transplants. In contrast, testosterone had no effect on tear free SC or IgA content in hypophysectomized rats with or without a pituitary transplant. These findings were evident irrespective of whether free SC and IgA data were expressed in terms of total amounts of concentration. In addition, results could not be accounted for by variations in tear volume between experimental groups.

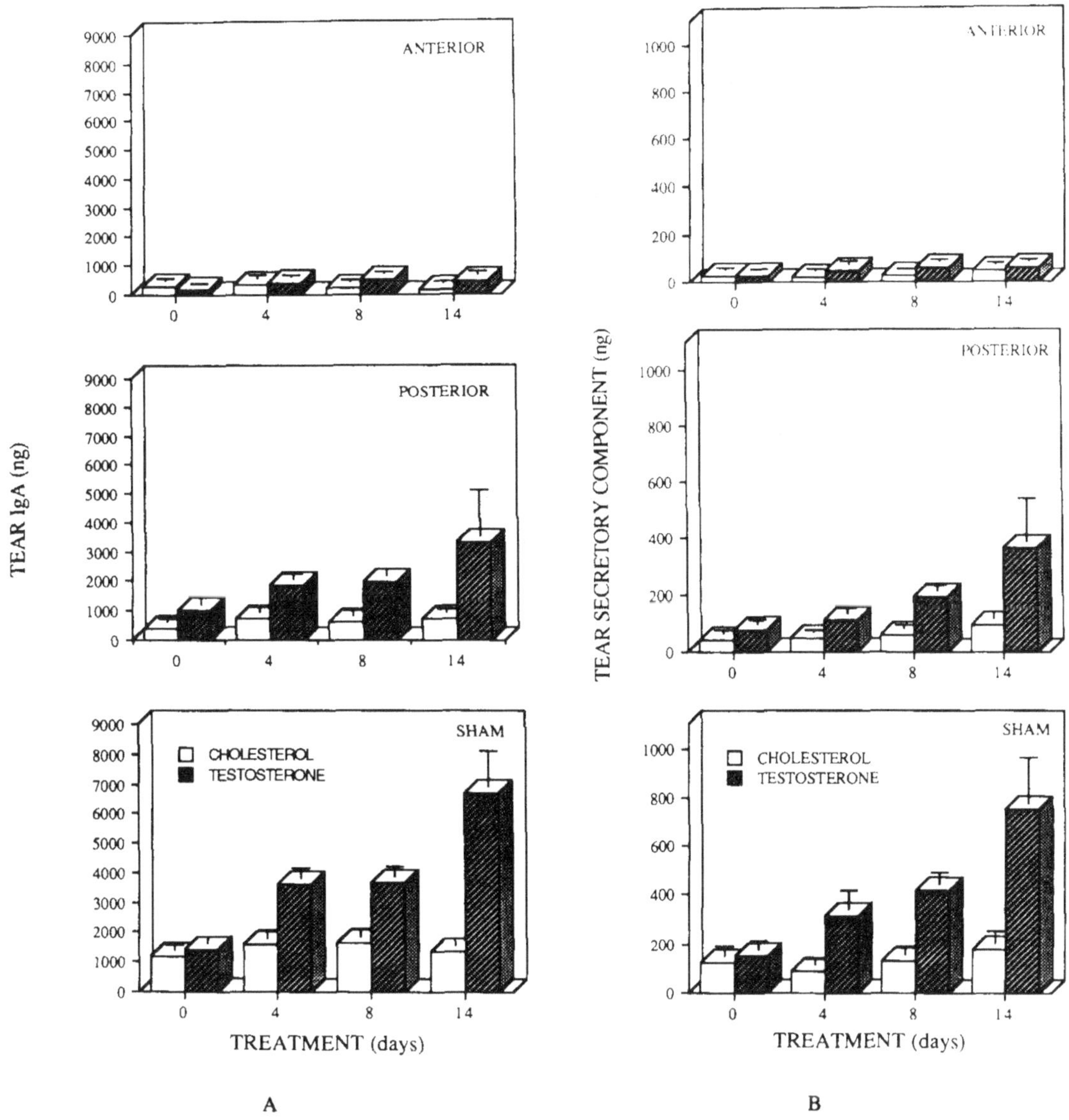


FIGURE 11. Impact of testosterone treatment on IgA (A) and free SC (B) levels in tears of orchiectomized rats with selective anterior, posterior or sham pituitary ablation. Data from Reference 190.

The second set of experiments evaluated whether the anterior or posterior pituitary modulates androgen action on mucosal immunity in the eye.[190] Orchiectomized rats underwent selective ablation of the anterior or posterior pituitary, or were sham-hypophysectomized, and 16 d later, received either placebo (50 mg) or testosterone (50 mg) pellets in the subscapular space. Tears were collected before pellet implantation and following 4, 8, and 14 d of hormone exposure. Testosterone administration to sham-hypophysectomized rats resulted in a significant and progressive rise in the levels of tear free SC and IgA (Figure 11A, B). In contrast, removal of the anterior pituitary significantly abrogated the action of testosterone on free SC and IgA in tears (Figure 11 A, B). Following 4 d of androgen exposure, the tear free SC and IgA response in these animals was completely inhibited. Continued testosterone administration for 8 and 14 d led to a slight rise in free SC and/or IgA content in tears; the magnitude of this increase, though, was markedly reduced relative to the hormone-induced effect in sham animals. Of interest, preliminary studies also demonstrated that the number of IgA-containing cells in lacrimal tissue, as well as tear IgG concentrations were substantially diminished in testosterone-treated rats without an anterior pituitary, compared to sham-operated controls.[367] With regard to the impact of posterior pituitary ablation, prolonged hormone treatment consistently elicited an increase in tear free SC and IgA, but the kinetics and extent of androgen action were highly variable (Figure 11 A, B). This variability may have been due to the regeneration of a functional neurohypophysis.[190] In the third series of experiments, the influence of prolactin, growth hormone, and α-melanocyte stimulating hormone, in the presence and/or absence of testosterone, was examined on tear IgA and free SC.[190] None of these peptide hormones, when given in physiological amounts for 4 d to hypophysectomized rats, modulated the ocular secretory immune system. These results were intriguing, because prolactin is a potent regulator of mucosal immunity in the mammary gland[156,174] and all three peptides regulate lacrimal gland structure and function[287,288,336,338] or systemic immunity.[1,2,5] The inability of prolactin to reconstitute the physiological environment necessary for testosterone action was consistent with the pituitary transplant data; this surgical procedure induces hyperprolactinemia.[341] In other studies (Sullivan, D. A., unpublished data, 1989), we have also found that administration of therapeutic maintenance doses of thyroxine and dexamethasone does not restore the capacity of the lacrimal gland in hypophysectomized rats to immunologically respond to testosterone exposure.

In summary, these results demonstrated that an intact, hypothalamic-pituitary axis either supports, mediates, or prevents inhibition of the androgen control of the secretory immune system of the eye. In this regard, it is of interest that the endocrine control of the hypothalamus and pituitary varies between males and females.[340] Further, gender appears to determine which pituitary hormones affect the lacrimal gland.[338] Thus, gender-related differences in the regulation of and output from the hypothalamic-pituitary axis may be the critical factors underlying the sexual dimorphism in ocular mucosal immunity.

## E. MECHANISMS INVOLVED IN THE HORMONAL CONTROL OF OCULAR MUCOSAL IMMUNITY

The precise mechanisms involved in the androgen regulation of the ocular secretory immune system have yet to be clearly elucidated. According to classical endocrine theories, androgen action on target tissues is initiated by hormone binding to specific, high affinity receptors in the cell cytoplasm, followed by a translocation of the androgen-receptor complex into the nucleus. This complex interacts with the genome in such a way to stimulate new or increased messenger RNA, and later, protein synthesis.[342,343] Androgen receptors, as noted above, have been identified in the lacrimal gland,[303] and testosterone has been shown to increase mRNA levels[246] and protein synthetic activity[290] in lacrimal tissue. However, to better understand androgen modulation of acinar cell SC production, the cellular location (i.e., epithelium, stroma) of receptors must be characterized. Both cell types in the lacrimal gland are influenced by androgens[247,289] and stromal cells in other tissues are known to modulate hormone-induced effects on the epithelium.[345] In the prostate, for example, testosterone is converted by the enzyme 5α-reductase-3α-hydroxysteroid dehydrogenase to dihydrotestosterone in epithelial cells; this hormone then acts through stromal fibroblast receptors to elaborate the secretion of cytokines, which modulate epithelial activity.[346] Furthermore, whether androgen-receptor complexes activate "SC" genes, initiate their transcription, or modify "SC" mRNA processing, translation, or degradation remains to be explored.

With regard to IgA, androgen receptors have not been found in lymphocytes,[311,312] but steroid hormones do influence the migration of IgA-positive cells to,[153,154,156] and IgA antibody synthesis in,[7] mucosal sites. Thus, androgens might possibly regulate the profile and function of lymphocytes in the lacrimal gland: such a hypothesis would be consistent with the known sexual dimorphism in the density of T cells[206] and IgA plasma cells,[132,213,214] as well as IgA production[166] in this tissue. However, androgen action does not appear to explain the gender-related differences in lymphocyte populations of the lacrimal gland in pre- or post-pubertal rats.[132,214] In addition, an androgen role may not account for the significant and increasing disparity between tear IgA levels of male and female rats from adulthood to senescence;[77,132] during this time interval, serum testosterone concentrations[347,348] and the quantity of androgen receptors in peripheral tissues[349] undergo a marked decrease.

One possible reason for these observations is that the hypothalamic-pituitary axis may be primarily responsible for androgen activity in the lacrimal gland; interruption of this axis completely inhibits androgen-induced effects on tear IgA and SC[169,189,190] and reduces the number of tissue IgA-containing cells.[367] It might be that a combination of hypothalamic-hypophyseal hormones and/or their target organ hormones modulates androgen action in the eye, either directly[320] or through steroid receptor regulation,[350] as has been found in other peripheral tissues. The nature of this hypothalamic-pituitary involvement, though, may be complex. Extirpation of the pituitary results in lacrimal gland atrophy[337] and a significant alteration in acinar area and density.[369] Nevertheless, hypophysectomy does not appear to entirely interfere with androgen effects on lacrimal gland histology;[282] this suggests that the hypothalamic-pituitary axis may have different modes of influence on immunity and morphology in the lacrimal gland.

In an attempt to better unravel the specific endocrine circuits and hormonal mechanisms responsible for the hypothalamic-pituitary-gonadal control of the secretory immune system of the eye, we have recently developed an effective procedure for the isolation and maintenance of lacrimal gland acinar cells in long term, serum-free cultures.[351] The rationale for this experimental approach is threefold: (1) acinar cell cultures will permit, under controlled conditions, analysis of the effect of multiple hormone combinations on the expression, secretion, and function of SC; (2) these hormonally-exposed cultures may indicate the precise endocrine pathways regulating the ocular secretory immune system *in vivo,* given that hormone action on SC tends to predict endocrine activity on IgA in the lacrimal gland;[6] and (3) these cultures, under coculture conditions, will allow assessment of the hormonal impact on acinar, stromal, and lymphocytic cell interactions.

At present, culture studies have focused upon the maintenance of homogeneous populations of acinar cells and their response to hormones. The isolation procedure involves serial enzymatic digestion of minced lacrimal tissue with EDTA, followed by treatment with collagenase, hyaluronidase and DNase. Dispersed cells are passed through nylon filters, then centrifuged on Ficoll gradients (2, 3, and 4%) to yield a single acinar cell suspension (>90% viability).[351] Functionally, freshly isolated cells show the presence of SC (by immunofluorescence microscopy) and specific receptors for $\alpha$-melanocyte-stimulating hormones (equilibrium binding). To permit proliferation in, and to allow maintenance of, long term cell cultures (e.g., 3.5 weeks), acinar cells are cultured on reconstituted basement membranes (Matrigel) in modified Eagle's medium containing dexamethasone, carbamyl choline, putrescine, L-ascorbic acid, insulin, transferrin, selenous acid, epidermal growth factor, reduced glutathione, fibroblast growth factor, and high-density lipoprotein.[351] The presence of an extracellular matrix support appears to be essential for cell attachment, metabolism, growth, morphological and functional differentiation, long-term maintenance, as well as hormonal control of epithelial cells.[352-358] In addition, the inclusion of supplements replaces the necessity for serum, which in certain culture systems may depress hormone action, disrupt metabolic activities, limit secretory function, and permit fibroblast overgrowth.[356,359] Under serum-free incubation conditions, acinar cells retain viabilities between 76.6 to 98.1%, express the typical "cobblestone" morphology, and, by transmission electron microscopy, exhibit classical acinar cell characteristics, with numerous secretory granules, Golgi and mitochondria, extensive rough endoplasmic reticulum, and many tonofilaments. Zonula

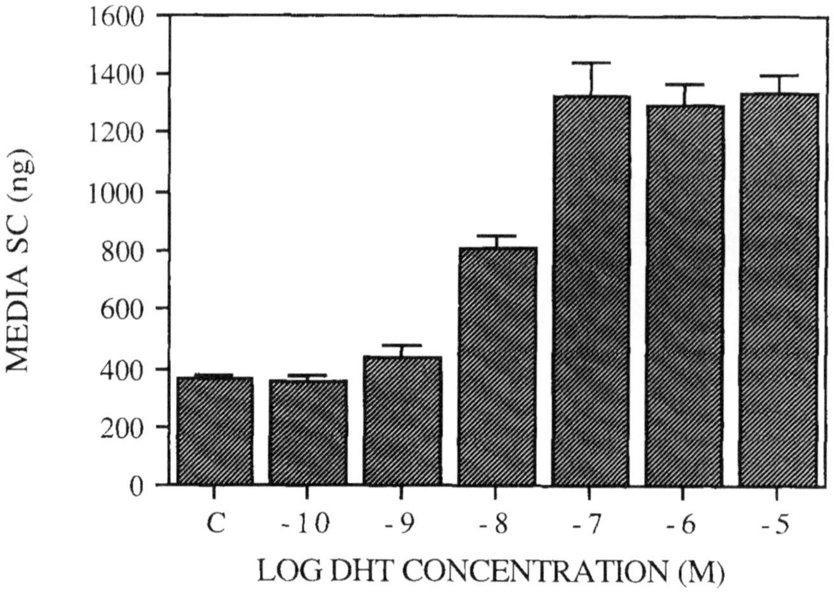


FIGURE 12. Effect of varying amounts of dihydrotestosterone on free SC accumulation in incubation media of cultured acinar cells from the rat lacrimal gland. Free SC levels equal combined amounts released into media during days 5 to 8 of cell culture. Data from Reference 366.

occludens and zonula adherens are also evident at the apical portion of cellular contact, and a high desmosome density is apparent at lateral regions of cell attachment.[351]

As concerns possible hormone-induced stimulation of SC output by cultured cells, experiments have been conducted following androgen exposure *in vivo* or *in vitro*.[366] First, short-term secretion (24 h) of SC was significantly enhanced if acinar cells were obtained from testosterone-treated, as compared to placebo-exposed rats. This SC release was reduced by cycloheximide inclusion in the incubation medium. Second, long-term studies have been performed *in vitro* with 4- to 21-d cell cultures in the presence or absence of testosterone or dihydrotestosterone in the media. Results demonstrated that both steroids significantly increased SC production by a two- to fourfold factor, relative to that of control cultures, without altering cell proliferation.[366] This hormone action was time- and dose-dependent (Figure 12) and specific for androgens; estradiol, progesterone, and cortisol had no effect on acinar cell SC output. Third, elevation of the insulin, transferrin and selenous acid (''ITS'') concentrations in incubation media from 1.25 to 5 μg/ml had a tremendous impact on androgen-induced SC production by acinar cells *in vitro* (Table 3); during a 4-d culture period, SC levels were increased by a 20-fold amount in enhanced ''ITS'' cultures.[370] These findings showed that androgens may act directly on acinar cells to regulate SC synthesis and secretion. Moreover, the results with insulin and transferrin suggest one possible mechanism by which the hypothalamic-pituitary axis modulates androgen control of the ocular secretory immune system *in vivo*. Decreased insulin availability dramatically impairs potential cellular responses to androgens[360] and following hypophysectomy or hypopituitarism, serum insulin[361-363] as well as transferrin[364] concentrations are significantly diminished. Thus, as observed in cultures with ''low'' insulin levels, androgen-induced production of acinar cell SC was decreased; the possible potentiation of hormone effects by insulin has been documented for other acinar cells.[365] Whether additional hormones modulated or secreted by the hypothalamic-pituitary axis also modify androgen action is currently being explored.

**TABLE 3**
**Influence of Insulin (I), Transferrin (T) and Selenous Acid (S) on Androgen-Induced SC Secretion by Lacrimal Gland Acinar Cells**

| | Medium SC (ng) | |
|---|---|---|
| ITS (μg/ml) | Control | Dihydrotestosterone |
| 1.25 | 6 ± 6 | 86 ± 23 |
| 5.00 | 156 ± 23 | 1529 ± 92 |

*Note:* Lacrimal gland acinar cells from male rats were plated at identical densities on Matrigel in 35 mm culture dishes containing 1.5 ml serum-free media. Media contained modified Eagle's medium, dexamethasone, carbamyl choline, putrescine, L-ascorbic acid, epidermal growth factor, reduced glutathione, fibroblast growth factor, high-density lipoprotein, varying concentrations of insulin, transferrin and selenous acid, and either dihydrotestosterone ($10^{-6}$ *M*) or equivalent amounts of vehicle (ethanol). Cultures were exposed to hormone from day 0 and the experimental period for media collection ranged from day 7 to 11. Cell proliferation was approximately fourfold greater in cultures with the higher ITS concentration. Numbers represent the mean ± SE. Data from Reference 370.

# IV. SUMMARY

Androgens control the secretory immune system of the eye.[6] These hormones, but not estrogens, progestins, or glucocorticoids, appear to stimulate the synthesis and/or secretion of SC and IgA by lacrimal tissue[166,168] and dramatically increase the tear levels of SC and IgA.[166-169,188,189] This androgen action, which is unique to the eye,[188,189] may explain the pronounced sexual dimorphism observed in the ocular mucosal immune system.[77,132,166-168,206,213,214,219] In fact, androgen influence is believed to account for the striking, gender-related differences in the morphology, biochemistry, immunology, and genetics of the lacrimal gland in a variety of species, including human.[167-169,240-251,282-290] The processes involved in the androgen regulation of the ocular secretory immune system might include binding to specific receptors in lacrimal tissue,[303] modulation of genomic activity[246] and protein synthesis[290] in this gland, as well as modulation of T[64-67] and IgA-positive[7] lymphocyte function. However, androgen action may also be supported or mediated by other hormones. Recently, we have demonstrated that the hypothalamic-pituitary axis plays an essential role in the androgen regulation of ocular mucosal immunity.[169,189,190] Disruption of this axis by hypophysectomy,[169] pituitary transplantation to the kidney capsule[189,190] or selective anterior pituitary ablation[190] completely abrogates the effect of androgens on the secretory immune system of the eye. Moreover, these endocrine manipulations have tremendous impact on the morphology[337,369] and tear production[291,369] of the lacrimal gland. These findings indicate that hypothalamic-pituitary hormones, which are known to influence lacrimal tissue structure and function,[282,287,288,335-339] may control either directly or indirectly (e.g., insulin levels) both androgen effects on, and the associated sexual dimorphism in, immunity in the eye. Indeed, the hypothalamic-hypophyseal axis may also mediate the known therapeutic action of androgens on the lacrimal gland in patients with

Sjogren's syndrome.[6,300-302] Clearly, further research is required to examine the elaborate endocrine regulation of the ocular secretory immune system. Such studies should significantly advance our basic understanding of immune defense in the eye and have potentially major clinical importance.

## REFERENCES


1. **Berczi, I. and Kovacs, K., Eds.**, *Hormones and Immunity*, MTP Press, Lancaster, England, 1987.
2. **Berczi, I., Ed.**, *Pituitary Function and Immunity*, CRC Press, Boca Raton, FL, 1986.
3. **Cotman, C. W., Brinton, R. E., Galaburda, A., McEwen, B., and Schneider, D. M.**, *The Neuro-Immune-Endocrine Connection*, Raven Press, New York, 1987.
4. **Comsa, J., Leonhardt, H., and Wekerle, H.**, Hormonal coordination of the immune response, *Rev. Physiol. Biochem. Pharmacol.*, 92, 115, 1982.
5. **Jancovik, B. D., Markovic, B. M., and Spector, N. H.**, *Neuroimmune Interactions*, The New York Academy of Sciences, New York, 1987.
6. **Sullivan, D. A.**, Hormonal control of the ocular secretory immune system, in *Hormones and Immunity*, Berczi, I. and Kovacs, K., Eds., MTP Press, Lancaster, England, 1987, 54.
7. **Cohn, D. A.**, High sensitivity to androgen as a contributing factor in sex differences in the immune response, *Arthritis Rheum.*, 22, 1218, 1979.
8. **Grossman, C. J.**, Regulation of the immune system by sex steroids, *Endocr. Rev.*, 5, 435, 1984.
9. **Munck, A., Guyre, P. M., and Holbrook, N. J.**, Physiological functions of glucocorticoids in stress and their relation to pharmacological actions, *Endocr. Rev.*, 5, 25, 1984.
10. **Johnson, H. M., Smith, E. M., Torres, B. A., and Blalock, J. E.**, Regulation of the in vitro antibody response by neuroendocrine hormones, *Proc. Natl. Acad. Sci. U.S.A.*, 79, 4171, 1982.
11. **Berczi, I.**, The effects of growth hormone and related hormones on the immune system, in *Pituitary Function and Immunity*, Berczi, I., Ed., CRC Press, Boca Raton, FL, 1986, 133.
12. **Berczi, I. and Nagy, E.**, Prolactin and other lactogenic hormones, in *Pituitary Function and Immunity*, Berczi, I., Ed., CRC Press, Boca Raton, FL, 1986, 161.
13. **Cannon, J. G., Tatro, J. B., Reichlin, S., and Dinarello, C. A.**, alpha-Melanocyte stimulating hormone inhibits immunostimulatory and inflammatory actions of interleukin 1, *J. Immunol.*, 137, 2232, 1986.
14. **Snow, E. C.**, Insulin and growth hormone function as minor growth factors that potentiate lymphocyte activation, *J. Immunol.*, 135, 776, 1985.
15. **Johnson, H. M. and Torres, B. A.**, Regulation of lymphokine production by arginine vasopressin and oxytocin: modulation of lymphocyte function by neurohypophyseal hormones, *J. Immunol.*, 135, 773, 1985.
16. **Payan, D. G. and Goetzl, E. J.**, Modulation of lymphocyte function by sensory neuropeptides, *J. Immunol.*, 135, 783, 1985.
17. **Stanisz, A. M., Befus, D., and Bienenstock, J.**, Differential effects of vasoactive intestinal peptide, substance P, and somatostatin on immunoglobulin synthesis and proliferation by lymphocytes from Peyer's patches, mesenteric lymph nodes, and spleen, *J. Immunol.*, 136, 152, 1986.
18. **Scicchitano, R., Stanisz, A. M., Payan, D. G., Kiyono, H., McGhee, J. R., and Bienenstock, J.**, Expression of substance P and somatostatin receptors on a T helper cell line, *Adv. Exp. Med. Biol.*, 216, 185, 1987.
19. **Ernst, P. B., Stanisz, A. M., Scicchitano, R., Paraskevas, F., Payan, D., and Bienenstock, J.**, Novel regulatory mechanisms of IgA synthesis: respective roles of neuropeptides and cells of the anti-suppressor circuit, *Adv. Exp. Med. Biol.*, 216, 57, 1987.
20. **O'Dorisio, M. S., Wood, C. L., and D'Orisio, T. M.**, Vasoactive intestinal peptide and neuropeptide modulation of the immune response, *J. Immunol.*, 135, 792, 1985.
21. **Berczi, I.**, The pituitary thyroid axis, in *Pituitary Function and Immunity*, Berczi, I., Ed., CRC Press, Boca Raton, FL, 1986, 213.
22. **Felten, D. L., Felten, S. Y., Carlson, S. L., Olschowka, J. A., and Livnat, S.**, Noradrenergic and peptidergic innervation of lymphoid tissue, *J. Immunol.*, 135, 755, 1985.
23. **Payan, D. G., McGillis, J. P., Renold, F. K., Mitsuhashi, M., and Goetzl, E. J.**, Neuropeptide modulation of leukocyte function, *Ann. N.Y. Acad. Sci.*, 496, 182, 1987.
24. **Jankovic, B. D. and Maric, D.**, Enkephalins and immunity. I. In vivo suppression and potentiation of humoral immune response, *Ann. N.Y. Acad. Sci.*, 496, 115, 1987.
25. **Maric, D. and Jankovic, B. D.**, Enkephalins and immunity. II. In vivo modulation of cell-mediated immunity, *Ann. N.Y. Acad. Sci.*, 496, 126, 1987.

26. **Shah, P. C., Freier, S., Park, B. H., Lee, P. C., and Lebenthal, E.,** Pancreozymin and secretin enhance duodenal fluid antibody levels to Cow's milk proteins, *Gastroenterology,* 83, 916, 1982.
27. **Freier, S., Eran, M., and Abrahamov, A.,** Cholecystokinin-induced release of IgA antibodies in rat intestine, *Adv. Exp. Med. Biol.,* 216, 413, 1987.
28. **Ahmed, S. A., Penhale, W. J., and Talal, N.,** Sex hormones, immune responses, and autoimmune diseases, *Am. J. Pathol.,* 121, 531, 1985.
29. **Blalock, J. E., Harbour-McMenamin, D., and Smith, E. M.,** Peptide hormones shared by the neuroendocrine and immunologic systems, *J. Immunol.,* 135, 858, 1985.
30. **Besedovsky, H. O., del Rey, A. E., and Sorkin, E.,** Immune-neuroendocrine interactions, *J. Immunol.,* 135, 750, 1985.
31. **Wira, C. R., Sullivan, D. A., and Sandoe, C. P.,** Estrogen-mediated control of the secretory immune system in the uterus of the rat, *Ann. N. Y. Acad. Sci.,* 409, 534, 1983.
32. **Weisz-Carrington, P.,** Secretory immunology in the mammary gland, in *Hormones and Immunity,* Berczi, I. and Kovacs, K., Eds., MTP Press, Lancaster, England, 1987, 172.
33. **Underdown, B. J. and Schiff, J. M.,** Immunoglobulin A: strategic defense initiative at the mucosal surface, *Annu. Rev. Immunol.,* 4, 389, 1986.
34. **Brandtzaeg, P.,** Role of J chain and secretory component in receptor-mediated glandular and hepatic transport of immunoglobulins in man, *Scand. J. Immunol.,* 22, 111, 1985.
35. **Mestecky, J. and McGhee, J. R.,** Immunoglobulin A (IgA): molecular and cellular interactions involved in IgA biosynthesis and immune response, *Adv. Immunol.,* 40, 153, 1987.
36. **Mestecky, J., McGhee, J. R., Bienenstock, J., and Ogra, P. L.,** *Recent Advances in Mucosal Immunology,* Plenum Press, New York, 1987.
37. **Svandborg-Eden, C. and Svennerholm, A. M.,** Secretory immunoglobulin A and G antibodies prevent adhesion of Escherichia coli to human urinary tract epithelial cells, *Infect. Immun.,* 22, 790, 1978.
38. **Williams, R. C. and Gibbons, R. J.,** Inhibition of bacterial adherence by secretory immunoglobulin A: a mechanism of antigen disposal, *Science,* 177, 697, 1972.
39. **Porter, P. and Linggood, M. A.,** Novel mucosal anti-microbial functions interfering with the plasmid-mediated virulence determinants of adherence and drug resistance, *Ann. N.Y. Acad. Sci.,* 409, 564, 1983.
40. **Abraham, S. N. and Beachey, E. H.,** Host defenses against adhesion of bacteria to mucosal surface, in *Advances in Host Defense Mechanisms: Mucosal Immunology,* Vol. 4, Gallin J. I. and Fauci, A. S., Eds., Raven Press, New York, 1985, 63.
41. **Boedeker, E. C., Cheney, C. P., and Cantey, J. R.,** Inhibition of enteropathogenic *Escherichia coli* (strain RDEC-1) adherence to rabbit intestinal brush borders by milk immune secretory immunoglobulin A, *Adv. Exp. Med. Biol.,* 216, 919, 1987.
42. **Wright, P. F., Murphy, B. R., Kervina, M., Lawrence, E. M., Phelan, M. A., and Karzon, D. T.,** Secretory immunologic response after intranasal inactivated influenza A virus vaccinations: evidence for immunoglobulin A memory, *Infect. Immun.,* 40, 1092, 1983.
43. **Taylor, H. P. and Dimmock, N. J.,** Mechanism of neutralization of influenza virus by secretory IgA is different from that of monomeric IgA or IgG, *J. Exp. Med.,* 161, 198, 1985.
44. **Loftness, T. J., Erlandsen, S. J., Wilson, I. D., and Meyer, E. A.,** Occurrence of specific secretory immunoglobulin A in bile after inoculation of *Giardia lamblia* trophozoites into rat duodenum, *Gastroenterology,* 87, 1022, 1984.
45. **Lloyd, S. and Soulsby, E. J. L.,** The role of IgA immunoglobulins in the passive transfer of protection to Taenia taeniaeformis in the mouse, *Immunology,* 34, 939, 1978.
46. **Kaplan, B. S., Uni, S., Aikawa, M., and Mahoud, A. A. F.,** Effector mechanism of host resistance in murine Giardiasis: specific IgG and IgA cell-mediated toxicity, *J. Immunol.,* 134, 1975, 1985.
47. **Svennerholm, A. M., Lange, S., and Holmgren, J.,** Correlation between intestinal synthesis of specific immunoglobulin A and protection against experimental cholera in mice, *Infect. Immun.,* 21, 1, 1978.
48. **Tamaru, T. and Brown, W. R.,** IgA antibodies in rat bile inhibit cholera toxin-induced secretion in ileal loops in situ, *Immunology,* 55, 579, 1985.
49. **Fubara, E. S. and Freter, R.,** Protection against enteric bacterial infection by secretory IgA antibodies, *J. Immunol.,* 111, 395, 1973.
50. **Newby, T. J. and Stokes, C. R.,** *Local Immune Response of the Gut,* CRC Press, Boca Raton, FL, 1984.
51. **Dhar, R. and Ogra, P. L.,** Local immune response, *Br. Med. Bull.,* 41, 28, 1985.
52. **Ganguly, R. and Waldman, R. H.,** Local immunity and local immune response, *Progr. Allergy,* 27, 1, 1980.
53. **Hanson, L. A., Brandtzaeg, P., Bjorkander, J., Carlsson, B., Griffiss, J. M., Mellander, L., and Karlsson, G.,** IgA deficiency with and without mucosal infections, *Adv. Exp. Med. Biol.,* 216, 1449, 1987.
54. **Kaiserlian, D., Delacroix, D., and Bach, J. F.,** The wasted mutant mouse. I. An animal model of secretory IgA deficiency with normal serum IgA, *J. Immunol.,* 135, 1126, 1985.

55. **McGhee, J. R. and Mestecky, J.,** *The Secretory Immune System,* The New York Academy of Sciences, New York, 1983.
56. **Tomasi, T. B. and Plaut, A. G.,** Humoral aspects of mucosal immunity, in *Advances in Host Defense Mechanisms: Mucosal Immunology,* Vol. 4, Raven Press, New York, 1985, 31.
57. **Drew, P. A., La Brooy, J. T., and Shearman, D. J. C.,** In vitro immunoglobulin synthesis by human intestinal lamina propria lymphocytes, *Gut,* 25, 649, 1984.
58. **Denham, S., Spencer, J., Barfoot, R., and Hall, J. G.,** IgA antibodies in the bile of rats. IV. Synthesis of secretory antibodies in irradiated rats reconstituted with spleen cells: influence of microenvironment on the regulation of IgA production, *Immunology,* 51, 45, 1984.
59. **Campbell, D. and Vose, B. M.,** T-cell control of IgA production. I. Distribution, activation conditions and culture of isotype-specific regulatory helper cells, *Immunology,* 56, 81, 1985.
60. **Elson, C. O., Kagnoff, M. A., Fiocchi, C., Befus, A, D., and Targan, S.,** Intestinal immunity and inflammation: recent progress, *Gastroenterology,* 91, 746, 1986.
61. **Strober, W. and Jacobs, D.,** Cellular differentiation, migration, and function in the mucosal immune system, in *Advances in Host Defense Mechanisms: Mucosal Immunology,* Vol. 4, Raven Press, New York, 1985, 1.
62. **Elson, C. O., Heck, J. A., and Strober, W.,** T cell regulation of murine IgA synthesis, *J. Exp. Med.,* 149, 632, 1979.
63. **Bienenstock, J., Befus, D., McDermott, M., Mirski, S., and Rosenthal, K.,** Regulation of lymphoblast traffic and localization in mucosal tissues, with emphasis on IgA, *Fed. Proc.,* 42, 3213, 1983.
64. **Manning, L. S. and Parmely, M. J.,** Cellular determinants of mammary cell-mediated immunity in the rat. I. The migration of radioisotopically labeled T lymphocytes, *J. Immunol.,* 125, 2508, 1980.
65. **Talal, N., Dauphinee, M. J., and Wofsy, D.,** Interleukin-2 deficiency, genes, and systemic lupus erythematosus, *Arthritis Rheum.,* 25, 838, 1982.
66. **Michalski, J. P., McCombs, C. C., Roubinian, J. R., and Talal, N.,** Effect of androgen therapy on survival and suppressor cell activity in aged NZB/NZW F1 hybrid mice, *Clin. Exp. Immunol.,* 52, 229, 1983.
67. **Ahmed, S. A., Dauphinee, M. J., and Talal, N.,** The effects of short term administration of sex hormones in normal and autoimmune mice, *Fed. Proc.,* 43, 1612, 1984.
68. **Stimson, W. H. and Crilly, P. J.,** Effects of steroids on the secretion of immunoregulatory factors by thymic epithelial cell cultures, *Immunology,* 44, 401, 1981.
69. **Franklin, R. M., McGee, D. W., and Shepard, K. F.,** Lacrimal gland directed B cell responses, *J. Immunol.,* 135, 95, 1985.
70. **Mayer, L.,** The role of epithelial cells as accessory cells, *Adv. Exp. Med. Biol.,* 216, 209, 1987.
71. **Bland, P. W.,** Antigen presentation by gut epithelial cells: secretion by rat enterocytes of a factor with IL-1-like activity, *Adv. Exp. Med. Biol.,* 216, 219, 1987.
72. **Ahnen, D. J., Brown, W. R., and Kloppel, T. M.,** Secretory component: the polymeric immunoglobulin receptor, *Gastroenterology,* 89, 667, 1985.
73. **Solari, R. and Kraehenbuhl, J. P.,** The biosynthesis of secretory component and its role in the trans-epithelial transport of IgA dimer, *Immunol. Today,* 6, 17, 1985.
74. **Fisher, M. M., Nagy, B., Bazin, H., and Underdown, B. J.,** Biliary transport of IgA: role of secretory component, *Proc. Natl. Acad. Sci. U.S.A.,* 76, 2008, 1979.
75. **Sullivan, D. A. and Wira, C. R.,** Hormonal regulation of immunoglobulins in the rat uterus: uterine response to multiple estradiol treatments, *Endocrinology,* 144, 650, 1984.
76. **Jonard, P. P., Rambaud, J. C., Dive, C., Vaerman, J. P., Galian, A., and Delacroix, D. L.,** Secretion of immunoglobulins and plasma proteins from the jejunal mucosa, *J. Clin. Invest.,* 74, 525, 1984.
77. **Sullivan, D. A. and Allansmith, M. R.,** The effect of aging on the secretory immune system of the eye, *Immunology,* 63, 403, 1988.
78. **Allansmith, M. R. and Gillette, T. E.,** Secretory component in human ocular tissues, *Am. J. Ophthalmol.,* 89, 353, 1980.
79. **Sullivan, D. A. and Allansmith, M. R.,** Source of IgA in tears of rats, *Immunology,* 53, 791, 1984.
80. **Brandtzaeg, P.,** Immunohistochemical characterization of intracellular J-chain and binding sites for secretory component (SC) in human immunoglobulin (Ig)-producing cells, *Mol. Immunol.,* 20, 941, 1983.
81. **Delacroix, D. L., Hodgson, H. J. F., McPherson, A., Dive, C., and Vaerman, J. P.,** Selective transport of polymeric immunoglobulin A in bile. Quantitative relationships of monomeric and polymeric immunoglobulin A, immunoglobulin M, and other proteins in serum, bile, and saliva, *J. Clin. Invest.,* 70, 230, 1982.
82. **Delacroix, D. L. and Vaerman, J. P.,** Secretory component (SC): preferential binding to heavy (>11S) IgA polymers and IgM in serum, in contrast to predominance of its 11S and free SC forms in secretions, *Clin. Exp. Immunol.,* 49, 717, 1982.
83. **Goodman, M. R., Link, D. W., Brown, W. R., and Nakane, P. K.,** Ultrastructural evidence of transport of secretory IgA across bronchial epithelium, *Am. Rev. Respir. Dis.,* 123, 115, 1981.

84. **Delacroix, D. L. and Vaerman, J. P.,** Function of the human liver in IgA homeostasis in plasma, *Ann. N. Y. Acad. Sci.*, 409, 383, 1983.
85. **Bienestock, J.,** Bronchus-associated lymphoid tissue, *Int. Arch. Allergy Appl. Immunol.*, 76 (Suppl. 1), 62, 1985.
86. **Sheldrake, R. F., Husband, A. J., Watson, D. L., and Cripps, A. W.,** Selective transport of serum-derived IgA into mucosal secretions, *Immunology*, 132, 363, 1984.
87. **Brandtzaeg, P.,** The secretory immune system of lactating human mammary glands compared with other exocrine organs, *Ann. N.Y. Acad. Sci.*, 409, 353, 1983.
88. **Solari, R. and Kraehenbuhl, J. P.,** Biosynthesis of the IgA antibody receptor: a model for the transepithelial sorting of a membrane glycoprotein, *Cell*, 36, 61, 1984.
89. **Vuitton, D. A., Seilles, E., Claude, P., Sava, P., and Delacroix, D. L.,** Gall bladder: the predominant source of bile IgA in man?, *Clin. Exp. Immunol.*, 62, 185, 1985.
90. **Valnes, K., Brandtzaeg, P., Elgjo, K., and Stave, R.,** Specific and nonspecific humoral defense factors in the epithelium of normal and inflammed gastric mucosa, *Gastroenterology*, 86, 402, 1984.
91. **Crago, S. S., Kulhavy, R., Prince, S. J., and Mestecky, J.,** Secretory component in epithelial cells is a surface receptor for polymeric immunoglobulins, *J. Exp. Med.*, 147, 1832, 1978.
92. **Brown, W., Isobe, K., Nakane, P. K., and Pacini, B.,** Studies on translocation of immunoglobulins across intestinal epithelium. IV. Evidence for binding of IgA and IgM to secretory component in intestinal epithelium, *Gastroenterology*, 73, 1333, 1977.
93. **Vaerman, J. P., Heremans, J. F., Bazin, H., and Beckers, A.,** Identification and some properties of rat secretory component, *J. Immunol.*, 114, 265, 1975.
94. **Sullivan, D. A., Underdown, B. J., and Wira, C. R.,** Steroid hormone regulation of free secretory component in the rat uterus, *Immunology*, 49, 379, 1983.
95. **Sullivan, D. A. and Wira, C. R.,** Variations in free secretory component levels in mucosal secretions of the rat, *J. Immunol.*, 130, 1330, 1983.
96. **Murdock, A. J. M., Buckley, C. H., and Fox, H.,** Hormonal control of the secretory immune system of the human cervix, *Br. J. Obstet. Gynaecol.*, 82, 812, 1982.
97. **Mostov, K. E., Friedlander, M., and Blobel, G.,** Structure and function of the receptor for polymeric immunoglobulins, *Biochem. Soc. Symp.*, 51, 113, 1985.
98. **Mostov, K. E., Friedlander, M., and Blogel, G.,** The receptor for transepithelial transport of IgA and IgM contains multiple immunoglobulin-like domains, *Nature (London)*, 308, 37, 1984.
99. **Sztul, E. S., Howell, K. E., and Palade, G. E.,** Biogenesis of the polymeric IgA receptor in rat hepatocytes. II. Localization of its intracellular forms by cell fractionation studies, *J. Cell Biol.*, 100, 1255, 1985.
100. **Mostov, K. E., Kraehenbuhl, J. P., and Blobel, G.,** Receptor mediated transcellular transport of immunoglobulin: synthesis of secretory component as multiple and larger transmembrane forms, *Proc. Natl. Acad. Sci. U.S.A.*, 77, 7257, 1980.
101. **Schiff, J. M., Fisher, M. M., and Underdown, B. J.,** Secretory component as the mucosal transport receptor: separation of physiochemically analagous human IgA fractions with different receptor-binding capacities, *Mol. Immunol.*, 23, 45, 1986.
102. **Brandtzaeg, P. and Prydz, H.,** Direct evidence for an integrated function of J chain and secretory component in epithelial transport of immunoglobulins, *Nature (London)*, 311, 71, 1984.
103. **Mostov, K. E., de Bruyn Kops, A., and Deitcher, D. L.,** Deletion of the cytoplasmic domain of the polymeric immunoglobulin receptor prevents basolateral localization and endocytosis, *Cell*, 47, 359, 1986.
104. **Schiff, J. M., Fisher, M. M., and Underdown, B. J.,** Receptor-mediated biliary transport of immunoglobulin A and asialoglycoprotein: sorting and missorting of ligands revealed by two radiolabeling methods, *J. Cell Biol.*, 98, 79, 1984.
105. **Geuze, H. J., Slot, J. W., Stous, G. J. A. M., Peppard, J., von Figura, K., Hasilik, A., and Schwartz, A. L.,** Intracellular receptor sorting during endocytosis: comparative immunoelectron microscopy of multiple receptors in rat liver, *Cell*, 37, 195, 1984.
106. **Nakamura, T., Nagura, H., Komatsu, N., and Watanabe, K.,** Immunocytochemical and enzymecytochemical studies on the intracellular transport mechanism of secretory immunoglobulins A in human salivary glands, *Virchows Arch. [A]*, 406, 367, 1985.
107. **Mullock, B. M., Luzio, J. P., and Hinton, R. H.,** Preparation of a low-density species of endocytic vesicle containing immunoglobulin A, *Biochem. J.*, 214, 823, 1983.
108. **Mostov, K. E. and Deitcher, D. L.,** Polymeric immunoglobulin receptor expressed in MDCK cells transcytoses IgA, *Cell*, 46, 613, 1986.
109. **Limet, J. N., Schneider, Y. J., Vaerman, J. P., and Trouet, A.,** Binding, uptake and intracellular processing of polymeric rat immunoglobulin A by cultured rat hepatocytes, *Eur. J. Biochem.*, 125, 437, 1982.
110. **Hoppe, C. A., Connolly, T. P., and Hubbard, A. L.,** Transcellular transport of polymeric IgA in the rat hepatocyte: biochemical and morphological characterization of the transport pathway, *J. Cell Biol.*, 101, 2113, 1985.

111. **Limet, J. N., Quintart, J., Schneider, Y. J., and Courtoy, P. J.,** Receptor-mediated endocytosis of polymeric IgA and galactosylated serum albumin in rat liver, *Eur. J. Biochem.*, 146, 539, 1985.
112. **Gebhardt, R.,** Participation of microtubules and microfilaments in the transcellular biliary secretion of immunoglobulin A in primary cultures of rat hepatocytes, *Experientia*, 40, 269, 1984.
113. **Goldman, I. S., Jones, A. L., Hradek, G. T., and Huling, S.,** Hepatocyte handling of immunoglobulin A in the rat: the role of the microtubules, *Gastroenterology*, 85, 130, 1983.
114. **Deitcher, D. L., Neutra, M. R., and Mostov, K. E.,** Functional expression of the polymeric immunoglobulin receptor from cloned cDNA in fibroblasts, *J. Cell Biol.*, 102, 911, 1986.
115. **Kloppel, T. M. and Brown, W. R.,** Rat liver membrane secretory component is larger than free secretory component in bile: evidence for proteolytic conversion of membrane form to free form, *J. Cell. Biochem.*, 24, 307, 1984.
116. **Solari, R., Racine, L., Tallichet, C., and Kraehenbuhl, J. P.,** Distribution and processing of the polymeric immunoglobulin receptor in the rat hepatocyte: morphological and biochemical characterization of subcellular fractions, *J. Histochem. Cytochem.*, 34, 17, 1986.
117. **Mullock, B. M., Jones, R. S., and Hinton, R. H.,** Movement of endocytic shuttle vesicles from the sinusoidal to the bile canalicular face of hepatocytes does not depend on occupation of receptor site, *FEBS Lett.*, 113, 201, 1980.
118. **Nagura, H., Nakane, P. K., and Brown, W. R.,** Secretory component in immunoglobulin deficiency: an immunoelectron microscopic study of intestinal epithelium, *Scand. J. Immunol.*, 12, 359, 1980.
119. **Altamarino, G. A., Barranco-Acosta, C., van Roost, E., and Vaerman, J. P.,** Isolation and characterization of secretory IgA (sIgA) and free secretory component (FSC) from rat bile, *Mol. Immunol.*, 17, 1525, 1980.
120. **Dietcher, D. L. and Mostov, K. E.,** Alternate splicing of rabbit polymeric immunoglobulin receptor, *Mol. Cell. Biol.*, 6, 2712, 1986.
121. **Eiffeut, H., Quentin, E., Decker, J., Hillemeir, S., Hufschmidt, M., Klingmuller, D., Weber, M., and Hilschmann, N.,** Die Primarstruktur der menschlichen freien Sekretkompnente und die Anordnung der Disulfidbrucken, *Hoppe-Seylers Z. Physiol. Chem.*, 365, 1489, 1984.
122. **Delacroix, D. L., Malburny, G. N., and Vaerman, J. P.,** Hepatobiliary transport of plasma IgA in the mouse: contribution to clearance of intravascular IgA, *Eur. J. Immunol.*, 15, 893, 1985.
123. **Brown, T. A., Russell, M. W., and Mestecky, J.,** Elimination of intestinally absorbed antigen into the bile by IgA, *J. Immunol.*, 132, 780, 1984.
124. **Russell, M. W., Brown, T. A., Claflin, J. L., Schroer, K., and Mestecky, J.,** Immunoglobulin A-mediated hepatobiliary transport constitutes a natural pathway for disposing of bacterial antigens, *Infect. Immun.*, 42, 1041, 1983.
125. **Stokes, C. R., Swarbrick, E. T., and Soothill, J. F.,** Immune elimination and enhanced antibody responses: functions of circulating IgA, *Immunology*, 40, 455, 1980.
126. **Stafford, H. A., Knight, K. L., and Fanger, M. W.,** Receptors for IgA on rabbit lymphocytes. II. Characterization of their binding parameters for IgA, *J. Immunol.*, 128, 2201, 1982.
127. **South, M. A.,** Discussion section, in *Immunology of Breast Milk*, Ogra, P. L. and Dayton, D. H., Eds., Raven Press, New York, 1979, 193.
128. **Crago, S. S., Kulhavy, R., Prince, S. J., and Mestecky, J.,** Inhibition of the pokeweed mitogen-induced response of normal peripheral blood lymphocytes by humoral components of colostrum, *Clin. Exp. Immunol.*, 45, 386, 1981.
129. **Sollid, L. M., Kvale, D., Brandtzaeg, P., Markussen, G., and Thorsby, E.,** Interferon-$\gamma$ enhances expression of secretory component, the epithelial receptor for polymeric immunoglobulins, *J. Immunol.*, 138, 4303, 1987.
130. **Kosaka, T., Asahina, T., and Kobayashi, N.,** Differential quantification of sIgA and SC by two-dimensional rocket method, *Immunology*, 40, 597, 1980.
131. **Takemura, T. and Eishi, Y.,** Distribution of secretory component and immunoglobulins in the developing lung, *Am. Rev. Respir. Dis.*, 131, 125, 1985.
132. **Sullivan, D. A., Hann, L. E., and Allansmith, M. R.,** The influence of age on the ocular secretory immune system of the rat, *Adv. Exp. Med. Biol.*, 216, 1395, 1987.
133. **Iwase, T., Moro, I., and Mestecky, J.,** Immunohistochemical study of the ontogeny of the secretory immune system, *Adv. Exp. Med. Biol.*, 216, 1359, 1987.
134. **Stanley, P. J. and Cole, P. J.,** The concentrations of IgA and free secretory piece in the nasal secretions of patients with recurrent respiratory infections, *Clin. Exp. Immunol.*, 59, 197, 1985.
135. **Dura, W. T. and Bernatowska, E.,** Secretory component, $\alpha_1$-antitrypsin and lysozyme in IgA deficient children, *Histopathology*, 8, 747, 1984.
136. **Scott, H., Brandtzaeg, P., Solheim, B. G., and Thorsby, E.,** Relation between HLA-DR-like antigens and secretory component (SC) in jejunal epithelium of patients with coeliac disease or dermatitis herpetaformis, *Clin. Exp. Immunol.*, 44, 223, 1981.

137. **Cooper, M. D., Dever, C., Tempel, K., Moticka, E. J., Hindman, T., and Stephens, D. S.,** Characterization of lymphoid cells from the human fallopian tube mucosa, *Adv. Exp. Med. Biol.*, 216, 387, 1987.
138. **Strober, W., Krakaver, R., Klaeveman, H. L., Reynolds, H. Y., and Nelson, D. L.,** Secretory component deficiency: a disorder of the IgA immune system, *N. Engl. J. Med.*, 294, 351, 1976.
139. **Ogra, P. L., Ogra, S. S., and Coppola, P. R.,** Secretory component and sudden-infant-death syndrome, *Lancet*, 2, 387, 1975.
140. **Dura, W. T., Gladkowska-Dura, M. J., and Johnson, W. W.,** Non-Hodgkin's lymphoma in the first two decades. Morphologic and immunocytochemical study, *Virchows Arch. [Pathol. Anat.]*, 390, 23, 1981.
141. **Elson, C. O. and Ealding, W.,** Generalized systemic and mucosal immunity in mice after mucosal stimulation with cholera toxin, *J. Immunol.*, 132, 2736, 1984.
142. **Dahlgren, U. I. H., Ahlstedt, S., and Hanson, L. A.,** Origin and kinetics of IgA, IgG and IgM milk antibodies in primary and secondary responses of rats, *Scand. J. Immunol.*, 23, 273, 1986.
143. **Cox, D. S. and Taubman, M. A.,** Oral induction of the secretory antibody response by soluble and particular antigens, *Int. Arch. Allergy Appl. Immunol.*, 75, 126, 1984.
144. **Keren, D. F., McDonald, R. A., Scott, P. J., Rosner, A. M., and Strubel, E.,** Effect of antigen form on local immunoglobulin A memory response of intestinal secretions to *Shigella flexneri*, *Infect. Immun.*, 47, 123, 1985.
145. **Nicklin, S. and Miller, K.,** Local and systemic immune responses to intestinally presented antigens, *Int. Arch. Allergy Appl. Immunol.*, 72, 87, 1983.
146. **Jeurissen, S. H. M., Claassen, E., van Rooijen, N., and Kraal, G.,** Intraintestinal priming leads to antigen-specific IgA memory cells in peripheral lymphoid organs, *Immunology*, 56, 417, 1985.
147. **Husband, A. J. and Dunkley, M. L.,** Lack of site of origin effects on distribution of IgA antibody-containing cells, *Immunology*, 54, 215, 1985.
148. **Dunkley, M. L. and Husband, A. J.,** The induction and migration of antigen-specific helper cells for IgA responses in the intestine, *Immunology*, 57, 379, 1986.
149. **Husband, A. J. and Gowans, J. C.,** The origin and antigen-dependent distribution of IgA-containing cells in the intestine, *J. Exp. Med.*, 148, 1146, 1978.
150. **Husband, A. J.,** Kinetics of extravasation and redistribution of IgA-specific antibody-containing cells in the intestine, *J. Immunol.*, 128, 1355, 1982.
151. **Weisz-Carrington, P. C. and Grimes, S. R.,** Specific attachment of stable murine monoclonal MN hybrids to the endothelium of the gut lamina propria, *Fed. Proc.*, 46, 1220, 1987.
152. **Ottaway, C. A. and Greenberg, G. R.,** Interaction of vasoactive intestinal peptide with mouse lymphocytes: specific binding and the modulation of mitogen responses, *J. Immunol.*, 132, 417, 1984.
153. **Wira, C. R., Hyde, E., Sandoe, C. P., Sullivan, D., and Spencer, S.,** Cellular aspects of the rat uterine IgA response to estradiol and progesterone, *J. Steroid Biochem.*, 12, 451, 1980.
154. **Canning, M. B. and Billington, W. D.,** Hormonal regulation of immunoglobulins and plasma cells in the mouse uterus, *J. Endocrinol.*, 97, 419, 1983.
155. **Rachman, F., Casimiri, V., Psychoyos, A., and Bernard, O.,** Immunoglobulins in the mouse uterus during the estrous cycle, *J. Reprod. Fertil.*, 69, 17, 1983.
156. **Weisz-Carrington, P., Roux, M. E., McWilliams, M., Phillips Quagliata, J. M., and Lamm, M. E.,** Hormonal induction of the secretory immune system in the mammary gland, *Proc. Natl. Acad. Sci. U.S.A.*, 75, 2928, 1978.
157. **Czinn, S. J. and Lamm, M. E.,** Selective chemotaxis of subsets of B lymphocytes from gut-associated lymphoid tissue and its implications for the recruitment of mucosal plasma cells, *J. Immunol.*, 136, 3607, 1986.
158. **Weisz-Carrington, P., Roux, M. E., McWilliams, M., Phillips Quagliata, J. M., and Lamm, M. E.,** Organ and isotype distribution of plasma cells producing specific antibody after oral immunization: evidence for a generalized secretory immune system, *J. Immunol.*, 123, 1705, 1979.
159. **Weisz-Carrington, P., Grimes, S. R., Jr., and Lamm, M. E.,** Gut-associated lymphoid tissue as a source of IgA immune response in respiratory tissues after oral immunization and intrabronchial challenge, *Cell Immunol.*, 106, 132, 1987.
160. **Tseng, J.,** Repopulation of the gut lamina propria with IgA-containing cells by lymphoid cells isolated from the gut lamina propria, *Eur. J. Immunol.*, 14, 420, 1984.
161. **McDermott, M. R. and Bienenstock, J.,** Evidence for a common immunologic system. I. Migration of B immunoblasts into intestinal, respiratory and genital tissues, *J. Immunol.*, 122, 1892, 1979.
162. **Tagliabue, A., Boraschi, D., Villa, L., Keren, D. F., Lowell, G. H., Rappuoli, R., and Nencioni, L.,** IgA-dependent cell-mediated activity against enteropathogenic bacteria: distribution, specificity and characterization of the effector cells, *J. Immunol.*, 133, 988, 1984.
163. **Barbeau, J., Bourget, J., and Deslauriers, N.,** RFc alpha-bearing cells and IgA-mediated phagocytosis in the mouse oral mucosa, *Adv. Exp. Med. Biol.*, 216, 563, 1987.

164. **Mirski, S., Befus, A. D., and Bienenstock, J.,** Effects of sex differences and gonadal hormone alterations on the accumulation of mesenteric lymphoblasts in the small intestine, *Ann. N.Y. Acad. Sci.*, 409, 845, 1983.
165. **McDermott, M. R., Clark, D. A., and Bienenstock, J.,** Evidence for a common mucosal immunologic system. II. Influence of the estrous cycle on B immunoblast migration into genital and intestinal tissues, *J. Immunol.*, 124, 2536, 1980.
166. **Sullivan, D. A. and Allansmith, M. R.,** Hormonal influence on the secretory immune system of the eye: androgen modulation of IgA levels in tears of rats, *J. Immunol.*, 134, 2978, 1985.
167. **Sullivan, D. A., Bloch, K. J., and Allansmith, M. R.,** Hormonal influence on the secretory immune system of the eye: androgen regulation of secretory component levels in rat tears, *J. Immunol.*, 132, 1130, 1984.
168. **Sullivan, D. A., Bloch, K. J., and Allansmith, M. R.,** Hormonal influence on the secretory immune system of the eye: androgen control of secretory component production by the rat exorbital gland, *Immunology*, 52, 234, 1984.
169. **Sullivan, D. A. and Allansmith, M. R.,** Hormonal influence on the secretory immune system of the eye: endocrine interactions in the control of IgA and secretory component levels in tears of rats, *Immunology*, 60, 337, 1987.
170. **Wira, C. R., Stern, J. E., and Colby, E.,** Estradiol regulation for secretory component in the uterus of the rat: evidence for involvement of RNA synthesis, *J. Immunol.*, 133, 2624, 1984.
171. **Wira, C. R. and Colby, E. M.,** Regulation of secretory component by glucocorticoids in primary cultures of rat hepatocytes, *J. Immunol.*, 134, 1744, 1985.
172. **Wira, C. R. and Sullivan, D. A.,** Estradiol and progesterone regulation of immunoglobulin A and G and secretory component in cervicovaginal secretions of the rat, *Biol. Reprod.*, 32, 90, 1985.
173. **Wira, C. R., Sullivan, D. A., and Sandoe, C. P.,** Epithelial cell involvement in the estradiol-stimulated accumulation of IgA in the rat uterus, *J. Steroid Biochem.*, 19, 469, 1983.
174. **Weisz-Carrington, P., Emancipator, S., and Lamm, M. E.,** Binding and uptake of immunoglobulins by mouse mammary gland epithelial cells in hormone-treated cultures, *J. Reprod. Immunol.*, 6, 63, 1984.
175. **Rowson, L. E., Lamming, G. E., and Fry, R. M.,** Influence of ovarian hormones on uterine infection, *Nature (London)*, 171, 749, 1953.
176. **Wira, C. R. and Sandoe, C. P.,** Specific IgA and IgG antibodies in the secretions of the female reproductive tract: effects of immunization and estradiol on expression of this response in vivo, *J. Immunol.*, 138, 4159, 1987.
177. **Forslin, L., Danielsson, D., and Falk, V.,** Adherence in vitro of *Neisseria gonorrhoeae, Escherichia coli* and group B Streptococci to vaginal epithelial cells of post-menopausal women, *Gynecol. Obstet. Invest.*, 11, 341, 1980.
178. **Botta, G. A.,** Hormonal and type-dependent adhesion of group B streptococci to human vaginal cells, *Infect. Immun.*, 25, 1084, 1979.
179. **Forslin, L., Danielsson, D., and Falk, V.,** Variations in attachment of *Neisseria gonorrhoeae* to vaginal epithelial cells during the menstrual cycle and early pregnancy, *Med. Microbiol. Immunol.*, 167, 231, 1979.
180. **Wira, C. R. and Merritt, K.,** Effect of estrous cycle, castration and pseudopregnancy on *E. coli* in the uterus and in uterine secretions of the rat, *Biol. Reprod.*, 17, 519, 1977.
181. **Baker, D. A. and Plotkin, S. A.,** Enhancement of vaginal infection in mice by *Herpes simplex* virus type II with progesterone, *Proc. Soc. Exp. Biol. Med.*, 158, 131, 1978.
182. **Buts, J. P. and Delacroix, D. L.,** Ontogenic changes in secretory component expression by villous and crypt cells of small rat intestine, *Immunology*, 54, 181, 1985.
183. **Sullivan, D. A. and Wira, C. R.,** Hormonal regulation of immunoglobulins in the rat uterus: uterine immunoglobulin response to a single estradiol treatment, *Endocrinology*, 112, 260, 1983.
184. **Martell, M., Oehninger, C., Scott, C., Delgado, L., Martinez, M., and Korc, I.,** Influence of glucocorticoid and betamimetic therapy on milk secretory IgA concentration produced by mothers delivering preterm infants, *J. Perinat. Med.*, 13, 61, 1985.
185. **Stanisz, A. M., Scicchitano, R., Dazin, P., Bienenstock, J., and Payan, D. G.,** Distribution of substance P receptors on murine spleen and Peyer's patch T and B cells, *J. Immunol.*, 139, 749, 1987.
186. **Wiedermann, C. J., Sertl, K., and Pert, C. B.,** Neuropeptides and the immune system: substance P receptors in bronchus-associated lymphoid tissue of the rat, *Ann. N.Y. Acad. Sci.*, 496, 205, 1987.
187. **Williams, J. A.,** Regulatory mechanisms in pancreas and salivary acini, *Annu. Rev. Physiol.*, 46, 361, 1984.
188. **Sullivan, D. A., Hann, L. E., and Vaerman, J. P.,** Selectivity, specificity and kinetics of the androgen regulation of the ocular secretory immune system, *Immunol. Invest.*, 17, 183, 1988.
189. **Sullivan, D. A. and Hann, L. E.,** Endocrine regulation of the ocular secretory immune system, *Adv. Exp. Med. Biol.*, 216, 765, 1987.
190. **Sullivan, D. A.,** Influence of the hypothalamic-pituitary axis on the androgen regulation of the ocular secretory immune system, *J. Steroid Biochem.*, 30, 429, 1988.

191. **Hann, L. E. and Sullivan, D. A.,** Isolation, characterization and hormonal response of lacrimal gland acinar cells, *Invest. Ophthalmol. Vis. Sci. Suppl.*, 28, 157, 1987.
192. **Weisz-Carrington, P., Dienstman, S., and Lamm, M. E.,** Surface IgA binding by mouse mammary gland epithelial cells in culture after treatment with hormones, *Fed. Proc.*, 39, 1153, 1980.
193. **Wira, C. R. and Sandoe, C. P.,** Sex steroid hormone regulation of IgA and IgG in rat uterine secretions, *Nature (London)*, 268, 534, 1977.
194. **Sullivan, D. A. and Wira, C. R.,** Estradiol regulation of secretory component in the female reproductive tract, *J. Steroid Biochem.*, 15, 439, 1981.
195. **Wira, C. R. and Sandoe, C. P.,** Hormonal regulation of immunoglobulins: influence of estradiol on immunoglobulins A and G in the rat uterus, *Endocrinology*, 106, 1020, 1980.
196. **Wira, C. R. and Sullivan, D. A.,** Effect of estradiol and progesterone on the secretory immune system in the female genital tract, *Adv. Exp. Med. Biol.*, 138, 99, 1982.
197. **Sullivan, D. A. and Wira, C. R.,** Estradiol regulation of secretory component in the rat uterus, *Ann. N.Y. Acad. Sci.*, 409, 882, 1983.
198. **Wira, C. R. and Stern, J. E.,** Estradiol regulation of the secretory immune system in the female reproductive tract: IgA in uterine and vaginal secretions of rats following portacaval anastomosis, *J. Steroid Biochem.*, 24, 33, 1986.
199. **Wira, C. R. and Sandoe, C. P.,** Origin of IgA and IgG antibodies in the female reproductive tract: regulation of the genital response by estradiol, *Adv. Exp. Med. Biol.*, 216, 403, 1987.
200. **Templeman, K. H. and Wira, C. R.,** Ultrastructural localization of IgA and IgG in uterine epithelial cells following estradiol administration, *Adv. Exp. Med. Biol.*, 216, 395, 1987.
201. **Gudmundsson, O. G., Sullivan, D. A., Bloch, K. J., and Allansmith, M. R.,** The ocular secretory immune system of the rat, *Exp. Eye Res.*, 40, 231, 1985.
202. **Allansmith, M. R., Kajiyama, G., Abelson, M. B., and Simon, M. A.,** Plasma cell content of main and accessory lacrimal glands and conjunctiva, *Am. J. Ophthalmol.*, 82, 819, 1976.
203. **Kett, K., Brandtzaeg, P., Radl, J., and Haiijman, J. J.,** Different subclass distribution of IgA-producing cells in human lymphoid organs and various secretory tissues, *J. Immunol.*, 136, 3631, 1986.
204. **Allansmith, M. R., Radl, J., Haaijman, J. J., and Mestecky, J.,** Molecular forms of tear IgA and distribution of IgA subclasses in human lacrimal glands, *J. Allergy Clin. Immunol.*, 76, 569, 1985.
205. **Crago, S. S., Kutteh, W. H., Moro, I., Allansmith, M. R., Radl, J., Haaijman, J. J., and Mestecky, J.,** Distribution of IgA1-, IgA2-, and J chain-containing cells in human tissues, *J. Immunol.*, 132, 16, 1984.
206. **Gudmundsson, O. G., Bjornsson, J., Olafsdottir, K., Bloch, K. J., Allansmith, M. R., and Sullivan, D. A.,** T cell populations in the lacrimal gland during aging, *Acta Ophthalmol.*, 66, 490, 1988.
207. **Montgomery, P. C., Peppard, J. V., and Skandera, C. A.,** Characterization of T-lymphocyte subpopulations from rat lacrimal glands, *Invest. Ophthalmol. Vis. Sci. Suppl.*, 28, 41, 1987.
208. **Ebersole, J. L., Taubman, M. A., Smith, D. J., and Steffen, M.,** Phenotype of lymphocytes in secretory tissues of aging rats, *Fed. Proc.*, 44, 979, 1985.
209. **Brandtzaeg, P., Gjeruldsen, S. T., Korsrud, F., Baklian, K., Berdal, P., and Ek, J.,** The human secretory immune system shows striking heterogeneity with regard to involvement of J chain-positive IgD immunocytes, *J. Immunol.*, 122, 503, 1979.
210. **Ebersole, J. L., Taubman, M. A., and Smith, D. J.,** Cellular and humoral IgA responses after single and multiple local injections of antigen, *Cell. Immunol.*, 77, 372, 1983.
211. **Montgomery, P. C., Ayyildiz, A., Lemaitre-Coelho, I. M., Vaerman, J. P., and Rockey, J. H.,** Induction and expression of antibodies in secretions, *Ann. N.Y. Acad. Sci.*, 409, 428, 1983.
212. **McGhee, D. W. and Franklin, R. M.,** Lymphocyte migration into the lacrimal gland is random, *Cell. Immunol.*, 86, 75, 1984.
213. **Sullivan, D. A., Colby, E. B., Hann, L. E., Allansmith, M. R., and Wira, C. R.,** Production and utilization of a mouse monoclonal antibody to rat IgA: identification of gender-related differences in the secretory immune system, *Immunol. Invest.*, 15, 311, 1986.
214. **Hann, L. E., Allansmith, M. R., and Sullivan, D. A.,** Impact of aging and gender on the Ig-containing cell profile of the lacrimal gland, *Acta Ophthalmol.*, 66, 87, 1988.
215. **Walcott, B., Sibony, P. A., Coyle, P. K., McKeon, C., and Keyser, K. T.,** Protein and immunoglobulin release from lacrimal gland fragments, *Invest. Ophthalmol. Vis. Sci. Suppl.*, 27, 25, 1986.
216. **Chandler, J. W. and Gillette, T. E.,** Immunologic defense mechanisms of the ocular surface, *Ophthalmology*, 90, 585, 1983.
217. **Franklin, R. M., Prendergast, R. A., and Silverstein, A. M.,** Secretory immune system of the rabbit ocular adnexa, *Invest. Ophthalmol. Vis. Sci.*, 18, 1093, 1979.
218. **Rebello, R., Green, F. H. Y., and Fox, H.,** A study of the secretory immune system of the female genital tract, *Br. J. Obstet. Gynaecol.*, 82, 812, 1975.
219. **Allansmith, M. R., Yee, L., and Sullivan, D. A.,** The influence of aging on the level of IgA and IgG in tears of rats, *Fed. Proc.*, 46, 1349, 1987.
220. **Gachon, A. M., Richard, J., and Dastugue, B.,** Human tears: normal protein pattern and individual protein determinations in adults, *Curr. Eye Res.*, 5, 301, 1983.

221. **Delacroix, D. L., Dive, C., Rambaud, J. C., and Vaerman, J. P.,** IgA subclasses in various secretions and serum, *Immunology,* 47, 383, 1982.
222. **Peppard, J. V. and Montgomery, P. C.,** Studies on the origin and composition of IgA in rat tears, *Immunology,* 62, 194, 1987.
223. **Allansmith, M. R., Gudmundsson, O. G., Hann, L. E., Keys, C., Bloch, K. J., and Sullivan, D. A.,** The immune response of the lacrimal gland to antigenic exposure, *Curr. Eye Res.,* 6, 921, 1987.
224. **Pederson, B., Andersen, S. M., Klauber, A., Ottavay, E., Prause, J. U., Zhong, C., and Norrild, B.,** Secretory IgA specific for herpes simplex virus in lacrimal fluid from patients with herpes keratitis — a possible diagnostic parameter, *Br. J. Ophthalmol.,* 66, 648, 1982.
225. **Finney, P. M. and Bushell, A. C.,** An enzyme-linked immunoassay (ELISA) for anti-chlamydial secretory immunoglobulin A in guinea pig tears, *J. Immunol. Methods,* 86, 71, 1986.
226. **Fox, P. D., Khaw, P. T., McBride, B. W., McGill, J. I., and Ward, K. A.,** Tear and serum antibody levels in ocular herpetic infection: diagnostic precision of secretory IgA, *Br. J. Ophthalmol.,* 70, 584, 1986.
227. **Taylor, H. R., Young, E., MacDonald, B., Schachter, J., and Prendergast, R. A.,** Oral immunization against Chlamydial eye infection, *Invest. Ophthalmol. Vis. Sci.,* 28, 249, 1987.
228. **Kahn, M., Barney, N. P., Briggs, R. M., Bloch, K. J., and Allansmith, M. R.,** Penetrating the conjunctival barrier: the role of molecular weight, *Invest. Ophthalmol. Vis. Sci.,* 28, 38, 1987.
229. **Mestecky, J., McGhee, J. R., Arnold, R. R., Michalek, S. M., Prince, S. J., and Babb, J. C.,** Selective induction of an immune response in human external secretions by ingestion of bacterial antigen, *J. Clin. Invest.,* 61, 731, 1978.
230. **Jackson, S. and Mestecky, J.,** Oral-parenteral immunization leads to the appearance of IgG auto-anti-idiotypic cells in mucosal tissues, *Cell. Immunol.,* 60, 498, 1981.
231. **Bergmann, K.Ch., Waldman, R. H., Tischner, H., and Pohl, W. D.,** Antibody in tears, saliva and nasal secretions following oral immunization of humans with inactivated influenza vaccine, *Int. Arch. Allergy Appl. Immunol.,* 80, 107, 1986.
232. **Kollock, C. W.,** Diseases and functional disorders of the eye, produced by normal and abnormal conditions of the sexual organs, *Trans. S. C. Med. Assoc.,* 97, 1888.
233. **Bettman, B.,** The relation between the eyes and disease of the female genital organs, *Am. J. Obstet. Dis. Women Child.,* 28, 498, 1893.
234. **Vance, R. A.,** The effect of menstrual disorders upon the vascularity and nutrition of the intra-ocular structures, *Boston Med. Surg. J.,* 9, 2310, 1872.
235. **McKay, R. J.,** Eye diseases dependent upon suppression of menses, *Am. J. Med. Sci.,* October, 382, 1882.
236. **Swanzy, H. R.,** The influence of the uterus in eye disease, *Obstet. J.,* 2, 118, 1878.
237. **Fitzgerald, D. E.,** Connexion between diseases of the eye and affections of the sexual organs in females, *Lancet,* March, 456, 1883.
238. **Ramsay, M.,** Diseases of the eye most frequently met with in gynaecological and obstetrical practice, *Lancet,* March, 854, 1893.
239. **Ludlam, R.,** The physiological and morbid relations existing between the uterus and the eye, *N.Y. Med. Times,* July, 99, 1893.
240. **Cavallero, C.,** Relative effectiveness of various steroids in an androgen assay using the exorbital lacrimal gland of the castrated rat, *Acta Endocrinol. (Copenhagen),* 55, 119, 1967.
241. **Hahn, J. D.,** Effect of cyproterone acetate on sexual dimorphism of the exorbital lacrimal gland in rats, *J. Endocrinol.,* 45, 421, 1969.
242. **Paulini, K., Beneke, G., and Kulka, R.,** Age- and sex-dependent changes in glandular cells. I. Histologic and chemical investigations on the glandular lacrimalis, glandular intraorbitalis, and glandula orbitalis external of the rat, *Gerontologia,* 18, 131, 1972.
243. **Paulini, K., Mohr, W., Beneke, G., and Kulka, R.,** Age- and sex-dependent changes in glandular cells. II. Cytomorphometric and autoradiographic investigations on the glandular lacrimalis, glandular intraorbitalis, and glandula orbitalis external of the rat, *Gerontologia,* 18, 147, 1972.
244. **Lauria, A. and Porcelli, F.,** Leucine aminopeptidase (LAP) activity and sexual dimorphism in rat exorbital gland, *Basic Appl. Histochem.,* 23, 171, 1979.
245. **Tier, H.,** Uper Zeilteilung and Kernklassen bildung in der Glandula orbitalis externa der Ratte, *Acta Pathol. Microbiol. Scand. Suppl.,* 50, 1, 1944.
246. **Shaw, P. H., Held, W. A., and Hastie, N. D.,** The gene family for major urinary proteins: expression in several secretory tissues of the mouse, *Cell,* 32, 755, 1983.
247. **Krawczuk-Hermanowiczowa, O.,** Effects of sexual glands on the lacrimal gland. II. Changes in rat lacrimal glands after castration, *Klin. Oczna,* 85, 15, 1983.
248. **Cornell-Bell, A. H., Sullivan, D. A., and Allansmith, M. R.,** Gender-related differences in the morphology of the lacrimal gland, *Invest. Ophthalmol. Vis. Sci.,* 26, 1170, 1985.

249. **Sullivan, D. A., Hann, L. E., Yee, L., and Allansmith, M. R.,** Age- and gender-related influence on the lacrimal gland and tears, submitted, 1989.

250. **Waterhouse, J. P.,** Focal adenitis in salivary and lacrimal glands, *Proc. R. Soc. Med.*, 56, 911, 1963.

251. **Sen, D. K., Sarin, G. S., Mathur, G. P., and Saha, K.,** Biological variation of immunoglobulin concentrations in normal human tears related to age and sex, *Acta Ophthalmol.*, 56, 439, 1978.

252. **Krzych, U., Thurman, G. B., Goldstein, A. L., Bressler, J. P., and Strausser, H. R.,** Sex-related immunocompetence of BALB/c mice. I. Study of immunologic responsiveness of neonatal, weanling, and young adult mice, *J. Immunol.*, 123, 2568, 1979.

253. **Mathur, S., Mathur, R. S., Goust, J. M., Williamson, H. O., and Fudenberg, H. H.,** Cyclic variations in white cell populations in the human menstrual cycle: correlations with progesterone and estradiol, *Clin. Immunol. Immunopathol.*, 13, 246, 1979.

254. **Sullivan, D. A., Richardson, G. S., MacLaughlin, D. T., and Wira, C. R.,** Variations in the levels of secretory component in human uterine secretions during the menstrual cycle, *J. Steroid Biochem.*, 20, 509, 1984.

255. **Tourville, D. R., Ogra, S. S., Lippes, J., and Tomasi, T. B., Jr.,** The human female reproductive tract: immunohistochemical localization of γA, γG, γM, secretory "piece," and lactoferrin, *Am. J. Obstet. Gynecol.*, 108, 1102, 1970.

256. **Hurliman, J., Dayal, R., and Gloor, E.,** Immunoglobulins and secretory component in endometrium and cervix. Influence of inflammation and carcinoma, *Virchow Arch. A*, 377, 211, 1978.

257. **Kelly, J. K. and Fox, H.,** The local immunological defense system of the human endometrium, *J. Reprod. Immunol.*, 1, 39, 1979.

258. **Tjokronegoro, A. and Sirisinha, S.,** Quantitative analysis of immunoglobulins and albumin in secretion of female reproductive tract, *Fertil. Steril.*, 26, 413, 1975.

259. **Chipperfield, E. J. and Evans, B. A.,** The influence of local infection on immunoglobulin in the human cervix, *Clin. Exp. Immunol.*, 11, 219, 1972.

260. **Roig de Vargas-Linares, C. E.,** Plasma cells in the hamster vagina: cyclical and experimental variations, *J. Reprod. Fertil.*, 15, 389, 1968.

261. **Zuckerman, H., Kahana, A., and Carmel, S.,** Antibacterial activity of human cervical mucus, *Gynecol. Invest.*, 6, 265, 1975.

262. **Waldman, R. H., Cruz, J. M., and Rowe, D. S.,** Immunoglobulin levels and antibody to *Candida albicans* in human cervicovaginal secretions, *Clin. Exp. Immunol.*, 9, 427, 1971.

263. **Jalanti, R. and Isliker, H.,** Immunoglobulins in human cervico-vaginal secretions, *Int. Arch. Allergy Appl. Immunol.*, 53, 402, 1977.

264. **Coughlan, B. M. and Skinner, G. R. B.,** Immunoglobulin concentrations in cervical mucus in patients with normal and abnormal cervical cytology, *Br. J. Obstet. Gynaecol.*, 84, 129, 1977.

265. **Schumacher, G. F. B.,** Soluble proteins in cervical mucus, *Biology of the Cervix*, Blandau, R. J. and Moghissi, K., Eds., University of Chicago Press, 1973, 201.

266. **Yang, S. L. and Schumacher, G. F. B.,** Immune response after vaginal application of antigens to the rhesus monkey, *Fertil. Steril.*, 32, 588, 1979.

267. **Hulka, J. F. and Omran, K. F.,** The uterine cervix as a potential local antibody secretor, *Am. J. Obstet. Gynecol.*, 104, 440, 1969.

268. **Schumacher, G. F. B.,** Hormonal immune factors in the female reproductive tract and their changes during the cycle, in *Immunological Aspects of Infertility and Fertility Regulation*, Schumacher, G. F. B. and Dhinsa, D. S., Eds., Elsevier/North-Holland, Amsterdam, 1980, 93.

269. **Allardyce, R. and Rademaker, M.,** Female genital tract immunity and infertility after oral vaccination with sperm antigens in mice, *Adv. Exp. Med. Biol.*, 216, 1807, 1987.

270. **Adamson, I. Y. and King, G. M.,** Sex differences in development of fetal rat lung. I. Autoradiographic and biochemical studies, *Lab. Invest.*, 50, 456, 1984.

271. **Adamson, I. Y. and King, G. M.,** Sex differences in development of fetal rat lung. II. Quantitative morphology of epithelial-mesenchymal interactions, *Lab. Invest.*, 50, 461, 1984.

272. **Manning, L. S. and Parmely, M. J.,** Migratory properties of mammary gland T lymphocytes in the rat, *Int. Conf. Reprod. Immunol.*, Banff, Alberta, Canada, 1981.

273. **McMichael, A. J. and Potter, J. D.,** Reproduction, endogenous and exogenous sex hormone, and colon cancer: a review and hypothesis, *J. Natl. Cancer Inst.*, 65, 1201, 1980.

274. **Aloe, L. and Levi-Montalcini, R.,** Comparative studies on testosterone and L-thyroxine effects on the synthesis of nerve growth factor in mouse submaxillary glands, *Exp. Cell Res.*, 125, 15, 1980.

275. **Chao, J. and Margolis, H.,** Differential effects of testosterone, thyroxine and cortisol on rat submandibular gland versus renal kallikrien, *Endocrinology*, 113, 2221, 1983.

276. **Stokes, C. R., Newby, T. J., and Bourne, F. J.,** The influence of oral immunization on local and systemic immune responses to heterologous antigens, *Clin. Exp. Immunol.*, 52, 399, 1983.

277. **Lim, T. S., Messiha, N., and Watson, R. R.,** Immune components of the intestinal mucosae of aging and protein deficient mice, *Immunology*, 43, 401, 1981.

278. **Alford, R. H.,** Effects of chronic bronchopulmonary disease and aging on human nasal secretion IgA concentrations, *J. Immunol.*, 101, 984, 1968.
279. **Schmucker, D. L., Gilbert, R., Jones, A. L., Hradek, G. T., and Bazin, H.,** Effect of aging on the hepatobiliary transport of dimeric immunoglobulin A in the male Fisher rat, *Gastroenterology*, 88, 436, 1985.
280. **Smith, D. J., Ebersole, J. L., and Taubman, M. A.,** Local and systemic immune response in aged hamsters, *Immunology*, 50, 407, 1983.
281. **Haaijman, J. J. and Hijmans, W.,** Influence of age on the immunological activity and capacity of the CBA mouse, *Mech. Ageing Dev.*, 3, 375, 1978.
282. **Cavallero, C.,** The influence of various steroids on the Lowenthal lachrymal glands of the rat, *Acta Endocrinol. (Copenhagen) Suppl.*, 51, 861, 1960.
283. **Cavallero, C., Chiappino, G., Milani, F., and Casella, E.,** Uptake of $^{35}$S-labelled sulfate in the exorbital lacrymal glands of adult and newborn rats under different hormonal treatment, *Experientia*, 16, 429, 1960.
284. **Cavallero, C. and Morera, P.,** Effect of testosterone on the nuclear volume of exorbital lacrimal glands of the white rat, *Experientia*, 16, 285, 1960.
285. **Cavallero, C. and Offner, P.,** Relative effectiveness of various steroids in an androgen assay using the exorbital lacrimal gland of the castrated rat. II. $C_{19}$-steroids of the 5 alpha-androstane series, *Acta Endocrinol. (Copenhagen)*, 55, 131, 1967.
286. **Quintarelli, G. and Dellovo, M. C.,** Activation of glycoprotein biosynthesis by testosterone propionate on mouse exorbital glands, *J. Histochem. Cytochem.*, 13, 361, 1965.
287. **Ebling, F. J., Ebling, E., Randall, V., and Skinner, J.,** The effects of hypophysectomy and bovine growth hormone on the responses to testosterone of prostate, preputial, harderian and lachrymal glands and of brown adipose tissue in the rat, *J. Endocrinol.*, 66, 401, 1975.
288. **Ebling, F. J., Ebling, E., Randall, V., and Skinner, J.,** The synergistic action of alpha-melanocyte stimulating hormone and testosterone on the sebaceous, prostate, preputial, harderian and lachrymal glands, seminal vesicles and brown adipose tissue in the hypophysectomized-castrated rat, *J. Endocrinol.*, 66, 407, 1975.
289. **Dzierzykay-Rogalska, I., Chodynicki, S., and Wisniewski, L.,** The effect of gonadectomy on the parotid salivary gland and Loeventhal's gland in white mice, *Acta Med. Pol.*, 4, 221, 1963.
290. **Martinazzi, M. and Baroni, C.,** Controllo ormonale delle ghiandola lacrimale extraorbitale nel topo con nanismo ipofisario, *Folia Endocrinol.*, 16, 123, 1963.
291. **Sullivan, D. A. and Allansmith, M. R.,** Hormonal modulation of tear volume in the rat, *Exp. Eye Res.*, 42, 131, 1986.
292. **Sullivan, D. A.,** Influence of the hypothalamic-pituitary-gonadal axis on tear volume and protein content, *Invest. Ophthalmol. Vis. Sci. Suppl.*, 28, 157, 1987.
293. **Liao, S.,** Cellular receptors and mechanisms of action of steroid hormones, *Int. Rev. Cytol.*, 41, 87, 1975.
294. **Laine, M. and Tenovuo, J.,** Effect on peroxidase activity and specific binding of the hormone 17β-oestradiol and rat salivary glands, *Arch. Oral Biol.*, 28, 847, 1983.
295. **Jacobs, M., Buxton, D., Kramer, P., Lubkin, V., Dunn, M., Herp, A., Weinstein, B., Southren, A. L., and Perry, H.,** *Invest. Ophthalmol. Vis. Sci. Suppl.*, 27, 25, 1986.
296. **Feldman, F., Bain, J., and Matuk, A. R.,** Daily assessment of ocular and hormonal variables throughout the menstrual cycle, *Arch. Ophthalmol.*, 96, 1835, 1978.
297. **Frankel, S. H. and Ellis, P. P.,** Effect of oral contraceptives on tear production, *Ann. Ophthalmol.*, 10, 1585, 1978.
298. **Friedman, B. and Gernand, H.,** Sjogren's syndrome-treated with stilbestrol, *Calif. West. Med.*, 64, 31, 1946.
299. **Prijot, E., Bazin, L., and Destexhe, B.,** Essai de traitment hormonal de la kerato-conjonctivite seche, *Bull. Soc. Belge Ophthalmol.*, 162, 795, 1972.
300. **Appelmans, M.,** La keratoconjonctivite seche de Gougerot-Sjogren, *Arch. Ophthalmol.*, 81, 577, 1948.
301. **Bruckner, V. R.,** Uber einem erfolgreich mit perandren behandelten fall von Sjogren'schem symptomen komplex, *Ophthalmologica*, 110, 37, 1945.
302. **Radnot, M. and Nemeth, B.,** Testosteronkeszitmenyek hatasa a konnymirigyre, *Orv. Hetil.*, 95, 580, 1954.
303. **Ota, M., Kyakumoto, S., and Nemoto, T.,** Demonstration and characterization of cytosol androgen receptor in rat exorbital lacrimal gland, *Biochem. Int.*, 10, 129, 1985.
304. **Porcelli, F. and Aureli, G.,** Variazioni dell'area e del contenuto nucleare in DNA nella ghiandola di Loewenthal di ratto in relazione all'eta' e al dimorfismo sessuale, *Riv. Istoch. Norm. Pat.*, 18, 130, 1974.
305. **Gelfand, J. A.,** Exploiting sex for therapeutic purposes, *N. Engl. J. Med.*, 308, 1417, 1983.
306. **Barbieri, R. L., Lee, H., and Ryan, K. J.,** Danazol binding to rat androgen, glucocorticoid, progesterone, and estrogen receptors: correlation with biologic activity, *Fertil. Steril.*, 31, 182, 1979.
307. **Gelfand, J. A., Sherins, R. J., Alling, D. W., and Frank, M. M.,** Treatment of hereditary angioedema with danazol: reversal of clinical and biochemical abnormalities, *N. Engl. J. Med.*, 295, 1444, 1976.

308. **Ahn, Y. S., Harrington, W. J., Simon, S. R., Mylvagnam, R., Pall, L. M., and So, A. G.,** Danazol for the treatment of idiopathic thrombocytopenic purpura, *N. Engl. J. Med.*, 308, 1396, 1983.
309. **Gralnick, H. R. and Rick, M. E.,** Danazol increases factor VIII and factor IX in classic hemophilia and Christmas disease, *N. Engl. J. Med.*, 308, 1393, 1983.
310. **Melez, K. A., Boegel, W. A., and Steinberg, A. D.,** Therapeutic studies in New Zealand mice. VII. Successful androgen treatment of NZB/NZW F1 females of different ages, *Arthritis Rheum.*, 23, 41, 1980.
311. **Neifeld, J. P., Lippman, M. E., and Tormey, D. C.,** Steroid hormone receptors in normal human lymphocytes: induction of glucocorticoid receptor activity by phytohemagglutinin stimulation, *J. Biol. Chem.*, 252, 2972, 1977.
312. **Cohen, J. H. M., Danel, L., Cordier, G., Saez, S., and Revillard, J. P.,** Sex steroid receptors in peripheral T cells: absence of androgen receptors and restriction of estrogen receptors to OKT8-positive cells, *J. Immunol.*, 131, 2767, 1983.
313. **Scott, B. L. and Pease, D. C.,** Electron microscopy of the salivary and lacrimal glands of the rat, *Am. J. Anat.*, 104, 115, 1959.
314. **Nikkinen, A., Lehtosalo, J. I., Uusitalo, H., Palkama, A., and Panula, P.,** The lacrimal glands of the rat and the guinea pig are innervated by nerve fibers containing immunoreactivities for substance P and vasoactive intestinal polypeptide, *Histochemistry*, 81, 23, 1984.
315. **Lehtosalo, J., Uusitalo, H., Mahrberg, T., Panula, P., and Palkama, A.,** Enkephalin-like immunoreactive nerve fibers in the lacrimal glands of the guinea pig, *Invest. Ophthalmol. Vis. Sci.*, 28, 23, 1987.
316. **Ruskell, G. L.,** Nerve terminals and epithelial cell variety in the human lacrimal gland, *Cell Tiss. Res.*, 158, 121, 1975.
317. **Wilson, C. M., Griffin, J. E., Reynolds, R. C., and Wilson, J. D.,** The interaction of androgen and thyroid hormones in the submandibular gland of the genetically hypothyroid (hyt/hyt) mouse, *Endocrinology*, 116, 2568, 1985.
318. **Carriere, R. M.,** The influence of thyroid and testicular hormones on the epithelium of the crypts of Lieberkuhn in the rat's intestine, *Anat. Rec.*, 156, 423, 1966.
319. **Carriere, R. M. and Buschke, M.,** The influence of thyroid and testicular hormones on globule leucocytes in the rat duodenal crypt epithelium, *Anat. Rec.*, 192, 407, 1978.
320. **Sica, G., Natoli, C., Marchetti, P., and Iacobelli, S.,** Effect of thyroid hormones on androgen responsiveness in a mammary tumor cell line, *J. Steroid Biochem.*, 15, 415, 1981.
321. **Motwani, N. M., Unakar, N. J., and Roy, A. K.,** Multiple hormone requirement for the synthesis of $\alpha 2\mu$-globulin by monolayers of rat hepatocytes in long-term primary culture, *Endocrinology*, 107, 1606, 1980.
322. **Steidler, N. E. and Reade, P. C.,** An immunohistochemical and histological study of the influence of the testes and adrenal glands on epidermal growth factor-containing cells in the submandibular salivary glands of male mice, *J. Anat.*, 135, 413, 1982.
323. **Roy, A. K. and Chaterjee, B.,** Sexual dimorphism in the liver, *Am. Rev. Physiol.*, 45, 37, 1983.
324. **Roy, A. K.,** Androgen-dependent synthesis of $\alpha 2\mu$-globulin in the rat: role of the pituitary gland, *J. Endocrinol.*, 56, 295, 1973.
325. **Norstedt, G. and Mode, A.,** On the primary site of action of estrogens and androgens in the regulation of hepatic prolactin receptors, *Endocrinology*, 111, 645, 1982.
326. **Norstedt, G., Mode, A., Hokfelt, T., Eneroth, P., Elde, R., Ferland, L., Labrie, F., and Gustafson, J.,** Possible role of somatostatin in the regulation of the sexually differentiated steroid metabolism and prolactin receptor in rat liver, *Endocrinology*, 112, 1076, 1983.
327. **Jeffery, S., Carter, N. D., and Wilson, C. A.,** Hypophysectomy abolishes sexual dimorphism of liver carbonic anhydrase. III, *FEBS Lett.*, 175, 129, 1984.
328. **Dupon, C. and Kim, M. H.,** Peripheral plasma levels of testosterone, androstenedione, and oestradiol during the rat oestrous cycle, *J. Endocrinol.*, 59, 653, 1973.
329. **Rush, M. E. and Blake, C. A.,** Serum testosterone concentrations during the 4-day estrous cycle in normal and adrenalectomized rats, *Proc. Soc. Exp. Biol. Med.*, 169, 216, 1982.
330. **Barbanel, G. and Assenmacher, I.,** Effects of thyroid hormones on the ontogeny of oestradiol-binding sites in the rat, *Mol. Cell. Endocrinol.*, 28, 247, 1982.
331. **Nover, A.,** Tierexperimentelle untersuchungen uber die regenerations und transplantation-sfahig keit der tranendruse uter lokalem und allgemeinem hormonreiz (thyrozin), *Graefes Arch. Ophthalmol. Bd.*, 156, 177, 1954.
332. **Crowe, J. P., Christensen, E., Butler, J., Wheeler, P., Doniach, D., Keenan, J., and Williams, R.,** Primary biliary cirrhosis: the prevalence of hypothyroidism and its relationship to thyroid autoantibodies and sicca syndrome, *Gastroenterology*, 78, 1437, 1980.
333. **Preslock, J. P.,** The pineal gland: basic implications and clinical correlations, *Endocrine Rev.*, 5, 282, 1984.
334. **Morel, G., Forest, M. G., and Dubois, P. M.,** Ultrastructural evidence for endogenous testosterone immunoreactivity in the pituitary gland of the rat, *Cell Tissue Res.*, 235, 159, 1984.

335. **Carriere, R.,** The influence of the thyroid gland on polyploid cell formation in the external orbital gland of the rat, *Am. J. Anat.*, 115, 1, 1964.
336. **Jahn, R., Padel, U., Porsch, P. H., and Soling, H. D.,** Adrenocorticotrophic hormone and alpha-melanocyte stimulating hormone induce secretion and protein phosphorylation in the rat lacrimal gland by activation of a cAMP-dependent pathway, *Eur. J. Biochem.*, 126, 623, 1982.
337. **Martinazzi, M.,** Effetti dell'ipofisectomia sulla ghiandola lacrimale extraorbitale del ratto, *Folia Endocrinol.*, 150, 120, 1962.
338. **Martinazzi, M.,** La ghiandola lacrimale extraorbitale o ghiandola di Loewenthal: sua morfologie e controllo ormonale, *Recenti Prog. Med.*, 34, 491, 1963.
339. **Wegilius, O. and Friman, C.,** Two different pituitary-controlled sulphation mechanisms in the rat, *Acta Med. Scand. Suppl.*, 412, 221, 1964.
340. **Bhatnagar, A. S., Ed.,** *The Anterior Pituitary Gland,* Raven Press, New York, 1983.
341. **Chen, C. L., Amenomori, Y., Lu, K. H., Voogt, J. L., and Meites, J.,** Serum prolactin levels in rats with pituitary transplants or hypothalamic lesions, *Neuroendocrinology,* 6, 220, 1970.
342. **Gorski, J. and Gannon, F.,** Current models of steroid hormone action: a critique, *Annu. Rev. Physiol.*, 38, 425, 1976.
343. **Clark, J. H., Schrader, W. T., and O'Malley, B. W.,** Mechanisms of steroid hormone action, in *Williams Textbook of Endocrinology,* Wilson, J. D. and Foster, D. W., Eds., W. B. Saunders, Philadelphia, 1985, 33.
344. **Kyakumoto, S., Kurokawa, R., Ohara-Nemoto, Y., and Ota, M.,** Sex difference in the cytosolic and nuclear distribution of androgen receptor in mouse submandibular gland, *J. Endocrinol.*, 108, 267, 1986.
345. **Cunha, G. R., Chung, L. W. K., Shannon, J. M., and Reese, B. A.,** Stomal-epithelial interactions in sex differentiation, *Biol. Reprod.*, 22, 19, 1980.
346. **Chang, S. M., Bell, C., Carol, and Chung, L. W. K.,** Androgen regulation of stromal-epithelial interaction in cultured prostatic cells, *Fed. Proc.*, 46, 525, 1987.
347. **Steiner, R. A., Bremner, W. J., Clifton, D. K., and Dorsa, D. M.,** Reduced pulsatile leuteinizing hormone and testosterone secretion with aging in the male rat, *Biol. Reprod.*, 31, 251, 1984.
348. **Swerdloff, R. S. and Heber, D.,** Effects of aging on male reproductive function, in *Endocrine Aspects of Aging,* Korenman, S. G., Ed., Elsevier, New York, 1982, 119.
349. **Roth, G. S.,** Hormone action during aging: alterations and mechanisms, *Mech. Ageing Dev.*, 9, 497, 1979.
350. **Sanborn, B. M., Wagle, J. R., and Steinberger, A.,** Control of androgen cytosol receptor concentrations in Sertoli cells: effect of androgens, *Endocrinology,* 114, 2388, 1984.
351. **Hann, L. E., Tatro, J. B., and Sullivan, D. A.,** Morphology and function of lacrimal gland acinar cells in primary culture, *Invest. Ophthalmol. Vis. Sci.*, 30, 145, 1989.
352. **Neville, M. C.,** Mammary cultures on floating gels: a model system for mammary function, *News Physiol. Sci.*, 2, 107, 1987.
353. **Wilde, C. J., Hasan, H. R., and Mayer, R. J.,** Comparison of collagen gels and mammary extracellular matrix as substrata for study of terminal differentiation in rabbit mammary epithelial cells, *Exp. Cell Res.*, 151, 519, 1984.
354. **Wicha, M. S., Liotta, L. A., Garbisa, S., and Kidwell, W. R.,** Basement membrane collagen requirements for attachment and growth of mammary epithelium, *Exp. Cell Res.*, 124, 181, 1979.
355. **Haeuptle, M. T., Suard, Y. L. M., Bogenmann, E., Reggio, H., Racine, L., and Kraehenbuhl, J. P.,** Effect of cell shape change on the function and differentiation of rabbit mammary cells in culture, *J. Cell Biol.*, 96, 1425, 1983.
356. **Suard, Y. M. L., Haeuptle, M. T., Farinon, E., and Kraehenbuhl, J. P.,** Cell proliferation and milk protein gene expression in rabbit mammary cell cultures, *J. Cell Biol.*, 96, 1435, 1983.
357. **Hadley, M. A., Byers, S. W., Suarez-Quian, C. A., Kleinman, H. K., and Dym, M.,** Extracellular matrix regulates Sertoli cell differentiation, testicular cord formation, and germ cell development in vitro, *J. Cell Biol.*, 101, 1511, 1985.
358. **Lee, E. Y. H. P., Lee, W. H., Kaetzel, C. S., Parry, G., and Bissell, M. J.,** Interaction of mouse mammary epithelial cells with collagen substrata: regulation of casein gene expression and secretion, *Proc. Natl. Acad. Sci. U.S.A.*, 82, 1419, 1985.
359. **Taub, M.,** Importance of hormonally defined, serum-free medium for in vitro studies concerning epithelial transport, in *Tissue Culture of Epithelial Cells,* Plenum Press, New York, 1985, 255.
360. **Jackson, F. L. and Hutson, J. C.,** Altered responses to androgen in diabetic male rats, *Diabetes,* 33, 819, 1984.
361. **Edwards, A. O. and Hazelwood, R. L.,** Effects of hypophysectomy on plasma levels of pancreatic polypeptide (APP) and insulin in adult chickens, *Proc. Soc. Exp. Biol. Med.*, 184, 510, 1987.
362. **Gause, I., Isaksson, O., Lindahl, A., and Eden, S.,** Effect of insulin treatment of hypophysectomized rats on adipose tissue responsiveness to insulin and growth hormone, *Endocrinology,* 116, 945, 1985.
363. **Heinze, E., Kleiner, W., and Voigt, K. H.,** Insulin release in rats 1 and 5 days after hypophysectomy, *Horm. Res.*, 14, 243, 1981.

364. **Matsubara, M., Nakagawa, K., and Akikawa, K.,** The significance of the measurement of plasma transferrin as a growth factor. I. The changes in hypophysial diseases before and after treatment, *Nippon Naibumpi Gakkai Zasshi,* 62, 1259, 1986.
365. **Williams, J. A., Sankaran, H., Korc, M., and Goldfine, I. D.,** Receptors for cholecystokinin and insulin in isolated pancreatic acini: hormonal control of secretion and metabolism, *Fed. Proc.*, 40, 2497, 1981.
366. **Sullivan, D. A. and Hann, L. E.,** Androgen regulation of secretory component synthesis by lacrimal gland acinar cells in vitro, submitted, 1989.
367. **Sullivan, D. A. and Hann, L. E.,** Hormonal influence on the secretory immune system of the eye: endocrine impact on the lacrimal gland accumulation and secretion of IgA and IgG, in press, 1989.
368. **Sullivan, D. A., Hann, L. E., Yee, L., Edwards, J. A., Soo, C., and Allansmith, M. R.,** Neural-immune interactions: role of ocular innervation in the secretory immune system of the lacrimal gland, *5th Int. Congr. Mucosal Immunol.*, in press, 1989.
369. **Sullivan, D. A., Block, L., and Hann, L. E.,** Endocrine regulation of the ocular secretory immune system, *Invest. Ophthalmol. Vis. Sci. Suppl.*, 30, 467, 1989.
370. **Hann, L. E., Kelleher, R. S., and Sullivan, D. A.,** Influence of culture conditions on the androgen control of secretory component synthesis by acinar cells from the lacrimal gland, submitted, 1989.
371. **Taubman, M. A., Halazonetis, T. D., Stack, W. E., Smith, D. J., Mandel, I. D., and Sullivan, D. A.,** Thyroid influence on the salivary secretory immune system, *5th Int. Congr. Mucosal Immunol.*, in press, 1989.

Chapter 16

# THE NEUROENDOCRINE-IMMUNE NETWORK IN THE GASTROINTESTINAL TRACT

**Serem Freier and Emanuel Lebenthal**

## TABLE OF CONTENTS

# I. INTRODUCTION

One of the problems that have to be overcome by multicellular forms of life is the dissemination of information to all relevant parts of the organism. This includes the recognition of sensory stimuli as well as providing the appropriate response—motor, secretory, and other. The nervous system, central, peripheral, and autonomic, has been created specifically for this purpose. This system has the advantage of speed. The blood vascular system, however, in many respects subserves a similar function. Although best known for its property of carrying nutrients, oxygen, and their metabolites in the bloodstream, it is also a vehicle for the transmission of steroid and peptide hormones. The action of these hormones is on those tissues having specific receptors for them. In addition, the immune system uses the bloodstream as well as the lymphatic channels for the conveyance of lymphocytes and their products. This phenomenon is of particular importance in transmitting information from the major sites of antigen sampling — the Peyer's patches in the intestine and the bronchus-associated lymphatic tissue in the lungs to all the mucosal sites of the body. This volume is devoted to the interactions of the immune, endocrine, and nervous systems in general, and this chapter deals with these interactions in the digestive tract in particular.

# II. THE ENTERIC NERVOUS SYSTEM AND ITS NEUROTRANSMITTERS

The enteric nervous system (ENS) is independent of the central nervous system.[1] This system is, therefore, autonomic in the true sense of the word. It must have its own sensory receptors and its primary afferent, integrative, and effector neurons.[2] It is for this reason that interruption of the extrinsic nerve supply of the gut does not affect intestinal motility. Although motor activity of the intestine is one of the major functions of the ENS, it is certainly not the only one. Secretion, including secretion of immunoglobulins, as this review will attempt to show, are important functions of the ENS. In the performance of these functions peptide hormones may serve as neurotransmitters.

Anatomically, the ENS is composed of two interconnected major plexuses of ganglion cells, as well as lesser, rather ill-defined groupings of nerve fibers derived from these ganglia.[3,4] The two ganglionated plexuses are found in the submucosa (Meissner's or submucosal plexus) and between the two muscle layers of the muscularis externa (Auerbach's or myenteric plexus). The neurons of the ENS differ widely in their morphology, as well as in their staining properties. The best known neurotransmitters of the ENS are acetylcholine (ACh) and norepinephrine (NE).[5-9] Segments of guinea pig ileum, *in vitro,* release relatively large amounts of ACh spontaneously,[10,11] and even more so when the gut is stimulated electrically.[12,13] As regards noradrenergic axon terminals, these can be demonstrated in the myenteric and submucosal plexuses. Of particular interest in the context of this chapter is the fact that noradrenergic axons ramify in the mucosa where they apparently innervate enteroendocrine cells and cause them to release serotonin.[14] There is, however, considerable evidence that some of the effects of exogenous NE, as well as those of sympathetic nerves, are indirect; they inhibit the release of ACh and thereby antagonize its effects.[15,16]

The noradrenergic innervation of the gut differs from the cholinergic innervation in this, that with rare exceptions, the cell bodies of the noradrenergic neurons reside entirely outside the gut itself.[17] Extrinsic denervation of the gut, therefore, will leave intact a cholinergic innervation only. Conversely, the gut is completely devoid of intrinsic neurons containing catecholeamine, epinephrine, dopamine, or NE.

Another substance which is believed to be a neurotransmitter in the gut is serotonin.[18-20] It is possible that this amine is synthesized locally in the intestinal neurons and is released as a result of nerve stimulation.[21] Serotonin in turn activates neurons that release ACh, as well as nonadrenergic, noncholinergic, and intrinsic inhibitory neurons.[21,25]

In recent years, research has centered on studying the role of peptide hormones as neurotransmitters in the ENS. First, it is of interest that 50% of mucosal neurons contain ACh in association with one or more peptides. The neuropeptides that have been located in the submucosal plexus of the intestine include vasoactive intestinal peptide (VIP),[26,27] and its homolog peptide histidine isoleucine,[27] substance P (SP),[28] somatostatin,[29-31] neuropeptide Y,[32,33] galanine,[34] and the C-terminal octapeptide of cholecystokinin (CCK-8).[31] Regarding CCK-8, this has been shown to be involved in immunoglobulin secretion of the gut.[35] With a few exceptions these neuropeptides have been found in the myenteric and in the submucosal plexuses. The following peptides have been found in the myenteric plexus only: bombesin, Met-enkephalin, derivatives of prepro-enkephalin, and Leu enkephalin derivatives of pre-prodynorphin.[30,36] Other peptides and related molecules that have been found at nerve endings in the gut include gastrin, neurotensin, angiotensin, and γ-aminobutyric acid.

It is of particular interest that of all the neurons supplying both the submucosal as well as the myenteric plexuses for any given peptide, there is a preponderance of submucosal over myenteric neurons.[37] These observations suggest that the submucosal neurons with their respective peptides may play an important role in secretory and absorptive processes in the gastrointestinal tract. Another point of interest is that more than one peptide may be present in the same neuron. This has been shown to be true for somatostatin, CCK-8, and neuropeptide Y in submucosal neurons.[27,32] In view of the participation of nervous and endocrine impulses in the release of immunoglobulins in the intestine the above brief discussion of the ENS seems relevant. The nerve supply to lymphatic tissue in the gut is discussed elsewhere in this volume.

## III. THE ENDOCRINE CELLS OF THE GUT

Cells storing serotonin, the so-called enterochromaffin (EC) cells because of their staining properties, have been known since the beginning of this century. In the last 20 years, it has transpired that there are peptide hormomone-containing non-EC cells in the intestine. Among these non-EC cells, several morphologically distinct cell types have been identified and correlated with the production of one or another member of the expanding family of gut endocrine peptides.[38-43] The fact that these cells shared certain histological, histochemical, and ultrastructural properties has prompted certain authors to group them together as members of the diffuse endocrine system (DES)[44] and the amine precursor uptake and decarboxylation (APUD) cell system.[45]

There is mounting evidence that the secretion of these cells, in addition to their neurocrine effects alluded to above and to their endocrine effects, may also express paracrine effects. Functionally, this is supported by the fact that one of the peptide hormones found in these cells is somatostatin, a hormone known to cause inhibition of a number of gut cells, including those releasing gastrin and cholecystokinin.[46-48] The endocrine-paracrine cells can be found throughout the gastric and small intestinal mucosa. The secretory granules are usually concentrated in the basal part of the cell. These cells have microvilli which reach up to the lumen, and resemble, superficially, those of the adjacent epithelial cells.[49] It appears that these villi may be the receptors for chemical stimuli from the lumen, triggering hormone release at the appropriate time, such as during digestion. The secretion of the granules takes place at the basolateral aspect of the cell into the interstitial spaces.[50] This observation also supports the concept of endocrine and paracrine effects of this hormone. Granule release at the lumenal aspect has never been observed, and secretions in the interstitial spaces would be prevented from reaching the lumen by the tight junction binding the enterocytes together near their apical surface. The presence of microtubules in the gut endocrine cells suggest that the secretory pathway is well defined.

The endocrine-paracrine cells may not produce a single peptide, but rather a family of related peptides. Sometimes a peptide and an amine, such as serotonin, may be produced in the same cell.

In addition to these peptide-producing cells in the intestinal tract, there are also extraintestinal sites for neuropeptide production and release. Thus VIP is produced in human neutrophils and eosinophils.[51] Virus infected lymphocytes synthesize adrenocorticotrophic hormone (ACTH) and thyroid stimulating hormone.[52,53] Monocytes and macrophage-like cells have been demonstrated to contain endorphins and bombesin.[54,55] Somatostatin-like peptides have recently been purified from basophils.[56] The ACTH isolated from lymphocytes appears to be functionally similar to that of the nervous system.[57] On the other hand, the somatostatin isolated from basophils appears to be different from the somatostatin-14 which is synthesized in neurons.[56,58]

## IV. THE EFFECT OF NEUROENDOCRINE HORMONES ON THE INTESTINAL IMMUNE SYSTEM

The three peptide hormones that have been studied most intensively with regard to their effect on the intestinal-immune tissue, are somatostatin, VIP, and SP. The interactions of these substances have been admirably reviewed in this volume by O'Dorisio. This chapter will therefore be mainly devoted to another of the neuropeptides, to wit, CCK-8 which has been studied by us during the last several years. In addition, we will discuss some actions of SP that have not been previously described. As the effects to be described involve the release of immunoglobulins into the lumen of the intestine, some introductory remarks on immunoglobulin transport from the lamina propria to the enterocyte, as well as transcytosis across the enterocyte are indicated. In this context some remarks on the hormonal control of the secretory component receptor are also relevant.

### A. THE LAMINA PROPRIA

The lamina propria is the lining of the gastrointestinal tract including the epithelium and the subjacent cells and the vascular and connective tissue down to the muscularis mucosa. The enterocyte is the cell involved in the digestion and absorption of food. Recently, however, it has become evident that in addition to its absorptive properties, this cell also has certain immune functions, such as the presentation of foreign antigens to the subjacent lymphocytes.[59] The immune cells lying within the lamina propria include the plasma cells and B and T lymphocytes, including the intraepithelial lymphocytes, mast cells, macrophages, and eosinophils. Some 80% of the plasma cells produce immunoglobulin A, (IgA) and the remainder produce IgM, IgE, IgD, and IgG.[40-62]

### B. PRODUCTION, TRANSPORT, AND SECRETION OF IMMUNOGLOBULINS

The plasma cells producing IgA in the lamina propria differ from those producing IgA elsewhere, for the IgA produced by them is in the dimeric form, whereas circulating IgA is in the monomeric form. The intestinal plasma cells also produce a polypeptide, the J or joining chain, which has a molecular weight of approximately 15 kDa and is joined to dimeric IgA by covalent bonds.[63-66] The only other immunoglobulin which is present in the polymeric form in the intestinal juice is IgM. It, too, is joined by J chain.[67-69] On the other hand, although J chain is also produced by plasma cells producing IgG and IgE, it does not complex with these immunoglobulins. The J chain is present in 80 to 100% of IgA immunocytes at mucosal sites,[70-72] but in the process of polymerization of IgA, the J chain may become buried within the molecule and therefore become immunologically obscured. One of the functions of J chain is believed to be the polymerization of immunoglobulins.[70-73] However, it is not essential for polymerization, nor does it produce polymerization of immunoglobulins other than IgA and IgM.

It is not known if the dimeric IgA leaves the plasma cells at a steady rate or if production at the mucosal site varies with different environmental circumstances. Most of the IgA leaving the lamina propria plasma cells diffuses towards the basolateral membrane of the enterocyte, where it is taken up by the secretory component (SC) receptor. A small amount is believed to enter the blood stream. Secretory component is a glycoprotein produced within the epithelial cells.[74-81] It is believed to move from its site of production in the rough endoplasmic reticulum to occupy sites along the basolateral membrane of the enterocyte up to the tight junction of the cell.[82] In this position SC serves as receptor for the two polymeric immunoglobulins, namely dimeric IgA and polymeric (usually pentameric) IgM. Secretory component is synthesized and expressed by glandular epithelial cells throughout the entire small intestine and colon. Secretory immunoglobulin A is, therefore, the product of two different kinds of cells: the plasma cells which produce the dimeric immunoglobulin, and the epithelial cells which produce the secretory component.[83] It is because of the diverse origin of SC and IgA, that SC in external fluids can be found in its free form or complexed with polymeric immunoglobulin.[84,85] Secretory component combines with the Fc fragment of immunoglobulin. The specificity of the SC receptor for polymeric immunoglobulins has been shown by electronmicroscopic studies. It will combine only with dimeric IgA or polymeric IgM. On the other hand, it will not combine with monomeric IgA nor with dimeric IgA which has already been complexed with SC.[86] After the immunoglobulin has been bound to its receptor, it is endocytosed and transported across the cell in vesicles. Ultimately, it is exocytosed at the apical surface of the cell.[86] The electron microscope has enabled us to follow the process of transcytosis. Within 5 min of the formation of the dimeric IgA-SC complex, endocytic vesicles containing the complex can be seen in the cytoplasm. *In vitro*, it takes about 30 min for the vesicles to transverse the cell, until ultimately they are exocytosed at the apical surface.

Chemical investigations have lent support to the concept of the specificity of SC for the polymeric immunoglobulins of IgA and IgM only.[87-92]

Although SC is justifiably regarded as a receptor, it has certain characteristics which distinguish it from other receptors. Thus, SC is not degraded within the cell nor is it recycled; it remains bound to its ligand and is secreted with it. Before secretion, SC is, however, proteolytically cleaved into a soluble protein. It is the soluble form of SC which appears in the intestinal fluid. Before this process of solubilization takes place, SC is referred to as membrane SC. In rabbits, it has been shown that secretory component varies in structure among different individuals — various allotypes may occur.[93] The structure of membrane SC in the rabbit has been deduced by means of cloning and sequencing cDNA.[94,95] Thus it has been shown that SC is not synthesized as such, but rather as a much larger transmembrane protein precursor. The complete primary structure of this polymer immunoglobulin receptor has now been established. In the rabbit it consists of 773 amino acids. Starting at the amino terminus the structure consists of an extracellular signal peptide of 18 amino acids. Following this comes the extracellular polymeric immunoglobulin receptor portion, the SC proper, containing 629 amino acids. It is the amino terminal first domain of SC containing residues 1 to 126 which is essential for dimeric IgA binding.[96] The membrane spanning segment is believed to possess 23 uncharged, predominantly hydrophobic, residues, while an intracellular tail has 103 amino acids. When the sequence of SC was compared with that of other known proteins, a significant degree of homology was observed between the above domain and the variable region of immunoglobulins. It may be, therefore, that the receptor is structurally related to a portion of its ligand. The production of SC occurs independently of that of the immunoglobulins. The factors controlling its production are not known completely, but there is evidence that hormonal control plays a part in regulating its synthesis. In the rat, for instance, it has been shown that uterine SC is present during proestrus, but not at diestrus.[97,98] The hormone responsible is estradiol, as in oophorectomized animals this hormone increased the rate of secretion of soluble SC into uterine fluids.[97,99] In the human female, too, SC concentrations in uterine secretions appear to be under hormonal control; they are highest during the secretory phase, reduced during the proliferative phase and lowest during menstruation.[100] In rat hepatocytes, on the other hand, it has been shown that the hormones which stimulate the production of SC are the glucocorticoids.[101] In the process of transcytosis, it is believed that the cytoplasmic tail of SC is instrumental in directing sIgA along the correct intracellular path.[102] The transport of the endocytic vesicle is determined by the microtubules of the cell.[86] This assumption stems from the fact that the addition of colchicine to the tissue culture causes the transport of the vesicles to be impeded. Within the cell the endocytic vesicles stay clear of lysozomes and thus escape degradation. This is further proof of the well-defined path taken by the endocytic vesicles.

## V. THE EFFECT OF NERVOUS IMPULSES AND OF CHOLECYSTOKININ ON THE RELEASE OF IMMUNOGLOBULINS IN THE INTESTINE

### A. THE EFFECT OF CHOLINERGIC STIMULATION AND INHIBITION

In a study of IgA secretion in perfused intestinal loops of anesthetized rats, it was demonstrated that the release of IgA is under autonomic control. This applies even to the baseline secretion of IgA.[103,104] The administration of the muscarinic blocking agent, atropine, resulted in a diminution of IgA secretion to 50% of the control value. On the other hand, the injection of the cholinergic agents pilocarpine, muscarine, and bethanechol resulted in a four- to eightfold increase in levels of IgA. It is of interest that the IgA obtained after muscarinic stimulation was of a heterogeneous molecular size distribution. A substantial quantity of the IgA was smaller than 11S secretory IgA. The significance of this finding may be that the classical, SC-mediated transport of IgA is not the only route taken by IgA after stimulation by cholinergic agents. In this context it may be relevant that cholinergic agents also stimulate the release of IgG.[104] The exact path taken by IgG in release from the intestinal lumen has not yet been defined, but the possibility must be considered that in the circumstances described here, IgG and some of the IgA leave the enterocyte by a common route.

As cholinergic impulses also mediate increased loss of chloride and diminish the absorption of sodium,[105] one is tempted to conclude that immunoglobulins and ion transport may be subject to the same impulses in the intestine. So far, there is not sufficient evidence to support such a contention, for it has been shown that the calcium channel blocker, verapamil, will inhibit the secretagogue action of pilocarpine,[106] but the release of IgA proceeds unhindered. The changes in cytosolic calcium after the administration of the cholinergic agent, carbechol, have also been demonstrated directly by measuring changes in the movement of $^{45}Ca$.[107] The possible mechanisms involved in secretagogue-induced immunoglobulin transport will be discussed more fully in the following section.

### B. THE EFFECT OF CHOLECYSTOKININ AND SECRETIN ON SPECIFIC ANTIBODY ACTIVITY IN HUMAN DUODENAL FLUID

The intravenous injection of cholecystokinin results in the human in a rise of specific antibody activity of immunoglobulins A, M, and E.[35,108] This activity reached a peak after 5 to 10 min, and thereafter returned to baseline values by 30 min. The activity of IgG was only slightly affected. There was a parallel rise of protein and enzymatic activity, suggesting that part, at least, of this antibody activity was derived from the pancreas.

Secretin, in spite of its dilutional effect on duodenal enzyme concentrations, also produced a rise in IgA and IgM antibody activity in duodenal fluid.

## C. THE EFFECT OF CHOLECYSTOKININ ON ANTIBODY ACTIVITY IN THE EXPERIMENTAL ANIMAL

### 1. The Known Physiological Effects of Cholecystokinin

Cholecystokinin exists in several molecular forms. Originally, the 33 amino acid peptide (CCK-33) was isolated from hog stomach. Later, larger peptides containing 39 and 58 amino acids were isolated.[109,110] The principal smaller form is the octapeptide (CCK-8) which is the form used in the experiments to be described later. The octapeptide is at least as potent as CCK-33.[111] The terminal pentapeptide of CCK is identical with that of gastrin.[112] The gene coding for CCK contains 345 cDNA bases and the CCK prepropeptide presumably contains 115 amino acids.[113,114]

The biological actions of CCK are listed in Table 1. The intracellular messengers activated in stimulus-secretion coupling of pancreatic acinar cells are changes in cytosolic calcium and in turnover of phosphoinositol.[115] In the rat intestine, intracellular calcium levels also appear to be altered by CCK. In experiments performed in our laboratory, CCK was found to provoke the release of antibodies into the gastrointestinal tract. The calcium channel blocker, verapamil, prevented this effect.[116] This would suggest, therefore, that changes in cytosolic calcium are involved in CCK-induced stimulus-secretion coupling in the rat small intestine.

In the pancreatic acinar cell, the action of CCK also opens a $K^+$ conductance pathway.[117] Receptors for CCK have been found on pancreatic acinar cells,[118-121] but have not so far been described on enterocytes. Of the biological actions of CCK, those on the intestine are particularly pertinent to this chapter. Cholecystokinin increases intestinal motility, presumably by an action on the smooth muscle, it promotes secretion of $Na^+$, $Cl^-$,[122,123] as well as of bicarbonate.[124] In addition, CCK increases intestinal lymph flow and augments antibody activity in the intestinal lumen. As already mentioned, cells containing CCK can be found along the mammalian duodenum and jejunum.[125,126] Cholecystokinin is released into the bloodstream during the process of digestion. Specifically, micelles containing fatty acids of nine or more carbon chain length, and amino acids are stimulants for CCK secretion.[127] Stimulants in the upper half of the intestine are more effective than those that are more distal.[128] In addition to the stimulus of food, one of the intestinal hormones, bombesin, has been found to be instrumental in releasing CCK.[129]

### 2. Antibody Release in the Intestine after CCK-OP

The experiments to be described in this section were invariably carried out on rats belonging to the Hooded-Lister strain weighing 160 to 180 g. The rats were immunized with ovalbumin. The experiments were performed under thiopentone anesthesia. A loop of small intestine 10cm distal to the pylorus and 10 cm long was severed at both ends and isolated *in situ*. The loop, therefore, did not receive any biliary or pancreatic secretions. After the loop was flushed for 30 min, a collection of specimens was begun. The loop was perfused with 0.9% saline at the rate of 0.2 ml/min. Aliquots of 2.5 min were collected — four for baseline determinations and the others after the intravenous injection of CCK-8. (Sincalide, Sqibb, Princeton, NJ). Antibody activity was measured by means of the ELISA technique.

Using this model we were able to show that the intravenous injection of CCK-OP provoked a significant ($p < 0.01$) rise of IgA antibody activity within 2.5 min of injection (Figure 1). By 12.5 min, the IgA antibodies were still significantly ($p < 0.05$) higher than in the controls. A similar phenomenon could be observed when we investigated the effect of CCK-8 on the release of IgG specific antibodies in the perfused intestine: there was a significant ($p < 0.01$) and maximal rise of IgG antibodies by 2.5 min and the response remained significant for 22.5 min (Figure 2). Antibodies of the IgM and IgE isotypes also showed significant but smaller rises.[116]

**TABLE 1**
**Biological Effects of Cholecystokinin**

| |
|---|
| **Pancreas** |
| Secretion of pancreatic enzymes, fluid and bicarbonate |
| Hypertrophy and hyperplasia of pancreas |
| **Esophagus** |
| Contraction of lower esophageal sphincter by direct action on sphincter muscle |
| Relaxation of lower esophageal sphincter by stimulation of inhibitory neurons |
| **Gall bladder** |
| Contraction of gall-bladder. Relaxation of Sphincter of Oddi |
| **Stomach** |
| Stimulation of gastric acidity. Reduction of gastric emptying rate |
| Contraction of antral smooth muscle |
| **Intestine** |
| Increase of intestinal motility |
| Increased secretion of $Na^+$ and $Cl^+$ |
| Increased intestinal bicarbonate secretion |
| Increased intestinal lymph flow |
| Increased antibody activity of G, A, M, and E isotypes |
| **Hormone release** |
| Increased insulin release |
| Raised somatostatin output |
| Augmentation of pancreatic polypeptide release |
| Increased gastric inhibitory peptide release |
| Release of calcitonin |
| **Appetite** |
| Decreased when given centrally |

### 3. Antibody Release after Administration of the CCK Antagonist Proglumide

The effect of the CCK antagonist, proglumide, (Milid. Rotta Research Laboratories, Milan, Italy) on jejunal antibodies is shown in Figures 3 and 4. There is a significant ($p < 0.01$) reduction of baseline IgA secretion starting 5 min after the injection of proglumide and lasting for 5 min. Similarly, a significant ($p < 0.05$) reduction in the release of IgG was noted after the injection of proglumide.

### 4. Antibody Release after the Oral Administration of a Protein Hydrolysate

The infant protein hydrolysate feeding formula Pregestimil (Mead Johnson, Evansville, IN) was instilled into the stomach in order to evaluate the possibility that food also can instigate a rise in specific antibody activity. It was found that the instillation of the protein hydrolysate formula, produced a significant rise of IgA and IgG antibody activity. In the case of IgA, this was five times higher, but rather more short-lived, than in the case of IgG. (Figures 5 and 6).

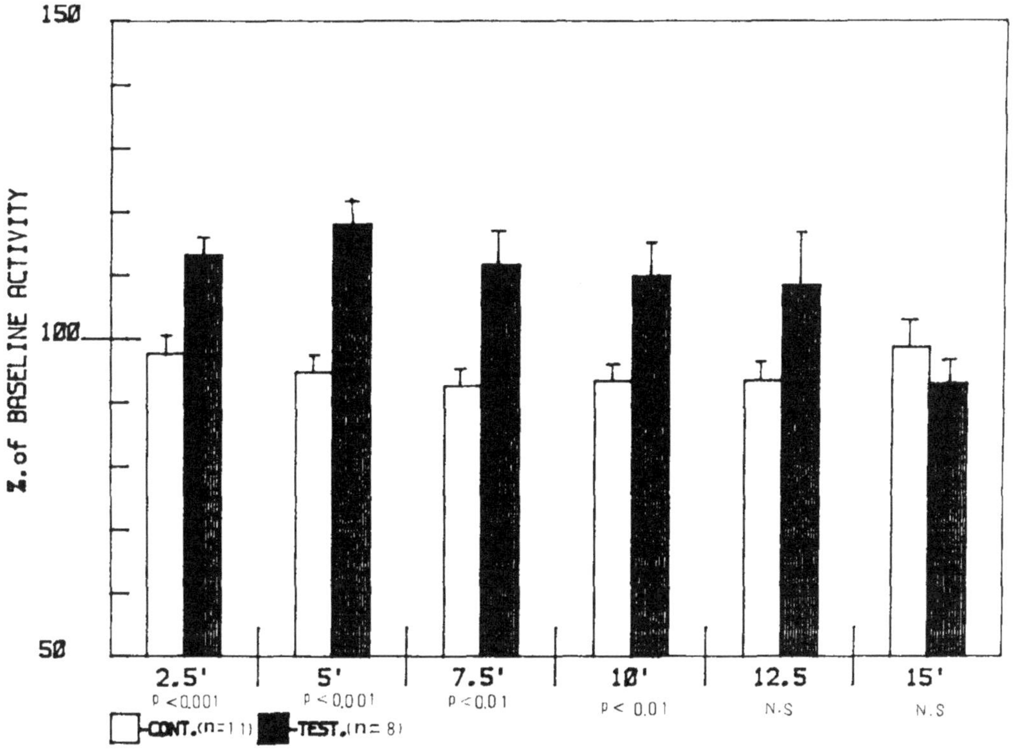


FIGURE 1. The injection of CCK-OP produces a rise of IgA antibody activity in the perfusion fluid which remains significant for 12.5 min. Values represent mean ± SEM.

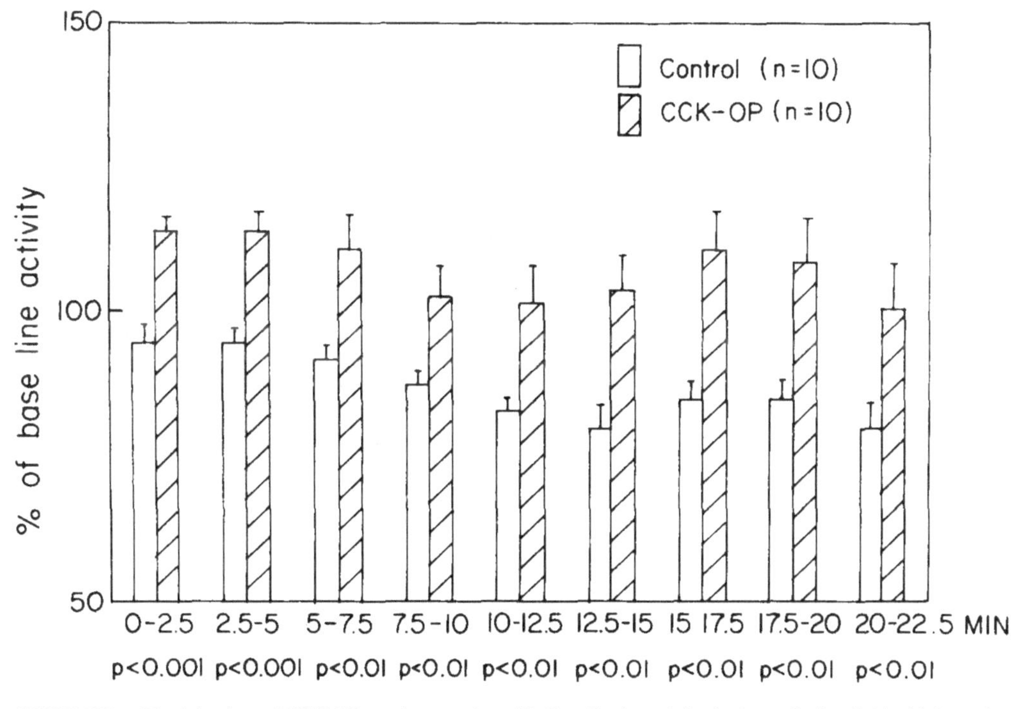


FIGURE 2. The injection of CCK-OP produces a rise of IgG antibody activity in the perfusion fluid which persists for at least 22.5 min. Values represent mean ± SEM. (Reprinted with permission from Freier, S., et al., Effect of cholecystokinin and of its antagonist, of atropine and of food on the release of immunoglobulin A and immunoglobulin G specific antibodies in the rat intestine, *Gastroenterology*, 93, (12), 1242, 1987, American Gastroenterological Association.)

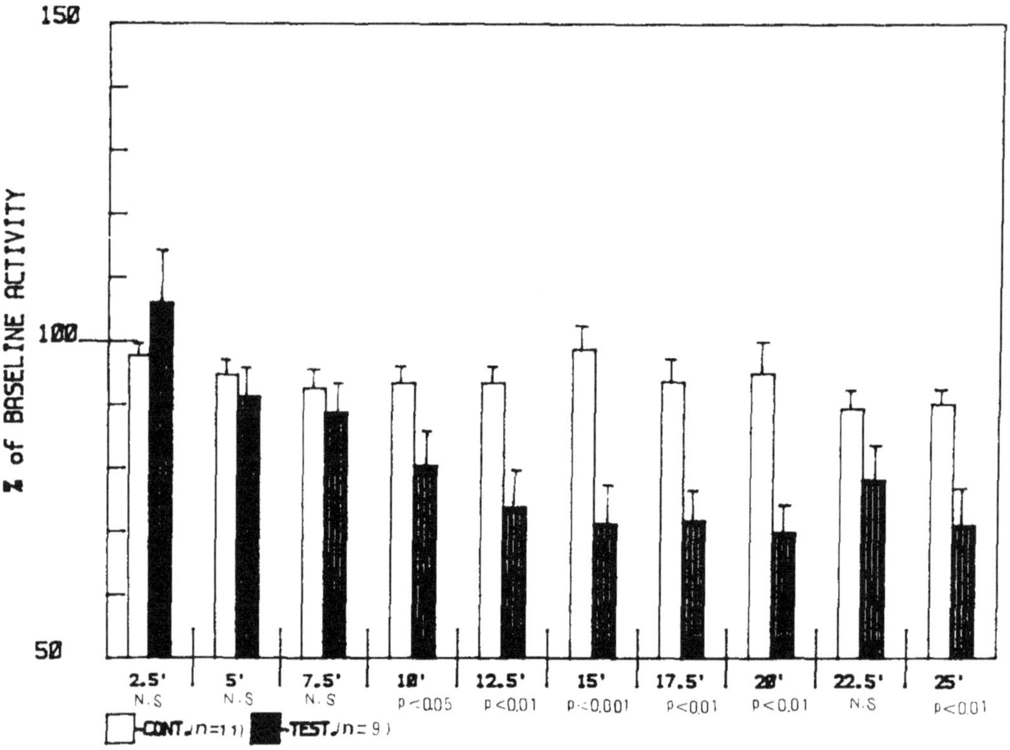


FIGURE 3. The CCK antagonist proglumide produces a drop in basal IgA antibody activity lasting at least 20 min. Values represent mean ± SEM.

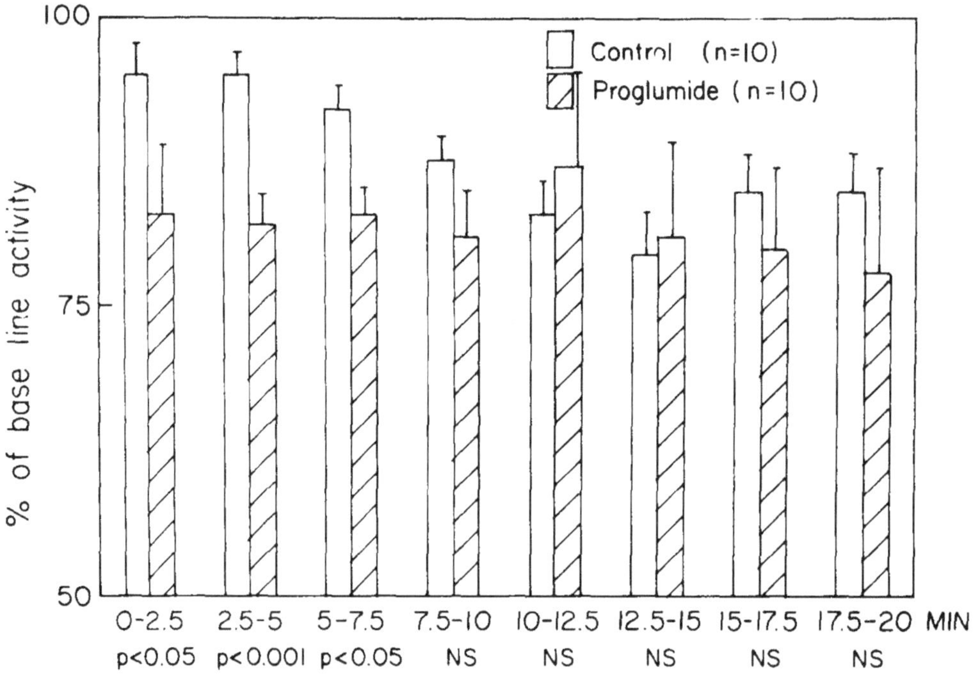


FIGURE 4. The CCK antagonist proglumide produces a drop in the basal IgG antibody activity which remains significant for 7.5 min, but lasts for 20 min. Values represent mean ± SEM. (Reprinted with permission as in Figure 2.)

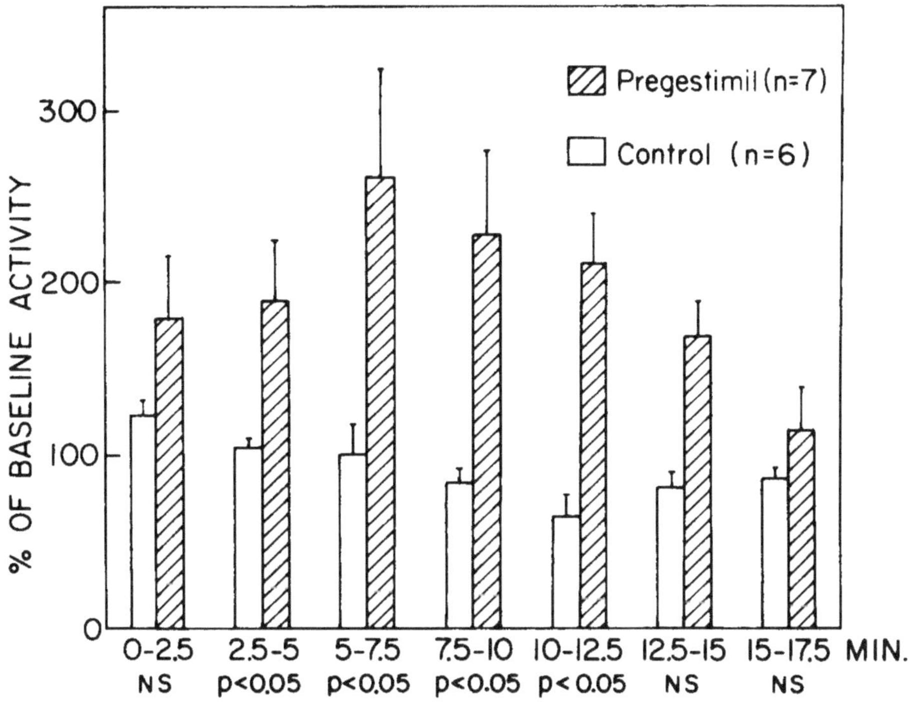


FIGURE 5. Effect of sham-feeding with Pregestimil on IgA antibody activity in perfusion fluid. Values represent mean ± SEM. (Reprinted with permission as in Figure 2).

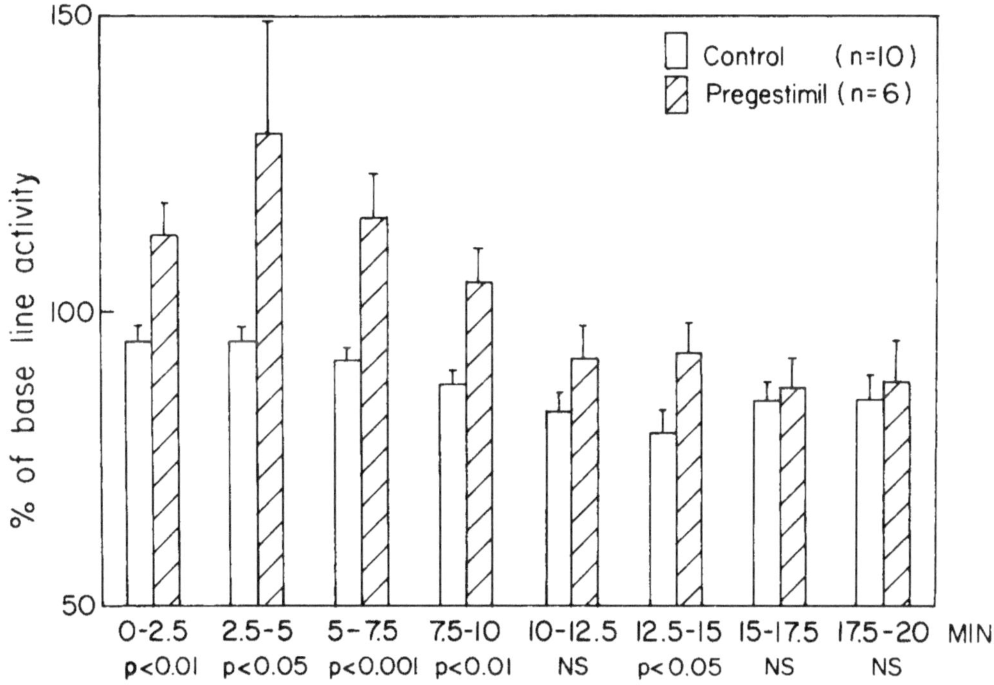


FIGURE 6. Effect of sham-feeding with Pregestimil on IgG antibody activity in perfusion fluid. Values represent mean ± SEM. (Reprinted with permission as in Figure 2).

Subsequent experiments showed that the effect of food was probably mediated by CCK, for the prior administration of the CCK antagonist, proglumide, inhibited the increase in antibody activity after the injection of CCK. We conclude, therefore, that CCK released by the ingestion of food is responsible for the rise in specific antibodies.

**5. Intracellular Messengers Activated by CCK and Cholinergic Agents**

The effect of CCK was observed in animals that were receiving an intravenous infusion of the calcium channel blocker, verapamil, before and during the injection of CCK. Verapamil completely cancelled the effect of CCK-induced release of IgA and IgG specific antibody activity, suggesting that a change in cytosolic calcium is an intracellular messenger involved in the stimulus secretion coupling of the antibody release. Possibly, it is not the only messenger involved, as the injection of verapamil did not reduce the baseline secretion of immunoglobulin release, although we had previously shown that CCK is also partly responsible for baseline release of IgA and IgG antibody activity. As mentioned above the injection of the cholinergic agent pilocarpine, also produces a rise of the IgA and IgG antibody specific activity. This effect was not prevented by the calcium channel blocker, suggesting that either a verapamil-insensitive calcium channel or other intracellular messengers are involved.

**6. Possible Modes of Action of CCK**

It is not certain if CCK has a direct mitogenic effect on lymphocytes. Recent *in vitro* experiments suggest that CCK has no stimulatory effects on rat spleen lymphocytes and, therefore, differs in this respect from substance P.[130] There is therefore no evidence to suggest that CCK acts at the level of lamina propria plasma cells. Furthermore, the time interval between the injection of CCK and the change in intestinal antibody activity is so rapid, that de novo synthesis and subsequent transenterocyte transport within the space of 2.5 min seems most unlikely.

A second possibility is a purely mechanical action. It has already been pointed out that CCK increases the motility of the intestine. It is conceivable, therefore, that antibodies that have already been released into the crypts could be expressed from them as a result of intestinal contraction induced by CCK. In support of this consideration, it must be stated that crypts stain more heavily for secretory component than do villi. The crypt, therefore, appears to be the main site of immunoglobulin release and theoretically could be a good storage depot of immunoglobulins until required.

An alternative mechanism that could be invoked would involve an accelerated transcytosis along the secretory pathway described above. This is an attractive hypothesis, but the observations of Wilson et al.[103] cast some doubt as to its applicability. These authors found that after stimulation with cholinergic agents, the IgA which was released into the lumen was of heterogeneous molecular size. Not all of it is, therefore, secretory IgA, and thus probably not transported via the classical endocytic vesicles.

In the light of the information which is at our disposal at the moment, a specific secretory mechanism is the most likely cause of the rise of antibody activity. This would fit with the observed phenomenon of increased release of IgG and IgE antibody activity. These two immunoglobulins are believed to pass into the lumen by diffusion and not by a definitive pathway such as that described for IgA and IgM. Furthermore, CCK is a secretagogue in the intestinal tract promoting the increased release of $Na^+$, $Cl^+$, and water.[122] We suggest, therefore, that initially, CCK impinges on a specific receptor at the basolateral membrane of the enterocyte. This, in turn, activates one or more intracellular messengers, one of which is a change in cytosolic calcium. The result is increased secretory activity of the cell, including the discharge into the lumen of electrolytes, water, and all of the four immunoglobulins which we have investigated. In order to find further support for this theory, it would be important to establish a number of additional factors. (1) It would be of interest to know if CCK acting on isolated enterocytes in tissue culture can also evoke a response of increased antibody activity. (2) The finding of specific receptors for CCK on enterocytes would fit in with our hypothesis. Staining of enterocytes by immunofluorescent or electronmicroscopic methods after exposure to CCK for the disposition of the immunoglobulins would also add to our understanding of the mechanisms involved in immunoglobulin release.

Another area for future research would be an inquiry into the effects of other neuropeptides in antibody release in the intestine. Preliminary studies in our laboratory have shown that substance P can also bring about the release of IgA and IgG specific antibodies. Indeed, if our concept is correct, one would expect that other secretory neuropeptides such as VIP would also have an effect on antibody release. On the other hand, neuropeptides which are antisecretory, such as somatostatin, endorphins and enkephalins may also inhibit the release of antibodies. Be this as it may, the neuroendocrine-immune network in the intestine proves to be a fruitful and exciting area for continued research.

## REFERENCES

1. **Trendelenburg, P.,** Physiologische und Pharmakologische Versuche uber die Dunndarm Peristaltick, *Naunyn-Schmiedebergs Arch. Exp. Pathol. Pharmakol.*, 81, 55, 1971.
2. **Gershon, M. D. and Erde, S. M.,** The nervous system of the gut, *Gastroenterology*, 80, 1571, 1981.
3. **Furness, J. B. and Costa, M.,** Types of nerves in the enteric nervous system, *Neuroscience*, 5, 1, 1980.
4. **Schofield, G. C.,** Anatomy of muscular and nervous tissues in the alimentary canal, in *Handbook of Physiology: Section 6, The Alimentary Canal,* Vol. 4, Code, C. F. and Heidel, W., Eds., Williams & Wilkins, Baltimore, 1968, 1579.
5. **Dale, H. H.,** Acetylcholine as a transmitter of the effects of nerve impulses. I. History of ideas and evidence of peripheral autonomic actions. Functional nomenclature of nerve fibres, *J. Mt. Sinai Hosp. N.Y.*, 4, 401, 1937.
6. **Dale, H. H.,** Acetylcholine as a transmitter of the effects of nerve impulses. II. Chemical transmission at ganglionic synapses and voluntary motor nerve endings. Some general considerations, *J. Mt. Sinai Hosp. N.Y.*, 4, 416, 1937.
7. **Finkelman, B.,** On the nature of inhibition in the intestine, *J. Physiol.*, 70, 145, 1930.
8. **von Euler, U. S.,** The nature of adrenergic nerve mediators, *Pharmacol. Rev.*, 3, 247, 1951.
9. **Gillespie, J. R. and McKenna, B. R.,** The inhibitor action of the sympathetic nerves on smooth muscle of the rabbit gut; its reversal by reserpine and restoration by catecholamines and by dopa, *J. Physiol.*, 156, 17, 1961.
10. **Feldberg, W. and Lin, R. C. Y.,** The effect of cocaine on the acetylcholine output of the intestinal wall, *J. Physiol. (London)*, 109, 475, 1949.
11. **Chujyo, N.,** Site of acetylcholine production in the wall of the intestine, *Am. J. Physiol.*, 174, 196, 1953.
12. **Schauman, W.,** Inhibition by morphine on the release of acetylcholine from the intestine of the guinea-pig, *Br. J. Pharmacol.*, 12, 115, 1957.
13. **Paton, W. D. M.,** The action of morphine and related substances on contraction and on acetylcholine output of coaxially stimulated guinea-pig ileum, *Br. J. Pharmacol. Chemother.*, 12, 119, 1957.
14. **Ahlman, H., Lundberg, J., Dahlstrom, A., and Kewenter, J.,** A possible vagal adrenergic release of serotonin from enterochromaffin cells in the gut, *Acta Physiol. Scand.*, 98, 366, 1976.
15. **Furness, J. B. and Costa, M.,** The adrenergic innervation of the gastrointestinal tract, *Ergeb. Physiol., Biol. Chem. Exp. Pharmakol.*, 69, 1, 1979.
16. **Cowie, L. A., Kosterlitz, H. W., and Waterfield, A. A.,** Factors influencing the release of acetylcholine from the myenteric plexus of the ileum of the guinea-pig and rabbit, *J. Pharmacol.*, 64, 565, 1978.
17. **Costa, M., Furness, J. B., and Gabella, G.,** Catecholamine containing nerve cells in the mammalian myenteric plexus, *Histochemie*, 25, 103, 1971.
18. **Juerio, A. V. and Gabella, G.,** Nor-adrenaline in the guinea-pig alimentary canal: regional distribution and sensitivity to denervation and reserpine, *J. Neurochem.*, 221, 851, 1974.
19. **Feldberg, W. and Toh, C. C.,** Distribution of 5-hydroxytrytamine (serotonin, enetramine) in the wall of the digestive tract, *J. Physiol. (London)*, 119, 352, 1953.

20. **Robinson R. and Gershon, M. D.,** Synthesis and uptake of 5-hydroxytryptamine by the myenteric plexus of the small intestine of the guinea-pig, *J. Pharmacol. Exp. Ther.*, 179, 29, 1971.
21. **Bulbring, E. and Gershon, M. D.,** 5-hydroxytryptamine participation in the vagal inhibitory innervation of the stomach, *J. Physiol. (London)*, 192, 823, 1967.
22. **Gershon, M. D.,** Effects of tetrodotoxin on innervated smooth muscle preparations, *Br. J. Pharmacol.*, 29, 259, 1967.
23. **Furness, J. B. and Costa, M.,** The nervous release and the action of substances which affect intestinal muscle through neither adrenoreceptors nor cholinoreceptors, *Philos. Trans. R. Soc. London Ser. B*, 265, 123, 1973.
24. **Drakontides, A. B. and Gershon, M. D.,** 5-HT receptors in the mouse duodenum, *Br. J. Pharmacol.*, 33, 480, 1968.
25. **Costa, M. and Furness, J. B.,** The sites of action of 5-HT in nerve muscle preparations from guinea-pig small intestine and colon, *Br. J. Pharmacol.*, 65, 237, 1979.
26. **Costa, M. and Furness, J. B.,** The origins, pathways and terminations of neurons with VIP-like immunoreactivity in the guinea-pig small intestine, *Neuroscience*, 76, 665, 1983.
27. **Makhlouf, G. M.,** Enteric neuropeptides: role in neuromuscular activity in the gut, *Trends Pharmacol. Sci.*, 6, 214, 1985.
28. **Costa, M., Cuello, A. C., Furness, J. B., and France, R.,** Distribution of enteric neurons showing immunoreactivity for substance P in the guinea-pig ileum, *Neuroscience*, 5, 323, 1980.
29. **Costa, M. and Furness, J. B.,** Neuronal peptides in the intestine, *Br. Med. Bull.*, 38, 247, 1982.
30. **Schultzberg, M., Hockfeld, T., Nilsson, G., Terenius, L., Rehfeld, J. F., Brown, M., Elde, R., Goldstein, M., and Said, S.,** Distribution of peptide- and catecholamine-containing neurons in the gastrointestinal tract of rat and guinea-pig: immunohistochemical studies with antisera to substance P, vasoactive intestinal peptide, enkephalins, somatostatin, *Neuroscience*, 5, 689, 1980.
31. **Keast, J. R., Furness, J. B., and Costa, M.,** Somatostatin in human enteric nerves. Distribution and characterization, *Cell Tissue Res.*, 237, 299, 1984.
32. **Furness, J. B., Costa, M., Emson, P. C., Hakanson, R., Moghimsadeh, E., Sundler, F., Taylor, I. L., and Chance, R. E.,** Distribution, pathways and reactions to drug treatment of nerves with neuropeptide Y and pancreatic polypeptide-like immunoreactivity in the guinea-pig, *Cell Tissue Res.*, 234, 71, 1983.
33. **Feher, E. and Burstock, G.,** Electronmicroscopic study of neuropeptide Y-containing nerve elements of the guinea-pig small intestine, *Gastroenterology*, 91, 965, 1986.
34. **Rokaens, A., Melander, T., Hockfeld, T., Lundberg, J. M., Tatemoto, J., Carquist, M., and Mutt, J.,** A galanine-like peptide in the central nervous system and intestine of the rat, *Neurosci. Lett.*, 47, 161, 1984.
35. **Shah, P. C., Freier, S., Park, B. H., Lee, P. C., and Lebenthal, E.,** IgE and IgD antibodies to cow milk and soy protein in duodenal fluid: effects of pancreozymin and secretin, *Immunology*, 83, 916, 1982.
36. **Furness, J. B., Costa, M., and Miller, R. J.,** Distribution and projections of nerves with enkephalin-like immunoreactivity in the guinea-pig small intestine, *Neuroscience*, 8, 653, 1983.
37. **Furness, J. B. and Costa, M.,** Types of nerves in the enteric nervous system, *Neuroscience*, 5, 1, 1980.
38. **Bussolati, G., Capella, C., Solcia, E., Vassalo, G., and Vazzadini, P.,** Ultrastructural and immunofluorescent investigations of the secretin cell in the dog intestinal mucosa, *Histochemie*, 26, 218, 1971.
39. **Capella, C., Solcia, E., and Vassalo, G.,** Identification of types of endocrine cells in the gastrointestinal mucosa of the rabbit, *Arch. Histol. Jpn.*, 30, 479, 1969.
40. **Forssman, W. G., Helmstedter, V., and Chance, R. E.,** Ultrastructural and immunohistochemical demonstration of pancreatic polypeptide-containing F cells in the stomach and pancreas of Toipia Belangeri, *Cell Tissue Res.*, 40, 692, 1977.
41. **Polak, J. M., Bloom, S., Coulling, I., and Pearse, A. G. E.,** Immunofluorescent localisation of enteroglucagon cells of the gastrointestinal tract of the dog, *Gut*, 12, 311, 1971.
42. **Solcia, E., Capella, C., Vassalo, G., and Buffa, R.,** Endocrine cells of the gastric mucosa, *Int. Rev. Cytol.*, 42, 223, 1975.
43. **Vassalo, G., Solcia, E., and Capella, C.,** Light and electron-microscopic identification of several types of endocrine cells in the gastrointestinal mucosa of the cat, *Z. Zellforsch. Mikrosk. Anat.*, 98, 933, 1969.
44. **Feyrter, F.,** *Über die Peripheren Endokrinen (Parakrinen) Drusen des Menschen*, W. Mandrich, Vienna, Austria, 1953, 1.
45. **Pearse, A. G. E. and Polak, J. M.,** The diffuse endocrine system and the APUD concept, in *Gut Hormones*, Bloom, S. R., Ed., Churchill Livingstone, Edinburgh, 1987, 33.
46. **Dubois, P. M. and Paulin, C.,** Gastrointestinal somatostatin cells in the human fetus, *Cell Tissue Res.*, 166, 179, 1976.
47. **Polak, J. M., Pearse, A. G. E., Grimelius, L., Bloom, S. R., and Avimura, A.,** Growth-hormone release-inhibiting hormone in gastrointestinal and pancreatic D cells, *Lancet*, 1, 1220, 1975.
48. **Pearse, A. G. E., Polak, J. M., and Bloom, S. R.,** The newer gut hormones. Cellular sources, physiology, pathology and clinical aspects, *Gastroenterology*, 72, 746, 1977.

49. **Solcia, E., Capella, C., Buffa, R., Fiocca, R., and Sessa, F.,** Endocrine cells of the digestive system, in *Physiology of the Gastrointestinal Tract,* Johnson, L. R., Ed., Raven Press, New York, 1987, 111.
50. **Kobayashi, S. and Sasagawa, T.,** Morphological aspects of the secretion of gastro-enteric hormones, in *Endocrine Gut and Pancreas,* Fujida, T., Ed., Elsevier, Amsterdam, 1976, 255.
51. **O'Dorisio, M. S., O'Dorisio, T. M., Cataland, S., and Balcersak, S. P.,** VIP as a biochemical marker for polymorphonuclear leucocytes, *J. Lab. Clin. Med.,* 96, 666, 1980.
52. **Blalock, J. E., Harbour-McMenamine, D., and Smith, E. M.,** Peptide hormones shared by the neuroendocrine and immunologic systems, *J. Immunol.,* 135, 858S, 1985.
53. **Lolait, S. J., Lim, A. T., Toh, B. H., and Funder, J. W.,** Immunoreactive β-endorphin in a subpopulation of mouse spleen macrophages, *J. Clin. Invest.,* 73, 277, 1984.
54. **Smith, E. M. and Blalock, J. E.,** Human lymphocyte production of corticotropin and endorphin-like substances: association with leukocyte interferon, *Proc. Natl. Acad. Sci. U.S.A.,* 78, 7530, 1981.
55. **Ruff, M. R. and Pert, C. B.,** Small cell carcinoma of lung: macrophage specific antigens suggest hemopoietic stem cell origin, *Science,* 225, 1034, 1984.
56. **Goetzl, E. J., Chernov-Rogan, T., Cooke, M. P., Renold, F., and Payan, D. G.,** Endogenous somatostatin-like peptides of rat basophilic leukemia cells, *J. Immunol.,* 135, 2707, 1985.
57. **Smith, E. M., Meyer, W. J., and Blalock, E. J.,** Virus-induced increases in corticosterone in hypophysectomized mice: a possible lymphoid adrenal axis, *Science,* 218, 1311, 1982.
58. **Goetzl, E. J., Chernov, T., Renold, F., and Payan, D. G.,** Neuropeptide regulation of the expression of immediate hypersensitivity, *J. Immunol.,* 135, 802S, 1985.
59. **Bland, P. W. and Warren, L. G.,** Antigen presentation by epithelial cells of the rat small intestine. Kinetics, antigens, specificity and blocking by anti-Ia antisera, *Immunology,* 58, 1, 1986.
60. **Perkkio, M. and Savilahti, E.,** Time of appearance of immunoglobulin containing cells of the mucosa of the neonatal intestine, *Pediatr. Res.,* 14, 953, 1980.
61. **Brown, W. R., Borthistle, B. K., and Chen, S. T.,** Immunoglobulin (IgE) containing cells in human gastrointestinal fluids and tissues, *Clin. Exp. Immunol.,* 20, 227, 1975.
62. **Tada, T. and Ishizaka, K.,** Distribution of τE forming cells in lymphoid tissues of the human and monkey, *J. Immunol.,* 27, 631, 1970.
63. **Morison, T. J. and Koshland, M. E.,** Characterization of the J chain from polymeric immunoglobulins, *Proc. Natl. Acad. Sci. U.S.A.,* 69, 124, 1972.
64. **Parkhouse, R. M. E.,** Biosynthesis of J chain in mouse IgA and IgM, *Nature (London),* 236, 9, 1972.
65. **Koshland, M. E.,** Structure and function of the J chain, *Adv. Immunol.,* 20, 41, 1975.
66. **Inman, F. P. and Mestecky, J.,** The J chain of polymeric immunoglobulins, *Contemp. Top. Mol. Immunol.,* 3, 111, 1974.
67. **Brandtzaeg, P., Fjellanger, I., and Gjeruldsen, S. T.,** Immunoglobulin M: local synthesis and selective secretion in patients with immunoglobulin A deficiency, *Science,* 160, 789, 1968.
68. **Brandtzaeg, P., Fjellanger, I., and Gjeruldsen, S. T.,** Human secretory immunoglobulins. I. Salivary secretions from individuals with normal or low levels of serum immunoglobulins, *Scand. J. Haematol. Suppl.,* 12, 1, 1970.
69. **Brandtzaeg, P.,** Human secretory immunoglobulins. II. Salivary secretions from individuals with selectively excessive or defective synthesis of serum immunoglobulins, *Clin. Exp. Immunol.,* 8, 69, 1971.
70. **Brandtzaeg, P.,** Immunohistochemical characterization of intracellular J chain and binding site for secretory component (SC) in the human immunoglobulin (Ig)-producing cells, *Mol. Immunol.,* 20, 941, 1983.
71. **Brandtzaeg, P.,** Presence of J chain in human immunocytes containing various immunoglobulin classes, *Nature (London),* 252, 418, 1974.
72. **Kosrud, F. R. and Brandtzaeg, P.,** Quantitative immunohistochemistry of immunoglobulin and J chain producing cells in human parotid and submandibular glands, *Immunology,* 39, 129, 1980.
73. **Parkhouse, R. M. E. and Della Corte, E.,** Biosynthesis of immunoglobulin A (IgA) and immunoglobulin M (IgM), *J. Biochem.,* 136, 607, 1973.
74. **Brandtzaeg, P.,** Structure, synthesis and external transfer of mucosal immunoglobulins, *Ann. Immunol. (Paris),* 124C, 417, 1973.
75. **Poger, M. E. and Lamm, M. E.,** Localisation of free and bound secretory component in human intestinal epithelial cells: a model for the assembly of secretory IgA, *J. Exp. Med.,* 139, 629, 1974.
76. **Brandtzaeg, P.,** Mucosal and glandular distribution of immunoglobulin components. Immunohistochemistry with a cold ethanol fixation technique, *Immunology,* 26, 1101, 1974.
77. **Tourville, D. R., Adler, R. H., Bienenstock, J., and Tomasi, T. B.,** The human secretory immunoglobulin system: immunohistological localisation of τA, secretory piece and lactoferrin in normal human tissues, *J. Exp. Med.,* 129, 411, 1969.
78. **Allen, W. D. and Porter, P.,** Localisation by immunofluorescence of secretory component and IgA in the intestinal mucosa of the young pig, *Immunology,* 24, 365, 1973.
79. **Comoglio, P. M. and Guglielmona, R.,** Immunohistochemical study of IgA transepithelial transfer into digestive tract secretions in the mouse, *Immunology,* 25, 71, 1973.

80. **Brandtzaeg, P.,** Mucosal and glandular distribution of immunoglobulin components: differential localisation of free and bound SC in secretory epithelial cells, *J. Immunol.*, 112, 1553, 1974.
81. **Brown, W. R. and Burns, F. A.,** Application of peroxidase-labelled antibody methods to identification of immunoglobulin-containing cells in gastrointestinal tissues, *J. Immunol. Methods*, 2, 303, 1973.
82. **Brown, W. R., Isobe, Y., and Nakane, P. K.,** Studies on translocation of immunoglobulins across intestinal epithelium. II. Immunoelectron-microscopic localization of immunoglobulins and secretory component in human intestinal mucosa, *Gastroenterology*, 71, 985, 1976.
83. **Ahnen, D. A., Brown, W. R., and Kloppel, T. M.,** Secretory component: the polymeric immunoglobulin receptor, *Gastroenterology*, 89, 667, 1985.
84. **South, M. A., Cooper, M. D., Wollheim, F. A., Hong, R., and Good, R. A.,** The IgA system. I. Studies of the transport and immunohistochemistry of IgA in the saliva, *J. Exp. Med.*, 123, 615, 1966.
85. **South, M. A., Warwick, W. J., Wollheim, F. A., and Good, R. A.,** The IgA system. III. IgA levels in the serum and saliva of pediatric patients — evidence for a local immunological system, *J. Pediatr.*, 71, 645, 1967.
86. **Nagura, N., Nakane, P. K., and Brown, W. R.,** Translocation of dimeric IgA through neoplastic colon cells in vitro, *J. Immunol.*, 123, 2359, 1979.
87. **Mach, J. P.,** In vitro combination of human and bovine free secretory component with IgA of various species, *Nature (London)*, 228, 1278, 1970.
88. **Radl, J., Klein, F., and van den Berg, P.,** Binding of secretory piece to polymeric IgA and IgM paraproteins in vitro, *Immunology*, 20, 843, 1971.
89. **Eskeland, T. and Brandtzaeg, P.,** Does J chain mediate the combination of 19S IgM and dimeric IgA with the secretory component rather than being necessary for their polymerization?, *Immunochemistry*, 11, 161, 1974.
90. **Weiker, J. and Underdown, B. J.,** A study of the association of human secretory component with IgA and IgM proteins, *J. Immunol.*, 114, 1337, 1975.
91. **Brandtzaeg, P.,** Complex formation between secretory component and human immunoglobulins related to their content of J chain, *Scand. J. Immunol.*, 5, 41, 1976.
92. **Lindh, E. and Bjork, I.,** Binding of secretory component to dimers of immunoglobulin A in vitro, *Eur. J. Biochem.*, 62, 263, 1976.
93. **Knight, K. L., Rosenzweig, M., Lichter, E. A., and Hanley, W. C.,** Rabbit secretory IgA: identification and genetic control of two allotypes of secretory component, *J. Immunol.*, 112, 877, 1974.
94. **Mostov, K. E., Kraehenbuehl, J.-P., and Blobel, G.,** Receptor-mediated transcellular transport of immunoglobulin: synthesis of secretory component as multiple and larger transmembrane forms, *Proc. Natl. Acad. Sci. U.S.A.*, 77, 7257, 1980.
95. **Mostov, K. E., Friedlander, M., and Blobel, G.,** The receptor for transepithelial transport of IgA and IgM contains multiple immunoglobulin-like domains, *Nature (London)*, 308, 37, 1984.
96. **Frutiger, S., Hughes, G. J., Hanley, C., Kingzette, M., and Jato, J.-C.,** The amino terminal domain of rabbit secretory component is responsible for non-covalent binding to immunoglobulin dimers, *J. Biol. Chem.*, 261, 16673, 1986.
97. **Wira, C. R. and Sandoe, C. P.,** Sex steroid hormone regulation of IgA and IgG in rat uterine secretions, *Nature (London)*, 268, 534, 1977.
98. **Wira, C. R., Sullivan, D. A., and Sando, C. P.,** Estrogen mediated control of the secretory immune system in the uterus of the rat, *Ann. N.Y. Acad. Sci.*, 83, 534, 1983.
99. **Sullivan, D. A. and Wira, C. R.,** Estradiol regulation of secretory component in the female reproductive tract, *J. Steroid Biochem.*, 15, 439, 1981.
100. **Sullivan, D. A., Richardson, G. S., MacLaughlin, I. D., and Wira, C. R.,** Variations in the level of secretory component in human uterine fluid during the menstrual cycle, *J. Steroid Biochem.*, 20, 509, 1984.
101. **Wira, C. A. and Colby, E. M.,** Regulation of secretory component by glucocorticoids in primary cultures of rat hepatocytes, *J. Immunol.*, 134, 1744, 1985.
102. **Mostov, K., de Bruyn Kops, A., and Deitcher, D. L.,** Deletion of cytoplasmic domain of the polymeric immunoglobulin receptor prevents basolateral localisation and endocytosis, *Cell*, 47, 359, 1986.
103. **Wilson, D. I., Soltis, R. D., Olson, R. E., and Erlandsen, S. L.,** Cholinergic stimulation of immunoglobulin A secretion in rat intestine, *Gastroenterology*, 83, 881, 1982.
104. **Freier, S., Eran, M., and Faber, J.,** Effect of cholecystokinin and of its antagonist, of atropine and of food on the release of immunoglobulin A and immunoglobulin G antibodies in rat intestine, *Gastroenterology*, 93, 1242, 1987.
105. **Dharmsathaporn, K., Pandol, S., and McRoberts, J. A.,** Chloride secretion induced by carbachol in a human colonic cell line: studies of the mechanism of action, *Gastroenterology* (Abstr.), 88, 1364, 1985.
106. **Donowitz, M. and Welch, M. J.,** $Ca^{2+}$ and cyclic AMP in regulation of intestinal sodium, potassium, and chloride transport, *Ann. Rev. Physiol.*, 48, 135, 1986.

107. **Donowitz, M., Fogel, R., Batesti, L., and Asarkof, N.,** The neurohumeral secretagogues carbechol, substance P and neurotensin increase $Ca^{2+}$ influx and calcium content in rabbit ileum: mechanisms of action?, *Life Sci.*, 31, 1929, 1982.
108. **Freier, S., Lebenthal, E., Freier, M., Shah, P. C., Park, B. H., and Lee, P. C.,** IgE and IgD antibodies to cow milk and soy protein in duodenal fluid: effects of pancreozymin and secretin, *Immunology*, 49, 69, 1983.
109. **Eysselein, V., Reeve, J., Jr., Shively, J., Hawke, D., and Walsh, J.,** Partial structure of a large canine cholecystokinin, (CCK58): amino acid sequence, *Peptides*, 3, 687, 1982.
110. **Mutt, V. and Jorpes, J.,** Structure of porcine cholecytokinin pancreozymin, *Eur. J. Biochem.*, 6, 156, 1968.
111. **Eyssellein, V., Deveney, C., Sankaran, H., Reeve, J., Jr., and Walsh, J.,** Biological activity of canine intestinal cholecystokinin-58, *Am. J. Physiol.*, 245, G313, 1983.
112. **Eysselein, V., Reeve, J., Shiveley, J., Miller, C., and Walsh, J.,** Isolation of a large cholecystokinin precursor from canine brain, *Proc. Natl. Acad. Sci. U.S.A.*, 81, 6565, 1984.
113. **Deschenes, R., Lorenz, L., Haun, R., Roos, B., Collier, K., and Dixen, J.,** Cloning and sequence analysis of cDNA encoding rat preprocholecystokinin, *Proc. Natl. Acad. Sci. U.S.A.*, 81, 726, 1984.
114. **Gubler, U., Chua, A., Hoffman, B., Colier, K., and Eng, J.,** Cloned cDNA to cholecystokinin mRNA predicts an identical preprocholecystokinin in pig brain and gut, *Proc. Natl. Acad. Sci. U.S.A.*, 81, 4307, 1984.
115. **Calderon, P., Furnelle, J., and Christophe, J.,** Phosphaditdyl inositol turnover and calcium movement in the rat pancreas, *Am. J. Physiol.*, 238, G247, 1980.
116. **Freier, S.,** unpublished observations.
117. **Maruyama, Y., Petersen, O., Flanagan, P., and Pearson, G.,** Quantification of $Ca^{2+}$-activated $K^+$-channels under hormonal control in pig pancreas acinar cells, *Nature (London)*, 305, 228, 1983.
118. **Williams, J., Sankaran, H., Roach, E., and Goldfine, I.,** Quantitative electron microscope autoradiographs of $^{125}$I-cholecystokinin pancreatic acini, *Am. J. Physiol.*, 243, G291, 1982.
119. **van Dijk, A., Richards, J., Treciak, A., Gillessen, D., and Mohler, H.,** Cholecystokinin receptors: biochemical demonstration and autoradiographical localization in rat brain and pancreas using [3H]cholecystokinin-8 and radioligand, *J. Neurosci.*, 4, 1021, 1984.
120. **Svoboda, M., Lambert, M., Furnelle, J., and Christophe, J.,** Specific photoaffinity crosslinking of $^{125}$I cholecystokinin to pancreatic plasma membranes. Evidence for a disulphide-linked M'76000 peptide in cholecystokinin receptors, *Regul. Pept.*, 4, 163, 1982.
121. **Rosenzweig, S., Miller, L., and Jamieson, J.,** Identification and localization of cholecystokinin-binding sites on rat pancreatic plasma membranes and acinar cells: a biochemical and autoradiographic study, *J. Cell Biol.*, 96, 1288, 1983.
122. **Pansu, D., Boshard, A., Dechelette, M. A., and Vague, M.,** Effect of pentagastrin, secretin and cholecystokinin on intestinal water and sodium absorption in the rat, *Digestion*, 20, 201, 1980.
123. **Allon, Y., et al.,** unpublished observations.
124. **Dockray, G.,** Immunoreactive component resembling cholecystokinin octapeptide in intestine, *Nature (London)*, 270, 359, 1977.
125. **Buffa, R., Solcia, E., and Go, V.,** Immunohistochemical identification of the cholecystokinin cell in the intestinal mucosa, *Gastroenterology*, 70, 528, 1976.
126. **Polak, J., Pearse, A., Bloom, S., Buchan, A., Rayford, P., and Thompson, J.,** Identification of cholecystokinin-secreting cells, *Lancet*, 2, 1016, 1975.
127. **Meyer, J. and Jones, R.,** Canine pancreatic responses to intestinally perfused fat and products of fat digestion, *Am. J. Physiol.*, 226, 1178, 1974.
128. **Konturek, S., Tasler, J., and Obtulowitz, W.,** Localization of cholecystokinin release in the intestine of the dog, *Am. J. Physiol.*, 222, 16, 1972.
129. **Jansen, J. and Lamres, C.,** Molecular form of cholecystokinin in human plasma during infusion of bombesin, *Life Sci.*, 33, 2197, 1983.
130. **Soder, O. and Hellstrom, P. M.,** Neuropeptide regulation of human thymocyte, guinea-pig T lymphocyte and rat B lymphocyte mitogenesis, *Int. Arch. Allergy. Appl. Immunol.*, 84, 205, 1987.

# INDEX

## A



## B

## E



## F



## G

## H



## I

## L

## M



## N

## O



## P

## Q



## R



## S

**Y**